Discrete Mathematics

Discrete Mathematics

Norman L. Biggs

Reader in Pure Mathematics
Royal Holloway College
University of London

CLARENDON PRESS · OXFORD
1985

Oxford University Press, Walton Street, Oxford OX2 6DP
Oxford New York Toronto
Delhi Bombay Calcutta Madras Karachi
Kuala Lumpur Singapore Hong Kong Tokyo
Nairobi Dar es Salaam Cape Town
Melbourne Auckland
and associated companies in
Beirut Berlin Ibadan Nicosia

Oxford is a trade mark of Oxford University Press

Published in the United States
by Oxford University Press, New York

British Library Cataloguing in Publication Data
Biggs, Norman L.
Discrete mathematics.
1. Mathematics—1961–
I. Title
510   QA39.2
ISBN 0-19-853252-0
ISBN 0-19-853266-0 Pbk

Library of Congress Cataloging in Publication Data
Biggs, Norman.
Discrete mathematics.
Includes index.
1. Electronic data processing—Mathematics.
1. Title.
QA76.9.M35B54 1985   512'.1   85-8964
ISBN 0-19-853252-0
ISBN 0-19-853266-0 (pbk.)

Filmset and printed in Northern Ireland by
The Universities Press (Belfast) Ltd.

To Christine and Juliet

Preface

This book is the result of many years experience of teaching Discrete Mathematics to students in the University of London. When I began lecturing on the subject there were only a few relevant courses, and they were very disjointed. Nowadays there are more courses, but still no general agreement about their shape or content. I hope that my book will contribute to the evolution of a coherent and comprehensive curriculum. It is intended to cover all the material needed by those students for whom mathematics is simply a tool, and also to provide a sound basis for advanced courses suitable for specialist mathematicians.

Although all the results in the book are part of the common mathematical heritage, the author fondly believes that there are some novel features in the method of presentation. Since such claims are notoriously ill-advised, they will not be elaborated, except to say that the proof of Theorem 4.6 is not due to the author, but to David Billington.

I am very grateful to those who have read and commented upon parts of the book, especially Lowell Beineke, Alan Boshier, Tony Gardiner, Gareth Jones, Keith Lloyd, Ann Penoyre, Fred Piper, John Shawe-Taylor, Anne Street, Art White, and Douglas Woodall. Most of the manuscript was typed, with great care and patience, by Marion Brooker.

Finally, I have to thank all those students at Royal Holloway College who have listened to my lectures and tried, with varying degrees of success, to solve my Exercises. Their work has contributed enormously to making the book a reality.

London   N. L. B.
June 1985

Contents

4. Subsets and designs

5. Partition, classification, and distribution

6. Modular arithmetic

PART II GRAPHS AND ALGORITHMS

7. Algorithms and their efficiency

8. Graphs

9. Trees, sorting, and searching

10. Bipartite graphs and matching problems

11. Digraphs, networks, and flows

12. Recursive techniques

PART III   ALGEBRAIC METHODS

13. Groups

14. Groups of permutations

15. Rings, fields, and polynomials

16. Finite fields and some applications

17. Error-correcting codes

18. Generating functions

19. Partitions of a positive integer

20. Symmetry and counting

Part I
Numbers and counting

In the first part of this book we study simple but fundamental ideas. Many of the concepts are familiar in a vague and intuitive way, because they arise frequently in the problems of everyday life. The aim here is to provide a sound mathematical foundation which can be used to analyse and solve problems more complex than the commonplace ones.

The background knowledge assumed is amply covered by the mathematics usually taught in schools and colleges to young people in the 13–16 age range. In particular, it is assumed that the reader is skilled in arithmetic and in the manipulation of algebraic expressions. Some familiarity with the language and notation of sets is also assumed and, for the sake of completeness, this will be reviewed briefly here.

A set S may be described by listing its members, thus

$$S = \{1, 2, 3, 4, 5, 6, 7, 8, 9\},$$

or by specifying a property which distinguishes its members from its non-members, thus

$$S = \{n \mid n \text{ is a whole number between 1 and 9}\}.$$

The latter notation is usually translated into ordinary language in the form 'S is the set of n such that n is a whole number between 1 and 9'. We use the notation $x \in S$ for the statement 'x is a member of S'.

A set B is a **subset** of a set A if every member of B is a member of A; this is written as $B \subseteq A$. Given two sets X and Y, their **union** $X \cup Y$ is the set whose members belong either to X or to Y (or both), and their **intersection** $X \cap Y$ is the set whose members belong to both X and Y. A set which has no members is said to be **empty**, and is denoted by $\varnothing$. Finally, when B is a subset of A the **difference** $A \setminus B$ is the set of members of A which are not in B.

When students begin to study mathematics at this level they are often confused by the logical links between the various statements which constitute a mathematical proof. The flow of an argument can sometimes be clarified by judicious use of the symbol $\Rightarrow$, corresponding to the word 'implies'. For example, it is correct to make such statements as

$$x = 2 \quad \Rightarrow \quad x^2 = 4,$$
$$x + 3 = 8 \quad \Rightarrow \quad x = 5.$$

The first of these can be read as '$x = 2$ *implies* $x^2 = 4$' or as '*if* $x = 2$ *then* $x^2 = 4$'. The direction of the arrow makes it clear that such one-way implications cannot, in general, be reversed. Thus we cannot say that if $x^2 = 4$ then $x = 2$, since there is the alternative possibility that $x = -2$. When the implication does work in both directions we use the symbol $\Leftrightarrow$, thus

$$x + 3 = 8 \quad \Leftrightarrow \quad x = 5.$$

This is often rendered into ordinary language as '$x + 3 = 8$ *if and only if* $x = 5$'.

Note on exercises

Before we begin in earnest, a few words about the plan of the book. Each section contains a number of exercises, most of which are designed to reinforce the ideas and techniques under discussion, rather than to test the reader's ingenuity. The student is advised to work through the majority of these exercises before going on to the next section. At the end of each chapter there is a selection of 'miscellaneous exercises'. Some of these are routine exercises, some of them are straightforward generalizations of the work covered in the chapter, and a few of them introduce important results which should be part of every mathematician's store of knowledge. Roughly speaking, the later exercises are the harder ones.

1
Integers

1.1 Arithmetic

Every reader of this book is familiar with the *integers*. At a very early stage in our lives we meet the positive integers, or 'whole numbers'

$$1, 2, 3, 4, 5, \text{ and so on.}$$

Later on we are introduced to 0 (zero), and the negative integers

$$-1, -2, -3, -4, -5, \text{ and so on.}$$

In mathematics we do not usually concern ourselves with logical and philosophical questions about the meaning of these objects, but we do need to know what properties they are assumed to possess. If everyone makes the same assumptions then everyone will arrive at the same results. The assumptions are known as *axioms*.

The point of view adopted in this book is, roughly speaking, as outlined above. We accept without question that there is a set of objects called **integers**, comprising the positive and negative integers, and zero, familiar from our early education and experience. The set of integers will be denoted by the special symbol $\mathbb{Z}$. The properties of $\mathbb{Z}$ will be given by a list of axioms, from which we shall be able to deduce all the results about integers needed in the subsequent discussions. We begin by listing those axioms concerned with the fundamental processes of addition and multiplication.

We adopt the usual notations: $a + b$ for the sum of two integers a and b, and $a \times b$ (or frequently just ab) for their product. We think of $+$ and $\times$ as *operations* which, when we are given a pair of integers a and b, define the integers $a + b$ and $a \times b$. The fact that $a + b$ and $a \times b$ are themselves integers, and not some alien objects like elephants, is our first assumption (Axiom **I1**). In the following list of axioms, a, b, c denote arbitrary integers, and 0 and 1 denote specific integers having the stated properties.

I1. $a + b$ and ab are in $\mathbb{Z}$.
I2. $a + b = b + a$; $ab = ba$.

I3. $(a+b)+c = a+(b+c)$; $(ab)c = a(bc)$.
I4. $a+0 = a$; $a1 = a$.
I5. $a(b+c) = ab + ac$.
I6. For each a in $\mathbb{Z}$ there is a unique integer $-a$ in $\mathbb{Z}$ such that $a+(-a) = 0$.
I7. If a is not equal to 0 and $ab = ac$, then $b = c$.

All the axioms correspond to familiar properties of the integers which we learn by rote at various stages of our mathematical education. From them most of the usual arithmetical rules for integers can be deduced. For example, we can *define* the operation of subtraction by saying that $a - b$ is to mean $a + (-b)$; and we can deduce the elementary rules for subtraction, such as the following.

Example Show that for any integers m and n

$$m - (-n) = m + n.$$

Solution By the definition of subtraction, $m - (-n)$ is the same as $m + (-(-n))$. Now Axiom **I6** tells us that $-(-n)$ is the unique integer which, when added to $-n$, gives zero. However, n itself is such an integer, since

$$(-n) + n = n + (-n) \qquad \text{(Axiom I2)}$$
$$= 0 \qquad \text{(Axiom I6)}$$

Hence $-(-n) = n$ and $m - (-n) = m + n$, as required. $\qquad \square$

Some similar results will be found in the following Exercises. Since we do not yet have all the axioms for integers, the results are not particularly exciting, but the important thing to remember is that they can be proved on the basis of the axioms alone.

Exercises 1.1

1 The following is a proof of the rule $x0 = 0$, using only the axioms stated above. Write out the proof in full explaining which axiom is being used at each step.

$$x(0+0) = x0$$
$$x0 + x0 = x0$$
$$-x0 + (x0 + x0) = -x0 + x0$$
$$(-x0 + x0) + x0 = 0$$
$$0 + x0 = 0$$
$$x0 = 0.$$

2 Construct a proof of the rule $(a+b)c = ac + bc$, explaining each step as in Ex. 1.

3 As usual, let x^2 denote xx. Prove that if any two integers a and b are given, then there is an integer c such that $(a+b)c = a^2 - b^2$.

4 Suppose there are two integers 0 and $0'$ with the property stated in Axiom **I4**, that is

$$a + 0 = a, \qquad a + 0' = a$$

for every $a \in \mathbb{Z}$. Show that this implies $0 = 0'$, so that 0 is in fact uniquely characterized by Axiom **I4**.

1.2 Ordering the integers

The natural order of the integers is at least as important as their arithmetical properties. From the very first we learn our numbers in the order 1, 2, 3, 4, 5, and the fact that 4 is 'bigger' than 3 soon becomes a matter of practical importance to us.

We express this idea formally by saying that there is an *ordering relation* on $\mathbb{Z}$, written

$$m \leqslant n \qquad (m, n \in \mathbb{Z}),$$

which is required to satisfy certain axioms. It turns out that just five axioms are needed to specify the basic properties of the symbol $\leqslant$, and they are listed below, with numbers continuing from the list in Section 1.1. As before, a, b, and c denote arbitrary integers.

I8. $a \leqslant a$.
I9. $a \leqslant b$ and $b \leqslant a \implies a = b$.
I10. $a \leqslant b$ and $b \leqslant c \implies a \leqslant c$.
I11. $a \leqslant b \implies a + c \leqslant b + c$.
I12. $a \leqslant b$ and $0 \leqslant c \implies ac \leqslant bc$.

These axioms hardly need to be learned, since they contain very familiar properties; the important point is that they enable us to deduce other, equally familiar facts. With this in mind, the following exercises should be solved using only the properties contained in Axioms **I1–I12**.

Exercises 1.2

1 Suppose $a \leqslant b$. By adding $-a$ and then $-b$ to both sides of the inequality, show that $-b \leqslant -a$. Deduce that if $a \leqslant b$ and $c \leqslant 0$ then $bc \leqslant ac$.

2 Show that $0 \leqslant x^2$ for any x in $\mathbb{Z}$, and deduce that $0 \leqslant 1$.

3 Deduce from the previous exercise that $n \leqslant n+1$ for all $n \in \mathbb{Z}$.

It is clear that we can define the other commonly used ordering symbols, $\geqslant$, $<$, and $>$, in terms of the symbol $\leqslant$. For example, $m > n$ must be defined as meaning that $n \leqslant m$ and $m \neq n$. We shall use these symbols as the need for them arises.

It might appear at first sight that we now have all the properties of $\mathbb{Z}$ that are required in mathematics, but, rather surprisingly, one vital axiom is missing. Suppose X is any subset of $\mathbb{Z}$; we say that the integer b is a **lower bound** for X if

$$b \leqslant x \quad \text{for all } x \in X.$$

Some subsets do not have lower bounds: for example, the set of negative integers -1, -2, -3, and so on, clearly has no lower bound. On the other hand, the set S denoted by the bold numbers in Fig. 1.1 has many lower bounds. A quick glance tells us that -40, for instance, is a lower bound, while a closer inspection reveals that -27 is the 'best' lower bound, since it actually belongs to S. In general, a lower bound for a set X which is itself a member of X is known as a **least member** for X.

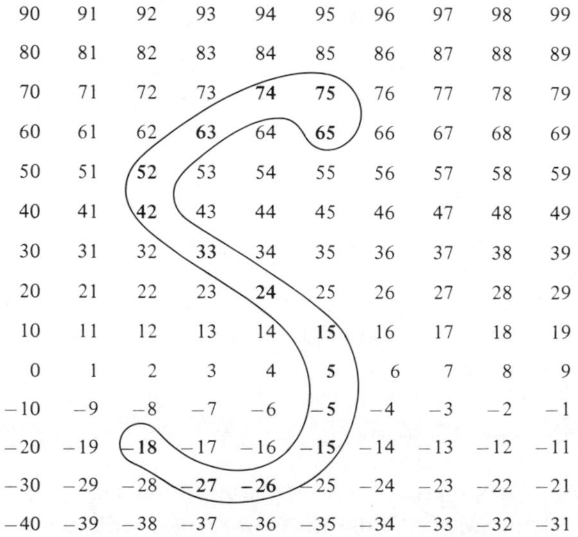

Fig. 1.1
The least member of S is -27.

Our final axiom for $\mathbb{Z}$ asserts what is (apparently) an obvious property.

> **I13.** If X is a subset of $\mathbb{Z}$ which is not empty and
> has a lower bound, then X has a least member.

Axiom **I13** is known as the **well-ordering axiom**. A good way to grasp its meaning is to consider a game in which two people alternately choose a member of X, subject to the rule that each number must be strictly less than the previous one. The axiom tells us that, when the numbers are required to be integers, the game will end; indeed the end occurs as soon as one of the players has the good sense to choose the least member. This apparently obvious property does *not* necessarily hold when we allow numbers which are not integers, because X might not have a least member even though it has a lower bound. For example, suppose X is the set of fractions $\frac{3}{2}, \frac{4}{3}, \frac{5}{4}$, and so on, having the general form $(n+1)/n$ $(n \geqslant 2)$. This set has a lower bound (1, for instance) but it has no least member, and so the players can go on playing for ever, choosing fractions closer and closer to 1.

The well-ordering axiom provides firm justification for our intuitive picture of the integers as a set of regularly spaced points on a straight line, extending indefinitely in either direction (Fig. 1.2). In particular, it says that we cannot get closer and closer to an integer without actually arriving, so that the picture in Fig. 1.3 is wrong.

Fig. 1.2
The correct intuitive picture of $\mathbb{Z}$.

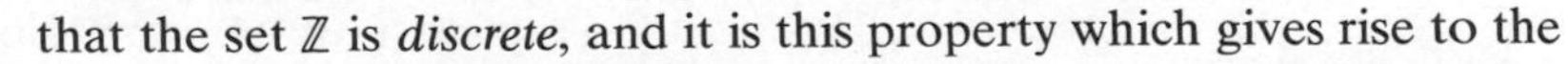

Fig. 1.3
An incorrect picture of $\mathbb{Z}$.

The fact that there are gaps between the integers leads us to say that the set $\mathbb{Z}$ is *discrete*, and it is this property which gives rise to the name 'discrete mathematics'. In calculus and analysis limiting processes are of fundamental importance, and it is essential to use number systems which are *continuous*, rather than discrete.

Exercises 1.2 (continued)

4 ' In each of the following cases say whether or not the set X has a lower bound, and if it has a lower bound, find its least member.

(i) $X = \{x \in \mathbb{Z} \mid x^2 \leqslant 16\}$.

(ii) $X = \{x \in \mathbb{Z} \mid x = 2y \text{ for some } y \in \mathbb{Z}\}$.

(iii) $X = \{x \in \mathbb{Z} \mid x^2 \leqslant 100x\}$.

5 A subset Y of Z is said to have an **upper bound** c if $c \geqslant y$ for all $y \in Y$. An upper bound which is also a member of Y is said to be a **greatest member** of Y. Use Axiom **I13** to show that if Y is not empty and it has an upper bound, then it has a greatest member. [Hint: apply the axiom to the set whose members are $-y$ ($y \in Y$).]

6 The integers n satisfying $1 \leqslant n \leqslant 25$ are arranged in a square array of five rows and five columns in an arbitrary way. The greatest member of each row is selected, and s denotes the least of these. Similarly, the least member in each column is selected, and t denotes the greatest of these. Show that $s \geqslant t$, and give an example in which $s \neq t$.

1.3 Recursive definitions

Let $\mathbb{N}$ denote the set of positive integers, that is

$$\mathbb{N} = \{n \in \mathbb{Z} \mid n \geqslant 1\},$$

and let $\mathbb{N}_0$ denote the set $\mathbb{N} \cup \{0\}$, that is

$$\mathbb{N}_0 = \{n \in \mathbb{Z} \mid n \geqslant 0\}.$$

If X is a subset of $\mathbb{N}$ (or $\mathbb{N}_0$) then it automatically has a lower bound, since each member x of X satisfies $x \geqslant 1$ (or $x \geqslant 0$). Thus in this case the well-ordering axiom takes the form

if X is a non-empty subset of $\mathbb{N}$ or $\mathbb{N}_0$
then X has a least member.

This is the form which is most often used in practice.

Our first use of the well-ordering axiom will be to justify a very common procedure. Frequently we encounter an expression of the form u_n, where n indicates any positive integer: for example, we might have $u_n = 3n + 2$, or $u_n = (n+1)(n+2)(n+3)$. In these examples u_n is given by an explicit formula and there is no difficulty in explaining how u_n is to be calculated when n is given a specific value. However, in many cases we do not know a formula for u_n; indeed our problem may be to find one. In such cases we may be given some values of u_n for small positive integers n, and a relationship between the general u_n and some of the u_r for $r < n$.

For example, suppose we are given that

$$u_1 = 1, \qquad u_2 = 2, \qquad u_n = u_{n-1} + u_{n-2} \qquad (n \geqslant 3).$$

To calculate the values of u_n for all n in $\mathbb{N}$ we could proceed as follows:

$$u_3 = u_2 + u_1 = 2 + 1 = 3,$$
$$u_4 = u_3 + u_2 = 3 + 2 = 5,$$
$$u_5 = u_4 + u_3 = 5 + 3 = 8,$$

and so on. This is an example of a *recursive definition*. It is plainly 'obvious' that the method will give a unique value of u_n for every positive integer n. But strictly speaking we need the well-ordering axiom to justify this conclusion, along the following lines.

Suppose there is a positive integer n for which u_n is not uniquely defined. Then, by the well-ordering axiom there is a least positive integer m with this property. Since u_1 and u_2 are specified explicitly, m is not 1 or 2, and the equation $u_m = u_{m-1} + u_{m-2}$ is applicable. By the definition of m, u_{m-1} and u_{m-2} are uniquely defined, and the equation gives a unique value of u_m, contrary to hypothesis. The contradiction arises from the assumption that u_n is not well-defined for some n, and hence this assumption must be false.

The reader should not be dismayed by the use of such contorted arguments to establish something which is 'obviously' true. In the first place, we shall not labour these points unduly, and in the second place, the fact that the result is 'obvious' simply means that we are working with the correct mental picture of the sets $\mathbb{N}$ and $\mathbb{Z}$. Once we have established that picture on a firm foundation we can set out to extend it and obtain results which may not be quite so 'obvious'.

The method of recursive definition will occur very often in the rest of the book. There are other forms of the procedure which are usually concealed by the notation. What do we mean by the following expressions?

$$\sum_{r=1}^{n} 2r - 1, \qquad 1 + 3 + 5 + \cdots + (2n - 1).$$

It is clearly not enough to say that each one means the same as the other, since each one contains a mysterious symbol, $\sum$ and $\cdots$, respectively. What we should say is that each of them is equal to the expression s_n given by the following recursive definition:

$$s_1 = 1, \qquad s_n = s_{n-1} + (2n - 1) \quad (n \geq 2).$$

This makes it clear that both mysterious symbols are really a form of shorthand for a recursive definition, and that consequently the expressions are properly defined for each n in $\mathbb{N}$.

Similar remarks apply to the definition of products such as $n!$ (spoken as n *factorial*). If we say that

$$n! = \prod_{i=1}^{n} i, \quad \text{or} \quad n! = 1 \times 2 \times 3 \times \cdots \times n,$$

then the meaning may be clear to everyone. But to be precise (and to make it clear to a computer) we should use the recursive definition

$$1! = 1, \qquad n! = n \times (n-1)! \quad (n \geq 2).$$

Exercises 1.3

1   In the following cases calculate (where possible) the values of u_1, u_2, u_3, u_4, and u_5 given by the equations. If you cannot calculate the values explain why the definition is faulty.

(i) $u_1 = 1, \qquad u_2 = 1, \qquad u_n = u_{n-1} + 2u_{n-2} \quad (n \geq 3).$

(ii) $u_1 = 1, \qquad u_n = u_{n-1} + 2u_{n-2} \quad (n \geq 2).$

(iii) $u_1 = 0, \qquad u_n = nu_{n-1} \quad (n \geq 2).$

2   Give a recursive definition of the 'nth power' 2^n for all $n \geq 1$.

3   Suppose u_n is defined by the equations

$$u_1 = 2, \qquad u_n = 2^{u_{n-1}} \quad (n \geq 2).$$

What is the least value of n for which it is not practicable to calculate u_n using a pocket calculator?

4   Write down explicit formulae for the expressions u_n defined by the following equations.

(i) $u_1 = 1, \qquad u_n = u_{n-1} + 3 \quad (n \geq 2).$

(ii) $u_1 = 1, \qquad u_n = n^2 u_{n-1} \quad (n \geq 2).$

1.4 The principle of induction

Suppose we are asked to prove the result

$$1+3+5+\cdots+(2n-1)=n^2.$$

In other words, we have to show that the expression defined recursively on the left is equal to that defined explicitly by the formula on the right, for all positive integers n. We might proceed as follows.

The formula is certainly correct when $n=1$, since $1=1^2$. Suppose it is correct for a specific value of n, say $n=k$, so that

$$1+3+5+\cdots+(2k-1)=k^2.$$

We can use this fact to simplify the left-hand side when $n=k+1$, as follows:

$$1+3+5+\cdots+(2k+1)=1+3+5+\cdots+(2k-1)+(2k+1)$$
$$=k^2+(2k+1)$$
$$=(k+1)^2.$$

So if the result is correct when $n=k$ then it is correct when $n=k+1$. Now we began by remarking that it is correct when $n=1$, hence it must be correct when $n=2$. By the same argument, since it is correct when $n=2$ it must be correct when $n=3$. Continuing in this way we see that it is correct for all positive integers n.

The essence of this argument is often referred to as the *principle of induction*. It is a powerful technique, easy to apply, and we shall use it frequently. But first we should examine its logical basis, and in order to do so, a more general formulation is needed.

Let S denote the subset of $\mathbb{N}$ for which the result is correct: of course, our aim is to prove that S is the whole set $\mathbb{N}$. The first step is to show that 1 is in S, and then we show that if k is in S so is $k+1$. Then we huff and puff (Fig. 1.4) and conclude that $S=\mathbb{N}$. Fortunately the huffing and puffing is not essential, because the principle of induction is a consequence of the axioms for $\mathbb{Z}$ and $\mathbb{N}$ that have been so carefully chosen. Specifically, it is a consequence of the well-ordering axiom.

Theorem 1.4   Suppose that S is a subset of $\mathbb{N}$ satisfying the conditions

    (i) $1\in S$,
    (ii) for each $k\in\mathbb{N}$, if $k\in S$ then $k+1\in S$.

Then it follows that $S=\mathbb{N}$.

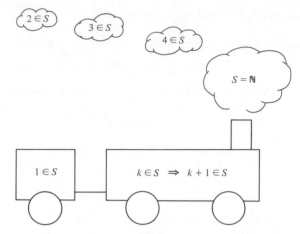

Fig. 1.4
The principle of induction.

Proof　If the conclusion is false, $S \neq \mathbb{N}$ and the complementary set $\bar{S}$ defined by

$$\bar{S} = \{r \in \mathbb{N} \mid r \notin S\}$$

is not empty. By the well-ordering axiom, $\bar{S}$ has a least member m. Since 1 belongs to S, $m \neq 1$. It follows that $m-1$ is in $\mathbb{N}$, and since m is the least member of $\bar{S}$, $m-1$ must be in S. Then putting $k = m-1$ in condition (ii) we conclude that m is in S, which contradicts the assertion that m is in $\bar{S}$. Thus the statement that $S \neq \mathbb{N}$ leads to an absurdity, so we must have $S = \mathbb{N}$. ☐

　　In practice, we usually present a 'proof by induction' in rather more descriptive terms. The fact that the result is true when $n = 1$ is called the *induction basis*, and the assumption that it is true when $n = k$ is called the *induction hypothesis*. When these terms are used there is no need to introduce the set S explicitly.

Example　The integer x_n is defined recursively by

$$x_1 = 2 \quad \text{and} \quad x_n = x_{n-1} + 2n \quad (n \geqslant 2).$$

Show that

$$x_n = n(n+1) \quad \text{for all } n \in \mathbb{N}.$$

Solution　(*Induction basis*) The result is true when $n = 1$ since $2 = 1 \times 2$.
(*Induction hypothesis*) Suppose the result is true when $n = k$, that is,

$x_k = k(k+1)$. Then

$$\begin{aligned} x_{k+1} &= x_k + 2(k+1) & \text{(by recursive definition)} \\ &= k(k+1) + 2(k+1) & \text{(by induction hypothesis)} \\ &= (k+1)(k+2). \end{aligned}$$

So the result is true when $n = k+1$, and by the principle of induction, it is true for all positive integers n. $\qquad\square$

There are several modified forms of the principle of induction. Sometimes it is convenient to take for the induction basis the value $n = 0$; on the other hand it may be appropriate to take a value like 2 or 3, since the first few cases may be exceptional. Each problem must be treated on its merits. Another useful modification is to take as induction hypothesis the assumption that the result is true for *all* relevant values $n \leqslant k$, rather than for $n = k$ alone. (This formulation is sometimes called the *strong* induction principle.) All these modifications can be justified by trivial changes in the proof of Theorem 1.4, as indicated in Ex. 1.4.6.

Exercises 1.4

1 Use the principle of induction to prove that

$$1^2 + 2^2 + \cdots + n^2 = \tfrac{1}{6} n(n+1)(2n+1)$$

for all positive integers n.

2 Make a table of values of

$$S_n = 1^3 + 2^3 + 3^3 + \cdots + n^3$$

for $1 \leqslant n \leqslant 6$. On the basis of your table suggest a formula for S_n. [Hint: the values of S_n are perfect squares.] Use the principle of induction to establish that the formula is correct for all $n \geqslant 1$. (If the method fails, your formula is wrong!)

3 Use the strong form of the principle of induction to show that if u_n is defined recursively by the rules

$$u_1 = 3, \qquad u_2 = 5, \qquad u_n = 3u_{n-1} - 2u_{n-2} \quad (n \geqslant 3),$$

then $u_n = 2^n + 1$ for all positive integers n.

4 Find the least positive integer n_0 for which it is true that $n! \geqslant 2^n$. Taking the case $n = n_0$ as the induction basis, show that the result holds for all $n \geqslant n_0$.

5   In the following cases find the appropriate value of n_0 for the induction basis and show that the statement is true for all $n \geqslant n_0$.

(i) $n^2 + 6n + 8 \geqslant 0$;   (ii) $n^3 \geqslant 6n^2$.

6   The following theorem incorporates all the modifications of the basic principle of induction outlined above.

Theorem 1.4*   Suppose n_0 is any integer (not necessarily positive), and let X denote the set of integers $n \geqslant n_0$. Let S be a subset of X satisfying the conditions

(i) $n_0 \in S$,
(ii) if $x \in S$ for all x in the range $n_0 \leqslant x \leqslant k$ then $k + 1 \in S$.

Then it follows that $S = X$.

Write out the proof of Theorem 1.4, and make the changes needed in order to prove Theorem 1.4*.

1.5   Quotient and remainder

As children we learn that when 6 'goes into' 27 the *quotient* is 4 and the *remainder* is 3, that is,

$$27 = 6 \times 4 + 3.$$

The important point is that the remainder must be less than 6. Although it is also true that, for instance

$$27 = 6 \times 3 + 9,$$

we are told that we must take the least value for the remainder, so that the amount 'left over' is as small as possible. The fact that the set of possible 'remainders' does have a least member is a consequence of the well-ordering axiom.

Theorem 1.5   If we are given integers a and b with $b \in \mathbb{N}$, then there are integers q and r such that

$$a = bq + r \quad \text{and} \quad 0 \leqslant r < b.$$

Proof   We shall apply the well-ordering axiom to the set of 'remainders'

$$R = \{x \in \mathbb{N}_0 \mid a = by + x \text{ for some } y \in \mathbb{Z}\}.$$

First we show that R is not empty. If $a \geqslant 0$ the identity

$$a = b0 + a$$

shows that $a \in R$, while if $a < 0$ the identity

$$a = ba + (1 - b)a$$

shows that $(1 - b)a \in R$. (In both cases it is necessary to check that the stated member of R is non-negative.)

Now, since R is a non-empty subset of $\mathbb{N}_0$, it has a least member r, and since r is in R it follows that $a = bq + r$ for some q in $\mathbb{Z}$. Furthermore

$$a = bq + r \quad \Rightarrow \quad a = b(q + 1) + (r - b)$$

so that if $r \geqslant b$ then $r - b$ is in R. But $r - b$ is less than r, contrary to the definition of r as the least member of R. Since the assumption $r \geqslant b$ leads to a contradiction we must have $r < b$, as required. $\square$

It is easy to see that the quotient q and the remainder r obtained in the theorem are unique. For suppose that q' and r' also satisfy the conditions, that is

$$a = bq' + r' \quad \text{and} \quad 0 \leqslant r' < b.$$

If $q' < q$ then $q - q' \geqslant 1$ so that we have

$$r' - a - bq' = (a - bq) + b(q - q') \geqslant r + b.$$

Since $r + b \geqslant b$, it follows that $r' \geqslant b$ contradicting the second property of r'. Hence the assumption $q' < q$ is false. The same argument with q and q' interchanged shows that $q < q'$ is false. So we must have $q = q'$, and consequently $r = r'$ also, since

$$r = a - bq = a - bq' = r'.$$

One important consequence of Theorem 1.5 is that it justifies our usual method of representing integers. Let $t \geqslant 2$ be a given integer, called the **base** for calculation. For any positive integer x we have, by repeated application of Theorem 1.5,

$$x = tq_0 + r_0$$
$$q_0 = tq_1 + r_1$$
$$\cdots$$
$$q_{n-2} = tq_{n-1} + r_{n-1}$$
$$q_{n-1} = tq_n + r_n.$$

Here each remainder r_i is one of the integers $0, 1, \ldots, t - 1$, and we stop when $q_n = 0$. Eliminating the quotients q_i we obtain

$$x = r_n t^n + r_{n-1} t^{n-1} + \cdots + r_1 t + r_0.$$

We have represented x (with respect to the base t) by the sequence of remainders, and we write $x = (r_n r_{n-1} \cdots r_1 r_0)_t$. Conventionally $t = 10$ is the base for calculations done by hand, and we omit the subscript, so we have the familiar notation

$$1984 = (1 \times 10^3) + (9 \times 10^2) + (8 \times 10) + 4.$$

This positional notation requires symbols only for the integers $0, 1, \ldots, t - 1$. When $t = 2$ it is particularly suited for machine calculations, since the symbols 0 and 1 can be represented physically by the absence or presence of a pulse of electricity or light.

Example   What is the representation in base 2 of $(109)_{10}$?

Solution   Dividing repeatedly by 2 we obtain

$$109 = 2 \times 54 + 1$$
$$54 = 2 \times 27 + 0$$
$$27 = 2 \times 13 + 1$$
$$13 = 2 \times 6 + 1$$
$$6 = 2 \times 3 + 0$$
$$3 = 2 \times 1 + 1$$
$$1 = 2 \times 0 + 1.$$

Hence

$$(109)_{10} = (1101101)_2. \qquad \square$$

Exercises 1.5

1   Find q and r satisfying Theorem 1.5 when

    (i) $a = 1001$, $b = 11$;   (ii) $a = 12345$, $b = 234$.

2   Find the representations of $(1985)_{10}$ in base 2, in base 5, and in base 11.

3   Find the usual (base 10) representations of

    (i) $(11011101)_2$;   (ii) $(4165)_7$.

1.6 Divisibility

Given any two integers x and y we say that y is a **divisor** of x, and write $y \mid x$, if

$$x = yq \quad \text{for some } q \in \mathbb{Z}.$$

We also say that y is a **factor** of x, that y **divides** x, that x is **divisible** by y, and that x is a **multiple** of y.

When $y \mid x$ we can use the symbol $\dfrac{x}{y}$ (or x/y) to denote the integer q such that $x = yq$. When y is not a divisor of x we have to assign a new meaning to the fraction x/y, since it is not an integer. The reader is undoubtedly familiar with rules for dealing with fractions, and we shall use those rules from time to time, but it is important to remember that fractions have not yet been formally defined within the framework of this book. It is even more important to remember that x/y is not a member of $\mathbb{Z}$ unless y divides x.

Example

Show that if c, d, and n are integers such that

$$d \mid n \quad \text{and} \quad c \left| \frac{n}{d} \right.$$

then

$$c \mid n \quad \text{and} \quad d \left| \frac{n}{c} \right..$$

Solution

Since $d \mid n$ there is an integer s such that $n = ds$, and n/d denotes the integer s. Since $c \mid n/d$ there is an integer t such that

$$s = \frac{n}{d} = ct.$$

It follows that

$$n = ds = d(ct) = c(dt)$$

so that $c \mid n$ and n/c denotes the integer dt. Finally, since $n/c = dt$ we have $d \mid n/c$, as required. $\qquad \square$

Exercises 1.6

1 Prove that $x \mid 0$ for every $x \in \mathbb{Z}$, but $0 \mid x$ only when $x = 0$.

2 Show that if $c \mid a$ and $c \mid b$, then $c \mid xa + yb$ for any integers x, y.

3 Show that if a and b are integers such that $ab = 1$ then $a = b = 1$ or $a = b = -1$. [Hint: either both a and b are positive or both are negative.] Deduce that if x and y are integers such that $x \mid y$ and $y \mid x$ then $x = y$ or $x = -y$.

4 Use the principle of induction to prove that, for all $n \geq 0$,

(i) $n^2 + 3n$ is divisible by 2; (ii) $n^3 + 3n^2 + 2n$ is divisible by 6.

1.7 The greatest common divisor

If a and b are integers we say that the integer d is a **greatest common divisor**, or **gcd**, of a and b if

(i) $d \mid a$ and $d \mid b$; (ii) if $c \mid a$ and $c \mid b$, then $c \mid d$.

Condition (i) says that d is a common divisor of a and b, and condition (ii) says that any common divisor of a and b is also a divisor of d. For example, 6 is a common divisor of 60 and 84, but it is *not* a greatest common divisor, since $12 \mid 60$ and $12 \mid 84$ but $12 \nmid 6$. (The symbol $\nmid$ means 'does not divide'.)

Conditions (i) and (ii) are not quite sufficient to ensure that two given integers have a unique gcd. For if d and d' both satisfy both conditions it follows that

$$d \mid d' \quad \text{and} \quad d' \mid d.$$

Hence, by Ex. 1.6.3, $d = d'$ or $d = -d'$. Thus, in order to get a unique gcd it is sufficient to impose a third condition:

(iii) $d \geq 0$.

We say that the unique integer d satisfying (i), (ii), and (iii) is *the* gcd of a and b, and write $d = \gcd(a, b)$. For example, $12 = \gcd(60, 84)$.

There is a very famous method for calculating the gcd of two given integers, based on the quotient and remainder technique. It depends on the fact that

$$a = bq + r \quad \Rightarrow \quad \gcd(a, b) = \gcd(b, r).$$

In order to prove this we remark that if d divides a and b then it surely divides $a - bq$; and $a - bq = r$, so d divides r. Thus any common divisor of a and b is also a common divisor of b and r. Conversely, if d divides b and r it also divides $a = bq + r$. Repeated application of this simple fact provides the method for calculating the gcd.

Example   Find the gcd of 2406 and 654.

Solution   We have

$$\begin{aligned}
\gcd(2406, 654) &= \gcd(654, 444) & \text{since} & \quad 2406 = 654 \times 3 + 444, \\
&= \gcd(444, 210) & \text{since} & \quad 654 = 444 \times 1 + 210, \\
&= \gcd(210, 24) & \text{since} & \quad 444 = 210 \times 2 + 24, \\
&= \gcd(24, 18) & \text{since} & \quad 210 = 24 \times 8 + 18, \\
&= \gcd(18, 6) & \text{since} & \quad 24 = 18 \times 1 + 6, \\
&= 6 & \text{since} & \quad 18 = 6 \times 3. \qquad \square
\end{aligned}$$

In general, in order to calculate the gcd of integers a and b (both ≥ 0) we define q_i and r_i recursively by the equations

$$\begin{aligned}
a &= bq_1 + r_1 & (0 \leq r_1 < b) \\
b &= r_1 q_2 + r_2 & (0 \leq r_2 < r_1) \\
r_1 &= r_2 q_3 + r_3 & (0 \leq r_3 < r_2) \\
&\cdots &
\end{aligned}$$

It is clear that the process must stop eventually, since each remainder r_i is strictly less than the preceding one. So the final steps are as follows:

$$\begin{aligned}
r_{k-4} &= r_{k-3} q_{k-2} + r_{k-2} & (0 \leq r_{k-2} < r_{k-3}) \\
r_{k-3} &= r_{k-2} q_{k-1} + r_{k-1} & (0 \leq r_{k-1} < r_{k-2}) \\
r_{k-2} &= r_{k-1} q_k,
\end{aligned}$$

where r_k vanishes, and the required gcd is r_{k-1}. This procedure is known as the **Euclidean algorithm**, after the Greek mathematician Euclid (c. 300 BC). It is extremely useful in practice, and has important theoretical consequences.

Theorem 1.7   Let a and b be integers with $b \geq 0$, and let $d = \gcd(a, b)$. Then there are integers m and n such that

$$d = ma + nb.$$

Proof   According to the calculation given above $d = r_{k-1}$, and using the penultimate equation we have

$$r_{k-1} = r_{k-3} - r_{k-2} q_{k-1}.$$

Thus d can be written in the form $m'r_{k-2} + n'r_{k-3}$, where $m' = -q_{k-1}$ and $n' = 1$. Substituting for r_{k-2} in terms of r_{k-3} and r_{k-4} we obtain

$$d = m'(r_{k-4} - r_{k-3}q_{k-2}) + n'r_{k-3}$$

which can be written in the form $m''r_{k-3} + n''r_{k-4}$, with $m'' = n' - m'q_{k-2}$ and $n'' = m'$. Continuing in this way we eventually obtain an expression for d in the required form.   □

For example, from the calculation used to find the gcd of 2406 and 654 we obtain

$$
\begin{aligned}
6 &= & & 24- \ 18\times 1 \ = & 1\ \times\ \ & 24+ & (-1)\times\ \ & 18 \\
&= & 24+\ (-1)\times\ & (210-\ 24\times 8)= & (-1)\times\ & 210+ & 9\ \times\ & 24 \\
&= & -210+\ \ 9\ \times\ & (444-210\times 2)= & 9\ \times\ & 444+ & (-19)\times & 210 \\
&= & 9\ \times 444+(-19)\times\ & (654-444\times 1)= & (-19)\times\ & 654+ & 28\ \times & 444 \\
&= (-19)\times 654+\ & 28\ \times\ & (2406-654\times 3)= & 28\ \times & 2406+ & (-103)\times & 654.
\end{aligned}
$$

Thus the required expression $d = ma + nb$ is

$$6 = 28\times 2406 + (-103)\times 654.$$

If $\gcd(a, b) = 1$ then we say that a and b are **coprime**, and in this case Theorem 1.7 asserts that there are integers m and n such that

$$ma + nb = 1.$$

This remark is very useful. For example, we are all familiar with the idea that a fraction can be reduced to its 'lowest terms', that is, to the form a/b with a and b coprime. The following *Example* establishes that this form is unique, and, as we shall see, the key fact in the proof is that we can express 1 as $ma + nb$.

Example   Suppose that a, a', b, b' are positive integers satisfying

(i) $ab' = a'b$;   (ii) $\gcd(a, b) = \gcd(a', b') = 1$.

Then $a = a'$ and $b = b'$.

(Condition (i) could be written as $a/b = a'/b'$, but we prefer to use a form which does not assume anything about fractions.)

Solution   Since $\gcd(a, b) = 1$ there are integers m and n such that $ma + nb = 1$. Consequently

$$b' = (ma + nb)b' = mab' + nbb' = (ma' + nb')b,$$

and so $b \mid b'$. By a similar argument using the fact that $\gcd(a', b') =$

1, we deduce that $b'|b$. Hence either $b = b'$ or $b = -b'$, and since b and b' are both positive we must have $b = b'$. Now (i) yields $a = a'$ and the result is proved. $\square$

1 Find the gcd of 721 and 448 and express it in the form $721m + 448n$ with $m, n \in \mathbb{Z}$.

2 Show that if there are integers m and n such that $mu + nv = 1$, then $\gcd(u, v) = 1$.

3 Use Theorem 1.7 and Ex. 2 to prove that if $\gcd(a, b) = d$, then

$$\gcd\left(\frac{a}{d}, \frac{b}{d}\right) = 1.$$

4 Let a and b be positive integers and let $d = \gcd(a, b)$. Prove that there are integers x and y which satisfy the equation $ax + by = c$ if and only if $d|c$.

5 Find integers x and y satisfying

$$966x + 686y = 70.$$

1.8 Factorization into primes

A positive integer p is said to be a **prime** if $p \geqslant 2$ and the only positive integers which divide p are 1 and p itself. Thus an integer $m \geqslant 2$ is not a prime if and only if we can write $m = m_1 m_2$, where m_1 and m_2 are integers strictly between 1 and m.

We remark that according to the definition 1 is *not* a prime. The first few primes are

$$2, 3, 5, 7, 11, 13, 17, 19, 23, 29, 31, 37, 41, 43, 47.$$

The reader is almost certainly familiar with the idea that any positive integer can be expressed as a product of primes; for example,

$$825 = 3 \times 5 \times 5 \times 11.$$

The existence of such a *prime factorization* for any positive integer $n \geqslant 2$ is a consequence of the well-ordering axiom. For let B be the set of positive integers $n \geqslant 2$ which do *not* have a prime factorization; if B is not empty then, by the well-ordering axiom, it has a least member m. If m were a prime p we should have the trivial prime factorization $m = p$; thus m is not a prime and $m = m_1 m_2$ where $1 < m_1 < m$ and $1 < m_2 < m$. Since m is supposed to be the least integer ($\geqslant 2$) which has no prime factorization, both m_1 and m_2 do have prime factorizations. But then the equation $m = m_1 m_2$ yields a prime factorization of m, contradicting the assumption that m is a member of B. Hence B must be empty, and the assertion is proved.

Exercises 1.8

1   Find all the primes p in the range $100 \leqslant p \leqslant 120$.

2   Write down prime factorizations of 201, 1001, and 201000.

3   Show that if p and p' are primes, and $p | p'$, then $p = p'$.

4   Show that if $n \geqslant 2$ and n is not prime then there is a prime p such that $p | n$ and $p^2 \leqslant n$.

5   Use the result of Ex. 4 to show that if 467 were not prime then it would have a prime divisor $p \leqslant 19$. Deduce that 467 is prime.

The ease with which we establish the existence of prime factorizations conceals two important difficulties. First, the problem of finding the prime factors is by no means straightforward; and secondly, it is not obvious that there is a *unique* prime factorization for any given integer $n \geqslant 2$. The next result is a key step in the proof of uniqueness.

Theorem 1.8.1

If p is a prime and $x_1, x_2, \ldots, x_n$ are any integers such that

$$p | x_1 x_2 \ldots x_n$$

then $p | x_i$ for some x_i $(1 \leqslant i \leqslant n)$.

Proof

We use the principle of induction. The result is manifestly true when $n = 1$ (induction basis). For the induction hypothesis, suppose it is true when $n = k$.

Suppose that $p | x_1 x_2 \ldots x_k x_{k+1}$, and let $x = x_1 x_2 \ldots x_k$. If $p | x$ then, by the induction hypothesis, $p | x_i$ for some x_i in the range $1 \leqslant i \leqslant k$. If $p \nmid x$ then (since p has no divisors except 1 and itself) we have $\gcd(p, x) = 1$. By Theorem 1.7 there are integers r and s

such that $rp + sx = 1$. Hence we have

$$x_{k+1} = (rp + sx)x_{k+1} = (rx_{k+1})p + s(xx_{k+1}),$$

and since p divides both terms it follows that $p \mid x_{k+1}$. Thus in either case p divides one of the x_i $(1 \leqslant i \leqslant k+1)$, and by the principle of induction the result is true for all positive integers n. $\qquad\square$

A very common error is to assume that Theorem 1.8.1 remains true when the prime p is replaced by an arbitrary integer. But that is clearly absurd: for example

$$6 \mid 3 \times 8 \quad \text{but} \quad 6 \nmid 3 \quad \text{and} \quad 6 \nmid 8.$$

Examples like this help us to understand that Theorem 1.8.1 expresses a very significant property of primes. Indeed, we shall see that this property plays a crucial part in the next result, which is sometimes called the *Fundamental theorem of arithmetic*.

Theorem 1.8.2   The prime factorization of a positive integer $n \geqslant 2$ is unique, apart from the order of the prime factors.

Proof   By the well-ordering axiom, if there is an integer for which the theorem is false, then there is a least such integer $n_0 \geqslant 2$. Suppose then that

$$n_0 = p_1 p_2 \ldots p_k \quad \text{and} \quad n_0 = p'_1 p'_2 \ldots p'_l,$$

where the p_i $(1 \leqslant i \leqslant k)$ are primes, not necessarily distinct, and the p'_j $(1 \leqslant j \leqslant l)$ are primes, not necessarily distinct. The first equation implies that $p_1 \mid n_0$, and the second equation implies that $p_1 \mid p'_1 p'_2 \ldots p'_l$. Hence, by Theorem 1.8.1, p_1 divides p'_j for some j $(1 \leqslant j \leqslant l)$. By re-ordering the second factorization we may assume that $p_1 \mid p'_1$, and since p_1 and p'_1 are primes, it follows that $p_1 = p'_1$ (Ex. 1.8.3). So, by Axiom **I7**, we may cancel the equal factors p_1 and p'_1, and obtain

$$p_2 p_3 \ldots p_k = p'_2 p'_3 \ldots p'_l = n_1, \text{ say.}$$

But the factorizations of n_0 were alleged to be different, and we have cancelled only the equal factors p_1 and p'_1, so n_1 has two different prime factorizations. This contradicts the definition of n_0 as the least such integer. Hence the theorem is true for all $n \geqslant 2$. $\qquad\square$

In practice we often collect equal primes in the factorization of n and write

$$n = p_1^{e_1} p_2^{e_2} \ldots p_r^{e_r},$$

where $p_1, p_2, \ldots, p_r$ are distinct primes and $e_1, e_2, \ldots, e_r$ are positive integers. For example, $7000 = 2^3 \times 5^3 \times 7$.

Example   Show that if m and n are integers such that $m \geqslant 2$ and $n \geqslant 2$, then $m^2 \neq 2n^2$.

Solution   Suppose that the prime factorization of n contains the prime 2 raised to the power x (where x is zero if 2 is not a prime factor of n). Then $n = 2^x h$, where h is a product of primes greater than 2, so

$$2n^2 = 2(2^x h)^2 = 2^{2x+1} h^2.$$

Thus 2 is raised to an *odd* power in the prime factorization of $2n^2$.

On the other hand, if $m = 2^y g$, where g is a product of primes greater than 2, then

$$m^2 = (2^y g)^2 = 2^{2y} g^2,$$

so 2 is raised to an *even* power (possibly zero) in the prime factorization of m^2. It follows that if $m^2 = 2n^2$ we should have two different prime factorizations of the same integer, contrary to Theorem 1.8.2. Hence $m^2 \neq 2n^2$. $\qquad\square$

It is clear that the conclusion of the *Example* holds if we allow either or both of m and n to be 1. So we may express the result by saying that there are no positive integers m and n such that

$$\left(\frac{m}{n}\right)^2 = 2$$

or equivalently, by saying that the square root of 2 cannot be expressed as a fraction m/n.

**Exercises
1.8
(continued)**

6   Let m and n be positive integers whose prime factorizations are

$$m = p_1^{e_1} p_2^{e_2} \ldots p_r^{e_r}, \qquad n = p_1^{f_1} p_2^{f_2} \ldots p_r^{f_r}.$$

Show that the gcd of m and n is $d = p_1^{k_1} p_2^{k_2} \ldots p_r^{k_r}$ where, for each i in the range $1 \leqslant i \leqslant r$, k_i is the smaller of e_i and f_i.

7   Show that if m and n are positive integers, such that $m \geqslant 2$, $n \geqslant 2$, and $m^2 = kn^2$, then k is the square of an integer.

8   Use the identity

$$2^{rs} - 1 = (2^r - 1)(2^{(s-1)r} + 2^{(s-2)r} + \cdots + 2^r + 1)$$

to show that if $2^n - 1$ is prime then n is prime.

9 Find the least value of n for which the converse of Ex. 8 is false: that is, n is prime but $2^n - 1$ is not.

1.9 Miscellaneous Exercises

1 Use the principle of induction to show that $2^n > n + 1$ for all integers $n \geqslant 2$.

2 Show that

$$1^4 + 2^4 + \cdots + n^4 = \tfrac{1}{30} n(n + 1)(2n + 1)(3n^2 + 3n + 1).$$

3 Show that $4^{2n} - 1$ is divisible by 15 for all integers $n \geqslant 1$.

4 Find the gcd of 1320 and 714, and express the result in the form $1320x + 714y \ (x, y \in \mathbb{Z})$.

5 Show that 725 and 441 are coprime and hence find integers x and y such that $725x + 441y = 1$.

6 Find a solution in integers to the equation

$$325x + 26y = 91.$$

7 The integer f_n is defined recursively by the equations

$$f_1 = 1, \qquad f_2 = 1, \qquad f_{n+1} = f_n + f_{n-1} \quad (n \geqslant 2).$$

Prove that $\gcd(f_{n+1}, f_n) = 1$ for all $n \geqslant 1$.

8 Let a and b be any two positive integers. Define the **least common multiple** of a and b to be the integer

$$l = \frac{ab}{\gcd(a, b)}.$$

Show that

(i) $a \mid l$ and $b \mid l$;
(ii) if m is a positive integer such that $a \mid m$ and $b \mid m$, then $l \mid m$.

9 Establish the following properties of the gcd.

(i) $\gcd(ma, mb) = m \gcd(a, b)$.
(ii) If $\gcd(a, x) = d$, and $\gcd(b, x) = 1$, then $\gcd(ab, x) = d$.

10 You have an unlimited supply of water, a drain, a large container, and two jugs which contain 7 litres and 9 litres respectively. How would you arrange to put one litre of water in the container? Explain the relationship between your method and Theorem 1.7.

11 By following the outline of the definition of gcd (a, b), frame a definition of the gcd of n integers $a_1, a_2, \ldots, a_n$. Prove that if $d =$
gcd $(a_1, a_2, \ldots, a_n)$ then there are integers $x_1, x_2, \ldots, x_n$ such that

$$d = x_1 a_1 + x_2 a_2 + \cdots + x_n a_n.$$

12 Let n be a positive integer with the following properties:

(i) n is square-free (that is, the prime factorization of n has no repeated
factors);
(ii) for all primes p, $p \mid n$ if and only if $p - 1 \mid n$.

Find the value of n.

13 The integer u_n is defined by the equations

$$u_1 = 2, \qquad u_{n+1} = u_n^2 - u_n + 1 \quad (n \geq 1).$$

Find the least value of n for which u_n is not prime and find the factors of
this u_n. Is u_6 prime?

14 Show that the integers defined in Ex. 13 satisfy

$$u_{n+1} = 1 + u_1 u_2 \ldots u_n.$$

Deduce that u_{n+1} has a prime factor which is different from any prime
factor of any one of $u_1, u_2, \ldots, u_n$. Hence show that the set of primes has no
greatest member.

15 Is 65537 a prime?

16 Prove that if n is a positive integer, none of the n consecutive integers
starting with $(n+1)! + 2$ is a prime.

17 Prove that there are no integers x, y, z, t for which

$$x^2 + y^2 - 3z^2 - 3t^2 = 0.$$

18 Prove that if gcd $(x, y) = 1$, and $xy = z^2$ for some integer z, then $x = n^2$
and $y = m^2$ for some integers m and n.

19 Show that if gcd $(a, b) = 1$ then gcd $(a + b, a - b)$ is either 1 or 2.

20 Show that it is possible to balance any integral weight from 1 to $2^n - 1$
grams if we are given weights of $1, 2, 4, \ldots, 2^{n-1}$ grams. Show that no other
set of n weights will do this.

2
Functions and counting

2.1 Functions

Suppose that X and Y are sets. We say that we have a **function** f **from** X **to** Y if for each x in X we can specify a unique element in Y, which we denote by $f(x)$. This situation is illustrated in Fig. 2.1; and from this picture we derive the standard notation $f: X \rightarrow Y$ for a function f from X to Y.

It is helpful to think of f as a rule which assigns to each object x in X a unique object $f(x)$ in Y. The object $f(x)$ is usually called the **value** of f at x. The important points are that $f(x)$ is defined for every x in X, and that there is just one such object for each x.

The most common functions in elementary mathematics are those for which X and Y are the sets $\mathbb{N}$ or $\mathbb{Z}$ or some other set of numbers. In this case the simplest method of specifying a function is by means of a formula. For example, the rule

$$f(n) = 3n + 4 \quad (n \in \mathbb{N})$$

defines the function f from $\mathbb{N}$ to $\mathbb{N}$ whose value at n is $3n + 4$. Some functions may require a split definition, such as the function g from $\mathbb{Z}$ to $\mathbb{Z}$ given by the rule

$$g(x) = \begin{cases} x & \text{if} \quad x \geq 0, \\ -x & \text{if} \quad x < 0. \end{cases}$$

This function assigns to each integer x its *absolute value*, usually written $|x|$. For instance, $|5| = |-5| = 5$.

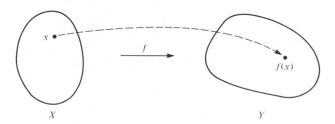

Fig. 2.1
A function $f: X \rightarrow Y$.

When $X = \mathbb{N}$ another way of specifying a function is the method of recursive definition outlined in Section 1.3. In that section we spoke (rather vaguely) of an expression u_n, defined for each n in $\mathbb{N}$. It would be more precise to say that u_n is just an alternative notation for $u(n)$, where u is a function from $\mathbb{N}$ to an appropriate set Y. For example, the equations

$$u(1) = 1, \qquad u(2) = 2, \qquad u(n) = u(n-1) + u(n-2) \quad (n \geqslant 3)$$

provide a recursive definition of a function u from $\mathbb{N}$ to $\mathbb{N}$ itself. We often refer to the list of values of such a function as a *sequence*; in this case the sequence is

$$1, 2, 3, 5, 8, 13, 21, \ldots$$

where the three dots at the end indicate that the list continues indefinitely. In general a sequence of members of a set Y is just another name for a function from $\mathbb{N}$ to Y. (Sometimes it is convenient to replace $\mathbb{N}$ by $\mathbb{N}_0$ or some other set of integers.) A sequence may be defined recursively, or by a formula, or in some other way, but in all cases we must have a uniquely defined element of Y for each relevant integer n.

A useful property of functions is that, in some circumstances, they can be combined as indicated in Fig. 2.2. Specifically, if we are given functions f from X to Y, and g from Y to Z, then there is a function from X to Z defined in the following way. For each x in X the value $f(x)$ is in Y, and the value of g at $f(x)$ is the element $g(f(x))$ of Z. Regarding this two-step operation as a single step, we have a function from X to Z which takes x to $g(f(x))$. It is called the **composite** of the functions $f \colon X \to Y$ and $g \colon Y \to Z$, written gf. Thus

$$(gf)(x) = g(f(x)).$$

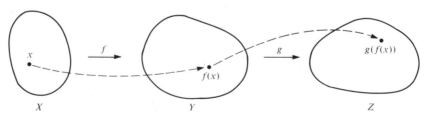

Fig. 2.2
The composite of $f \colon X \to Y$
and $g \colon Y \to Z$.

It is advisable to remember that (in this book) gf means 'first f, then g'. Although gf is a kind of product of g and f, it is misleading to think of it in that way. If X, Y, Z are different sets, then gf is defined as above, but fg has no meaning. Even if both gf and fg are defined, as when $X = Z$, there is no reason why they should be the same.

Exercises 2.1

1　The functions s and t from $\mathbb{Z}$ to $\mathbb{Z}$ are defined by

$$s(x) = x + 1, \qquad t(x) = 2x \quad (x \in \mathbb{Z}).$$

Show that $st \neq ts$.

2　Let $X = \{1, 2, 3, 4, 5\}$ and let $f: X \to X$ be the function defined by

$$f(1) = 2, \qquad f(2) = 2, \qquad f(3) = 4, \qquad f(4) = 4, \qquad f(5) = 4.$$

Show that $ff = f$. Find a function $g \neq f$ such that $gf = f$ and $fg = f$.

3　Let U denote the set of citizens of the state of Utopia. Which of the following statements correctly specify a function from U to U? (Any assumptions you make about the Utopian civilization should be stated explicitly.)

　(i)　$f(x)$ is the mother of x.　　(ii)　$g(x)$ is the daughter of x.
　　　　　　　　　(iii)　$h(x)$ is the wife of x.

4　Suppose that f, g, and h are functions such that the composite $h(gf)$ is defined. Show that $(hg)f$ is also defined, and $(hg)f = h(gf)$.

2.2 Surjections, injections, bijections

There are certain special kinds of functions which have special names. The diagrams in Fig. 2.3 illustrate the following definitions, and the reader is advised to sketch similar illustrations.

Fig. 2.3
A surjection, an injection, and a bijection.

Surjection　　　Injection　　　Bijection

Definition   The function f from X to Y is a **surjection** if every y in Y is a value $f(x)$ for *at least one* x in X. It is an **injection** if every y in Y is a value $f(x)$ for *at most one* x in X. It is a **bijection** if it is both a surjection and an injection, that is, if every y in Y is a value $f(x)$ for *exactly one* x in X.

Example   The following formulae define functions from $\mathbb{Z}$ to $\mathbb{Z}$. Which of them are surjections, which are injections, and which are bijections?

(i) $f(x) = x^2$;   (ii) $g(x) = 2x$;   (iii) $h(x) = x + 2$.

Solution   (i) Since $f(x) = x^2$ and x^2 is never negative, a negative integer such as -1 cannot be a value of f. Hence there is no integer x such that $f(x) = -1$ and f is not a surjection. Furthermore, there are some integers y for which there are two solutions to the equation $f(x) = y$; for example, taking $y = 4$ we have $f(2)$ and $f(-2)$ both equal to 4. Consequently f is not an injection.

(ii) Since $g(x) = 2x$ and $2x$ is even, an odd integer such as 3 cannot be a value of g. Hence g is not a surjection. On the other hand, g is an injection. In order to prove this, we suppose that there are two integers x and x' such that $g(x)$ and $g(x')$ take the same value y: that is, $y = 2x = 2x'$. Cancelling the factor 2 (according to Axiom **I7**) we get $x = x'$, which implies that there is at most one solution to the equation $g(x) = y$, and g is an injection.

(iii) If we are given an integer y, then taking $x = y - 2$ we get

$$h(x) = x + 2 = y.$$

So there is at least one integer x such that $h(x) = y$ and h is a surjection. If there were two such integers x and x' we should have

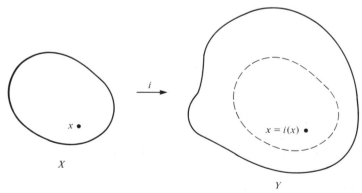

Fig. 2.4
The inclusion function
$i\colon X \to Y$.

$x + 2 = x' + 2$, which implies $x = x'$. Hence h is an injection, and indeed, a bijection. $\square$

It is worth remarking that the technique used above for proving that g and h are injections is the most convenient one in practice. In general, in order to prove that a function f is an injection we assume $f(x) = f(x')$ and deduce that $x = x'$.

One particular kind of injection has a special name. If X is a subset of Y then the **inclusion** function $i: X \rightarrow Y$ (Fig. 2.4), defined by $i(x) = x$ is clearly an injection. When $X = Y$ it is a bijection and in that case it is sometimes called the **identity** function on X.

The following theorem is very useful.

Theorem 2.2.1

If $f: X \rightarrow Y$ and $g: Y \rightarrow Z$ are injections, then so is the composite $gf: X \rightarrow Z$. If f and g are surjections so is gf, and if f and g are bijections so is gf.

Proof

Suppose $gf(x) = gf(x')$. Since $g(f(x)) = g(f(x'))$ and g is an injection we must have $f(x) = f(x')$, and since f is an injection $x = x'$. Hence gf is an injection.

Now if g is a surjection, then for any given z in $\mathbb{Z}$ we have $z = g(y)$ for some y in Y, and if f is a surjection there is some x in X for which $f(x) = y$. Thus $z = g(f(x)) = gf(x)$, and gf is a surjection.

The statement for bijections is a direct consequence of the two preceding ones. $\square$

The concept of a bijection is fundamental to the process of counting, as we shall see in the next section. It can also be formulated in a different way, as we shall now explain.

Definition

A function f from X to y has an **inverse** function g from Y to X if, for all x in X and y in Y,

$$(gf)(x) = x, \qquad (fg)(y) = y.$$

In other words, gf is the identity function on X and fg is the identity function on Y. (See Fig. 2.5.)

Roughly speaking, an inverse function g reverses the effect of f. The equations in the definition can be restated in the equivalent form

$$f(x) = y \quad \Leftrightarrow \quad g(y) = x,$$

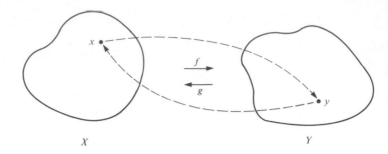

Fig. 2.5
Inverse functions.

which corresponds more closely to the intuitive notion of an inverse
function. For example, given the function f from $\mathbb{Z}$ to $\mathbb{Z}$ defined by
$f(x) = x + 3$, an inverse function g can be found by remarking that

$$x + 3 = y \quad \Leftrightarrow \quad y - 3 = x,$$

so that $g(y) = y - 3$.

Of course, not every function has an inverse. But if there is one,
then it is unique. For if $f\colon X \to Y$ is given, and g and g' both satisfy
the conditions for an inverse of f, then (in particular) $g'f$ is the
identity on X and fg is the identity on Y. So we have

$$g = (g'f)g = g'(fg) = g'.$$

Thus $g = g'$ and the inverse is unique.

This argument justifies our speaking of *the* inverse of f (if it
exists) and the use of the notation f^{-1} for the unique inverse
function. Moreover, the inverse of f^{-1} is f, so $(f^{-1})^{-1} = f$.

Theorem 2.2.2 A function has an inverse if and only if it is a bijection.

Proof Suppose that f is a bijection from X to Y. For each y in Y there is
precisely one x in X such that $f(x) = y$. The rule $g(y) = x$ defines a
function from Y to X which is an inverse of f.

Conversely, suppose f has an inverse f^{-1}. Given y in Y we know
that $f(f^{-1}(y)) = y$, so that putting $x = f^{-1}(y)$ gives $f(x) = y$. Thus f is
a surjection. In order to show that f is also an injection, suppose
$f(x) = f(x')$. Applying f^{-1} to both sides of this equation, we obtain
$x = x'$, as required. $\square$

Of course, it is an immediate consequence of this theorem that
the inverse of a bijection is also a bijection.

1   Which of the following functions from $\mathbb{Z}$ to $\mathbb{Z}$ are surjections, which of them are injections, and which of them are bijections?

$$\text{(i)} \quad f(x) = x^3, \qquad \text{(ii)} \quad g(x) = x - 3,$$
$$\text{(iii)} \quad h(x) = 3x + 1, \qquad \text{(iv)} \quad i(x) = x^2 - 1.$$

2   The function u from $\mathbb{N}$ to $\mathbb{N}$ is defined recursively by the rules

$$u(1) = 1, \qquad u(n+1) = \begin{cases} \frac{1}{2}u(n) & \text{if } u(n) \text{ is even;} \\ 5u(n) + 1 & \text{otherwise.} \end{cases}$$

Show that u is neither an injection nor a surjection.

3   Prove that if f and g are bijections, and gf is defined, then the inverse of gf is $f^{-1}g^{-1}$.

4   The function $f: X \to Y$ is said to have a **left inverse** $l: Y \to X$ if lf is the identity function on X. Show that

(i)   if f has a left inverse then it is a surjection;
(ii) if f is an injection then it has a left inverse.

5   Formulate and prove results about a *right* inverse of $f: X \to Y$ which correspond to those given in the previous exercise.

2.3   Counting

What do we mean when we say that a set has n members? One way to answer that question is to remember how we count simple sets. We say the words one, two, three, and so on, and point to the objects in turn. When each object has received a number we stop, and the last number uttered is the number of members of the set.

In order to translate this say-and-point technique into mathematical language we must first define, for each positive integer n, the set

$$\mathbb{N}_n = \{1, 2, 3, \ldots, n\}.$$

The say-and-point technique assigns to each member of $\mathbb{N}_n$ a member of the set X being counted; in other words, it determines a function f from $\mathbb{N}_n$ to X (Fig. 2.6). Furthermore it is clear that the function f is a *bijection* since, if we have counted correctly, every member of X receives just one number. So, if X is a set and n a

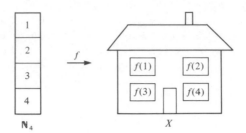

Fig. 2.6
Counting the windows.

positive integer, and if there is a bijection from $\mathbb{N}_n$ to X, then we shall say that X has n members.

It must be remarked at once that the definition does not explicitly exclude the possibility that a set may have both n members and m members, for $m \neq n$. Indeed, we have all had the experience of counting and re-counting some moderately large set of objects, such as the sheep in a field, and getting a different answer each time. The next theorem is the key step in the proof that this is due solely to practical ineptitude, and that there really is a correct answer. In other words, a set which has n members cannot also have m members, for $m \neq n$.

Theorem 2.3
If m and n are positive integers such that $m < n$, then there is no injection from $\mathbb{N}_n$ to $\mathbb{N}_m$.

Proof
Let S denote the set of positive integers n for which there is an injection from $\mathbb{N}_n$ to $\mathbb{N}_m$ for some $m < n$. If S is not empty it has a least member k, and since k is in S there is an injection i from $\mathbb{N}_k$ to $\mathbb{N}_l$ for some $l < k$. We cannot have $l = 1$, since any function from $\mathbb{N}_k$

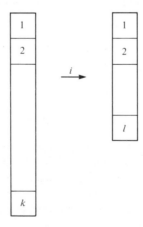

Fig. 2.7
The supposed injection
$i : \mathbb{N}_k \to \mathbb{N}_l$.

to $\mathbb{N}_1$ can take only the value 1, and cannot therefore be an injection defined on $\mathbb{N}_k$ for $k > 1$. Thus $l - 1$ is a positive integer, and the situation may be depicted as in Fig. 2.7.

If none of the values $i(1), i(2), \ldots, i(k-1)$ is equal to l, then restricting i to the subset $\mathbb{N}_{k-1}$ gives an injection from $\mathbb{N}_{k-1}$ to $\mathbb{N}_{l-1}$. On the other hand, if $i(b) = l$ for some b in the range $1 \le b \le k - 1$, then we must have $i(k) = c \ne l$, since i is an injection. In this case we can construct an injection i^* from $\mathbb{N}_{k-1}$ to $\mathbb{N}_{l-1}$ as illustrated in Fig. 2.8; that is

$$i^*(b) = c, \qquad i^*(r) = i(r) \quad (r \ne b).$$

In either case the existence of an injection from $\mathbb{N}_{k-1}$ to $\mathbb{N}_{l-1}$ contradicts the definition of k as the least member of S. Hence S must be empty, and the result is proved. $\qquad\square$

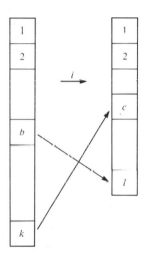 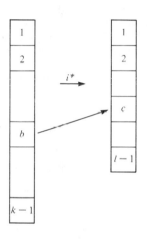

Fig. 2.8
Construction of i^* when
$i(b) = l$.

Suppose there is a set X which has n members, and also m members for some $m < n$. It follows that there are bijections

$$\beta : \mathbb{N}_n \to X, \qquad \gamma : \mathbb{N}_m \to X,$$

and, by the results established in the previous section, both γ^{-1} and $\gamma^{-1}\beta$ are also bijections. In particular, $\gamma^{-1}\beta$ is an injection from $\mathbb{N}_n$ to $\mathbb{N}_m$, contrary to Theorem 2.3. Thus the statement 'X has n members' can hold for at most one positive integer n.

When X has n members we write $|X| = n$, and say that the **cardinality** (or **size**) of X is n. For the empty set $\varnothing$ we make a special but eminently reasonable definition,

$$|\varnothing| = 0.$$

When $|X| = n$ we often find it convenient to put

$$X = \{x_1, x_2, \ldots, x_n\},$$

which is really just another way of saying that there is a bijection β from $\mathbb{N}_n \to X$, such that $\beta(i) = x_i$ $(1 \leqslant i \leqslant n)$.

Finally, a warning: many sets do not have a cardinality according to the definition made above. The set $\mathbb{N}$ itself is an example. We shall return to this question in Section 2.5.

Exercises 2.3

1 In each of the following cases find the appropriate value of n and write down a formula for a bijection $f : \mathbb{N}_n \to X$.

(i) $X = \{2, 4, 6, 8, 10\}$.
(ii) $X = \{-3, -8, -13, -18, -23, -28\}$.
(iii) $X = \{10, 17, 26, 37, 50, 65, 82, 101\}$.
(iv) $X = \{k \in \mathbb{N} \mid \text{the } k\text{th day of this month is a Monday}\}$.

2 Discuss any difficulties which might arise in using say-and-point method to count the following sets.

(i) The set of sheep in a field.
(ii) The set of your own ancestors.
(iii) The set of even positive integers.

3 Show that if $|X| = n$ and there is a bijection from X to Y then $|Y| = n$.

2.4 The pigeonhole principle

Theorem 2.3, which we proved in order to justify our definition of cardinality, can also be used in more practical ways. Suppose we have a set X, whose members we shall refer to as 'objects', and a set Y whose members are 'boxes', or 'pigeonholes'. A *distribution* of the objects into the boxes is simply a function f from X to Y: if the object x goes into box y, then $f(x) = y$. For example, the distribution in Fig. 2.9 corresponds to the function shown on the right.

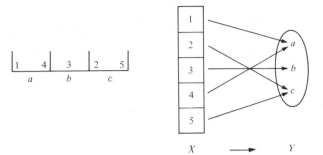

Fig. 2.9
A distribution and the
corresponding function.

In this model the function is a surjection if each box receives at least one object, and an injection if each box receives at most one object. Now it is clear that if there are more objects than boxes then some box must receive at least two objects; in other words, the function cannot be an injection. Formally, this is a consequence of Theorem 2.3. For suppose $|X| = m$ and $|Y| = n$, where $m > n$; then an injection from X to Y would yield an injection from $\mathbb{N}_m$ to $\mathbb{N}_n$, which is impossible according to the theorem. This observation is usually known as the pigeonhole principle:

> *if m objects are distributed into n boxes, and m > n,*
> *then at least one box receives at least two objects.*

There are many obvious applications of the pigeonhole principle, such as the following.

(i) In any set of 13 or more people, there are at least two whose birthdays fall in the same month.

(ii) In any set of 51 or more people born in the USA, there are at least two who were born in the same state.

(iii) In any set of one million people, there are at last two who have the same number of hairs on their heads.

There are also more subtle applications.

Example Show that if X is any set of people there are two members of X who have the same number of friends in X. (It is assumed that if x is a friend of x' then x' is a friend of x.)

Solution Consider the function f defined on X by the rule that, for each member x of X,

$$f(x) = \text{number of friends of } x \text{ in } X.$$

If $|X| = m$, the possible values of $f(x)$ are $0, 1, \ldots, m - 1$, since the friends of x can be any of the members of X except x. So f is a function from X to the set $Y = \{0, 1, \ldots, m-1\}$.

At this point we cannot immediately apply the pigeonhole principle, since Y has the same cardinality m as X. However, if there is a person x^* who has $m-1$ friends then everyone else is a friend of x^*, and consequently no one has no friends. In other words the numbers $m-1$ and 0 cannot both be values of f. Hence f is a function from a set of size m to a set of size $m-1$ (or less), and the pigeonhole principle tells us that there are at least two people x_1 and x_2 such that $f(x_1) = f(x_2)$, as required. □

Exercises 2.4

1   A blindfold man has a heap of 10 grey socks and 10 brown socks in a drawer. How many must be select in order to guarantee that, among them, there is a matching pair? How many must he select in order to guarantee a grey pair?

2   Suppose five points are chosen inside an equilateral traingle with side-length 1. Show that there is at least one pair of points whose distance apart is at most $\frac{1}{2}$. [Hint: divide the triangle into four suitable boxes.]

3   Show that in any set of 12 integers there are two whose difference is divisible by 11.

4   Let X be a subset of $\{1, 2, \ldots, 2n\}$, and let Y be the set of odd integers $\{1, 3, \ldots, 2n-1\}$. Define a function $f: X \to Y$ by the rule

$$f(x) = \text{greatest odd integer which divides } x.$$

Show that if $|X| \geq n+1$ then f is not an injection, and deduce that in this case X contains distinct integers x_1 and x_2 such that $x_1 | x_2$.

5   Show that it is possible to find a subset X of $\{1, 2, \ldots, 2n\}$ which has n members and is such that no member of X is a divisor of another member.

2.5   Finite or infinite?

We have been careful to avoid the use of the words 'finite' and 'infinite' so far, but now we are in a position to give a formal definition.

Definition   A set X is **finite** if it is empty or if $|X| = n$ for some positive integer n. A set which is not finite is said to be **infinite**.

According to the definition, X is infinite if it is not the empty set and there is no bijection from $\mathbb{N}_n$ to X for any positive integer n. Of course, the most obvious candidate for an infinite set is $\mathbb{N}$ itself, and we shall begin by convincing ourselves that $\mathbb{N}$ is indeed infinite on the basis of our definition.

Certainly $\mathbb{N}$ is not empty, since it contains 1, for instance. Suppose there were a bijection β from $\mathbb{N}_n$ to $\mathbb{N}$ for some positive integer n. Since $\mathbb{N}_{n+1}$ is a subset of $\mathbb{N}$, the inclusion $i: \mathbb{N}_{n+1} \to \mathbb{N}$ is an injection and the composite

$$\mathbb{N}_{n+1} \xrightarrow{i} \mathbb{N} \xrightarrow{\beta^{-1}} \mathbb{N}_n$$

is an injection also. But this contradicts Theorem 2.3, so $\mathbb{N}$ is infinite, as claimed.

A more exciting example of an infinite set is given in the following *Example*. The proof (attributed to Euclid) is considered to be one of the most elegant pieces of pure mathematical reasoning.

Example   The set P of primes is infinite.

Solution P is not empty, since 2 is a prime. Suppose P were finite, so that the primes, being in bijective correspondence with a set $\{1, 2, \ldots, n\}$, could be listed as $p_1, p_2, \ldots, p_n$. We shall show that such a list cannot possibly contain all the primes.

Consider the positive integer

$$m = p_1 p_2 \ldots p_n + 1.$$

None of the primes $p_1, p_2, \ldots, p_n$ divides m, but on the other hand, we know from Section 1.8 that m has a unique factorization into primes. Hence this factorization must contain primes other than $p_1, p_2, \ldots, p_n$, and our proposed list is incomplete. $\qquad\square$

Most people have the vague idea that, given an infinite set, we can continue the say-and-point counting technique indefinitely, without ever coming to an end. The next theorem expresses this idea more precisely.

Theorem 2.5   The non-empty set X is infinite if and only if there is an injection from $\mathbb{N}$ to X.

Proof   If X is infinite we can define a function f from $\mathbb{N}$ to X recursively, as follows. Take $f(1)$ to be any member of X; and if $f(1), \ldots, f(k)$ have been defined, take $f(k+1)$ to be any member of X except $f(1), \ldots, f(k)$. This means that no two values of f are equal, so f is an injection. Furthermore, the definition of $f(k+1)$ is always possible, since if there were no values available for $f(k+1)$ we should have $X = \{f(1), \ldots, f(k)\}$ and f would be a bijection from $\mathbb{N}_k$ to X, contrary to the hypothesis that X is infinite.

Conversely, suppose there is an injection $f: \mathbb{N} \to X$. If X were finite we should have a bijection $\beta: \mathbb{N}_n \to X$ for some positive integer n, and consequently a chain of injections

$$\mathbb{N}_{n+1} \overset{i}{\to} \mathbb{N} \overset{f}{\to} X \overset{\beta^{-1}}{\longrightarrow} \mathbb{N}_n$$

where i is the inclusion function. Now the composite of these injections is an injection from $\mathbb{N}_{n+1}$ to $\mathbb{N}_n$, contradicting Theorem 2.3 (yet again). Hence X must be infinite. ☐

According to Theorem 2.5, if we are given an infinite set X we can always attempt to 'count' it by constructing an injection f from $\mathbb{N}$ to X. When this is done, there are two possible outcomes. In some cases we shall be able to construct f in such a way that any given member of X eventually receives a number. If so, f is a surjection as well as an injection (and consequently a bijection), and in this case we say that X is **countable**. On the other hand, it may be impossible to construct a bijection from $\mathbb{N}$ to X, and in that case we say that X is **uncountable**.

Let us briefly reiterate the distinctions between the terms finite and infinite, countable and uncountable. If there is a counting process which terminates, the set is *finite*, or as we might say, *counted*. If there is a counting process which, although it does not terminate, will eventually reach any given member, then the set is *infinite*, but *countable*. If no counting process will eventually reach every member, then the set is *infinite* and *uncountable*. Roughly speaking, discrete mathematics is concerned with finite or countable sets, whereas calculus and analysis deal with uncountable sets such as the set $\mathbb{R}$ of real numbers.

We end this chapter with a word of warning. The properties of finite sets are very familiar to us, and for that reason our intuition is a fairly reliable guide for dealing with them. For example, we accept without question the assertion that if A is subset of a finite set B then A is finite and $|A| \leqslant |B|$. (In fact, it is quite tricky to prove this, but it can be done on the basis of our axioms and definitions using,

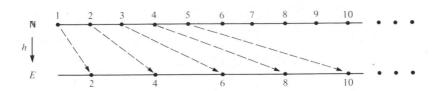

Fig. 2.10
The function $h: \mathbb{N} \to E$ is
a bijection!

once again, the ubiquitous Theorem 2.3.) On the other hand, intuition can be a very poor guide in our dealings with infinite sets. This is because the definitions which mathematicians have found to be logically consistent do not always correspond to our everyday experience of sets which are, necessarily, finite.

One example of the strange behaviour of infinite sets will suffice. Let E denote the subset of $\mathbb{N}$ consisting of the even integers, $E = \{2, 4, 6, \ldots\}$. It is obvious that the function h from $\mathbb{N}$ to E defined by the rule $h(n) = 2n$ is a *bijection* (Fig. 2.10), and that E is a *proper* subset of $\mathbb{N}$, that is, a subset which is not the whole of $\mathbb{N}$. Thus an infinite set can have a proper subset which is in bijective correspondence with the whole set.

Exercises 2.5

1   By constructing an injection from $\mathbb{N}$ to X show that each of the following sets X is infinite:

    (i) $\mathbb{Z}$,   (ii) $\{x \in \mathbb{Z} \mid x < 0\}$,   (iii) $\{n \in \mathbb{N} \mid n \geq 10^6\}$.

2   Prove that the function $f: \mathbb{N} \to \mathbb{Z}$ defined by

$$f(n) = \begin{cases} n/2 & \text{if } n \text{ is even,} \\ -(n-1)/2 & \text{if } n \text{ is odd,} \end{cases}$$

is a bijection. Use this result to exhibit another example of the 'strange behaviour' referred to in the final paragraph above.

3   Any prime except 2 and 3 is of the form $6m + 1$ or $6m + 5$ for some integer m. Use Euclid's method, as given in the *Example*, to show that the set of primes of the form $6m + 5$ is infinite. [Hint: replace the $+1$ in Euclid's construction by -1.]

4   Use Theorem 2.5 to prove that if X is a subset of Y, and X is infinite, then Y is infinite.

5   Let X be a non-empty subset of $\mathbb{Z}$ which has no least member. Show that we can choose a sequence $x_1, x_2, \ldots$ of members of X such that $x_n < x_{n-1}$ $(n \in \mathbb{N})$. Deduce that X is infinite.

6   According to Ex. 5 a non-empty *finite* subset S of $\mathbb{Z}$ must have a least member: it is often referred to as the **minimum** of S and written min S. Show that if S and T are non-empty finite subsets of $\mathbb{Z}$ then

$$\min (S \cup T) \leqslant \min S,$$
$$\min (S \cap T) \geqslant \min S.$$

Formulate a definition of the **maximum** of S (under suitable conditions) and discuss its properties.

2.6   Miscellaneous Exercises

1   Which of the following functions from $\mathbb{Z}$ to $\mathbb{Z}$ are surjections, which of them are injections, and which of them are bijections?

   (i) $f(x) = 1 + x^2$,   (ii) $g(x) = 1 + x^3$,   (iii) $h(x) = 1 + x^2 + x^3$.

2   Let X be a set with $|X| = 3$. How many different bijections f from X to X are there, and how many of them satisfy $f = f^{-1}$?

3   Show that if X is a finite set and the function $g: X \to X$ is an injection, then g is a bijection.

4   Show that if X is a finite set and $f: X \to X$ is a surjection then f is a bijection.

5   Let X be a finite set and $g: X \to X$ a function such that $g^2(x) = x$ for all $x \in X$. Show that g is a bijection.

6   Let us refer to the following subsets of $\mathbb{Z}$ as *blocks*:

   $\{1, 2, 4\}, \{2, 3, 5\}, \{3, 4, 6\}, \{4, 5, 7\}, \{1, 5, 6\}, \{2, 6, 7\}, \{1, 3, 7\}.$

Define
$$f(x, y) = \begin{cases} z & \text{if } x \neq y \text{ and } \{x, y, z\} \text{ is a block,} \\ x & \text{if } x = y. \end{cases}$$

Is f a function? Is it an injection?

7   Show that if any five points are chosen within a square of side-length 2, then there are two of them whose distance apart is at most $\sqrt{2}$.

8 Prove that if any ten points are chosen within an equilateral triangle of side-length 1, then there are two of them whose distance apart is at most $\frac{1}{3}$.

9 How many points must be chosen inside a square of side-length 2 in order to ensure that at least one pair are not more than distance $\sqrt{2}/n$ apart?

10 Show that in any set of 172 integers there must be a pair whose difference is divisible by 171. Is the result true if the word 'difference' is replaced by 'sum'?

11 Show that the set of positive integers whose digits (in the usual base 10 representation) are all different is finite. How many of them are there?

12 A *lattice point* in three-dimensional space is a point whose coordinates are integers. Show that if we are given nine lattice points then there is at least one pair for which the mid-point of the line segment joining them is also a lattice point.

13 A golfer has d days to prepare for a tournament and must practice by playing at least one round each day. In order to avoid staleness he should not play more than m rounds altogether. Show that if r satisfies $1 \leqslant r \leqslant 2d - m - 1$ then there is a sequence of consecutive days during which he plays exactly r rounds.

14 Let $x_1, x_2, \ldots, x_r$ be a sequence of distinct integers. For each i ($1 \leqslant i \leqslant r$) let m_i denote the length of the longest increasing subsequence starting at x_i and let n_i denote the length of the longest decreasing subsequence beginning at x_i. Show that the function f which assigns to i the pair (m_i, n_i) is an injection.
  Deduce that a sequence of $mn + 1$ distinct integers must contain either an increasing subsequence of length m or a decreasing subsequence of length n.

15 Prove that any subset of a countable set is either finite or countable.

16 Show that the union of any two countable sets is countable.

17 Show that the union of a countable collection of countable sets is countable.

18 Let X be any set and let Y be the set of all subsets of X. Show that there is no bijection from X to Y.

19 Show that the set of all subsets of $\mathbb{N}$ is an uncountable set.

20 Prove that if A is a subset of a finite set B then A is finite and $|A| \leqslant |B|$.

3
Principles of counting

3.1 The addition principle

A major theme of this course is the development of effective techniques for counting a finite set X. When X arises in a complex problem we may require sophisticated counting methods far removed from the say-and-point technique of constructing a bijection from $\mathbb{N}_n$ to X. In this chapter we shall begin to develop such methods.

Our first rule is so simple that it has been used in practice since the dawn of civilization. Only in recent times, and in the context of a strict mathematical development of the subject, has it been given a formal status.

Theorem 3.1

If A and B are non-empty finite sets, and A and B are disjoint (that is $A \cap B = \varnothing$, the empty set), then

$$|A \cup B| = |A| + |B|.$$

Proof

Since A and B are non-empty and finite, we can list A and B in the standard way:

$$A = \{a_1, a_2, \ldots, a_r\}, \qquad B = \{b_1, b_2, \ldots, b_s\}.$$

Since A and B are disjoint, $A \cup B$ can be listed in a similar way:

$$A \cup B = \{c_1, c_2, \ldots, c_r, c_{r+1}, \ldots, c_{r+s}\}$$

where

$$c_i = a_i (1 \leqslant i \leqslant r) \quad \text{and} \quad c_{r+i} = b_i \ (1 \leqslant i \leqslant s).$$

Hence $|A \cup B| = r + s = |A| + |B|$, as claimed. $\qquad \square$

It is clear that the rule is still valid if A, or B, or both A and B, are empty. Furthermore, the rule can be extended to the union of any number of disjoint sets $A_1, A_2, \ldots, A_n$ in the obvious way, that is:

$$|A_1 \cup A_2 \cup \cdots \cup A_n| = |A_1| + |A_2| + \cdots + |A_n|.$$

The proof is an easy exercise using the principle of induction (Ex. 3.1.3).

A simple application of this rule yields a slightly more general form of the pigeonhole principle than the one given in Section 2.4. Suppose that a number of objects are distributed into n boxes, and A_i denotes the set of objects which are in box i $(1 \leq i \leq n)$. Since the sets A_i are disjoint the total number of objects is $|A_1| + |A_2| + \cdots + |A_n|$, and if no box contains more than r objects this number is at most

$$r + r + \cdots + r = nr.$$

Putting this argument in reverse, we have the generalized pigeonhole principle:

> *if m objects are distributed into n boxes and m > nr,*
> *then at least one box contains at least r + 1 objects.*

Example Show that in any set of six people there are either three mutual friends or three mutual strangers.

Solution Let α be any one of the people, and distribute the other five people into two 'boxes', box 1 containing the people who are friends of α, and box 2 containing those who are strangers to α. Since $5 > 2 \times 2$, one of these boxes contains at least three people.

Suppose box 1 contains β, γ, and δ (and possibly other people). If any two of $\{\beta, \gamma, \delta\}$ are friends, say β and γ, then $\{\alpha, \beta, \gamma\}$ is a set of three mutual friends. On the other hand, if no pair of $\{\beta, \gamma, \delta\}$ are friends, then $\{\beta, \gamma, \delta\}$ is a set of three mutual strangers.

If it happens that box 2 contains three or more people, a parallel argument with friends and strangers interchanged leads to the same conclusions. □

Exercises 3.1

1 The rules for the University of Folornia five-a-side soccer competition specify that the members of each team must have birthdays in the same month. How many mathematics students are needed in order to guarantee that they can raise a team?

2 What is wrong with the following argument? Since one-half of the numbers n in the range $1 \leq n \leq 60$ are multiples of 2, 30 of them cannot be primes. Since one-third of the numbers are multiples of 3, 20 of them cannot be primes. Hence at most 10 of them are primes.

3   Write out a proof (using induction on n) of the fact that if $A_1, A_2, \ldots, A_n$ are disjoint finite sets, then

$$|A_1 \cup A_2 \cup \cdots \cup A_n| = |A_1| + |A_2| + \cdots + |A_n|.$$

4 Show that in any set of 10 people there are either four mutual friends or three mutual strangers.

3.2   Counting sets of pairs

Quite often we have to count things which can be described more naturally as pairs of objects, rather than as single objects. Suppose, for example, that the Mathematics Department at the University of Folornia is calculating its teaching load for the current year. In order to do this, Professor McBrain makes a table like Table 3.2.1.

Table 3.2.1

	Calculus	Discrete Math.	$\cdots$	Algebra
Angus	√	√		
Benjamin		√		√
Clare	√	√		
. . .				
Zoot	√	√		√

In the table, each row corresponds to a student, each column corresponds to a course, and if student x is taking course y then a mark is entered in the corresponding position (x, y) in the table. The total number of marks is the teaching load for the department. In other words, the problem is to count the set S of pairs (x, y) such that student x is taking course y. In general, given any sets X and Y we may define the **product set** $X \times Y$ to be the set of *all* ordered pairs (x, y), so that the general problem is to count a subset S of $X \times Y$.

It is clear from Professor McBrain's table that there are two ways of doing the calculation. We may count the number of courses taken by each student and add the results, or we may count the number of students taking each course and add those results. Naturally, we expect to get the same answer by either method.

We can make these ideas precise in the following way. We suppose that a subset S of $X \times Y$ (where X and Y are finite sets) is specified by the marks in a general form of McBrain's table, as in Table 3.2.2.

Table 3.2.2

·	·	·	·	y	$\cdots$	Row total
·		$\checkmark$	$\checkmark$		$\checkmark$	·
·			$\checkmark$	$\checkmark$		·
x	$\checkmark$	$\checkmark$		$\checkmark$	$\checkmark$	$r_x(S)$
·		$\checkmark$	$\checkmark$	$\checkmark$		·
·					$\checkmark$	·
·	$\checkmark$				$\checkmark$	·
Column total	·	·	·	$c_y(S)$	$\cdots$	$\lvert S \rvert$

The first method of counting is to count the marks in row x and find the row total $r_x(S)$, for each x in X. The grand total is then obtained by adding all the row totals, that is

$$\lvert S \rvert = \sum_{x \in X} r_x(S).$$

The second method is to count the marks in column y and find the column total $c_y(S)$, for each y in Y. In this case the grand total is obtained by adding all the column totals:

$$\lvert S \rvert = \sum_{y \in Y} c_y(S).$$

The fact that we have two different expressions for $\lvert S \rvert$ is often used in practice to check the arithmetic. The same fact can also be very helpful in theory, because we can sometimes derive quite unexpected results by equating the two expressions. But before we turn to applications, we shall formulate the principle and some of its immediate consequences as a theorem.

Theorem 3.2   Let X and Y be finite non-empty sets, and let S be a subset of $X \times Y$. Then the following results hold.

(i) The cardinality of S is given by

$$|S| = \sum_{x \in X} r_x(S) = \sum_{y \in Y} c_y(S),$$

where $r_x(S)$ and $c_y(S)$ are the row and column totals as described above.

(ii) If $r_x(S)$ is a constant r, independent of x, and $c_y(S)$ is a constant c, independent of y, then

$$r\,|X| = c\,|Y|.$$

(iii) (The multiplication principle) The cardinality of $X \times Y$ is given by

$$|X \times Y| = |X| \times |Y|.$$

Proof

(i) The 'set of marks in row x' can be formally defined as the set of pairs in S whose first component is x, so $r_x(S)$ is the cardinality of this set. Since these sets are disjoint, the fact that $|S|$ is equal to the sum of the numbers $r_x(S)$ follows from the addition principle. The result for the column totals is proved in the same way.

(ii) If $r_x(S) = r$ for all $x \in S$, then there are $|X|$ terms in the first expression for $|S|$ and each of them is equal to r. Hence

$$|S| = r\,|X|.$$

Similarly, $|S| = c\,|Y|$, and the result follows.

(iii) In the special case when $S = X \times Y$ the row total is $r_x(S) = |Y|$ for each x in X. Hence, by part (ii) $|X \times Y| = |X| \times |Y|$. $\square$

Example

Suppose that Professor McBrain has decreed that, for administrative convenience, every student must take exactly four of the seven courses available. The lecturers report that the numbers attending the courses are 52, 30, 30, 20, 25, 12, 18. What conclusion can be drawn?

Solution

Let n be the total number of students. Since each student is supposed to attend four courses, the total teaching load calculated by the 'row total' method is $4n$. On the other hand, the lecturers reports give the 'column totals', and we should have

$$4n = 52 + 30 + 30 + 20 + 25 + 12 + 18 = 187.$$

But this is impossible, because 187 is not divisible by 4. We must dismiss the suggestion that the lecturers have counted wrongly, so the only conclusion is that some students are skipping classes. Alas, it was ever so. $\square$

Exercises 3.2

1 In Dr Cynthia Angst's Calculus class, 32 of the students are boys. Each boy knows five of the girls in the class and each girl knows eight of the boys. How many girls are in the class?

2 Suppose we have a number of different subsets of $\mathbb{N}_8$, with the property that each one has four members, and each member of $\mathbb{N}_8$ belongs to exactly three of the subsets. How many subsets are there? Write down a collection of subsets which satisfies the conditions.

3 Is it possible to find a collection of subsets of $\mathbb{N}_8$ such that each one has three members and each member of $\mathbb{N}_8$ belongs to exactly five of the subsets?

4 If $X_1, X_2, \ldots, X_n$ are sets, the **product set** $X_1 \times X_2 \times \cdots \times X_n$ is defined to be the set of all ordered n-tuples $(x_1, x_2, \ldots, x_n)$, with $x_i \in X_i$ $(1 \leq i \leq n)$. Use the principle of induction to prove that

$$|X_1 \times X_2 \times \cdots \times X_n| = |X_1| \times |X_2| \times \cdots \times |X_n|.$$

5 In a simple language there are the usual 26 letters and all words have four letters. Any arrangement of the letters, including repetitions, is allowed. How many words are there? How many of them do not contain the letter b?

3.3 Euler's function

In this section we shall prove an important and useful theorem using only the most basic counting principles.

The theorem is concerned with the divisibility properties of integers. Recall that two integers x and y are *coprime* if $\gcd(x, y) = 1$; and for each $n \geq 1$ let $\phi(n)$ denote the number of integers x in the range $1 \leq x \leq n$ such that x and n are coprime. We can calculate the first few values of $\phi(n)$ by making a table (Table 3.3.1)

The function ϕ is called **Euler's function** after Leonhard Euler (1707–1783). When n is a prime, say $n = p$, each one of the integers $1, 2, \ldots, p-1$ is coprime to p, so we have

$$\phi(p) = p - 1 \quad (p \text{ prime}).$$

An explicit general formula for $\phi(n)$ will be obtained in Section 4.4.

Our present task is to prove a result concerning the sum of the

Table 3.3.1

n	Coprime to n	$\phi(n)$
1	1	1
2	1	1
3	1, 2	2
4	1, 3	2
5	1, 2, 3, 4	4
6	1, 5	2
7	1, 2, 3, 4, 5, 6	6
8	1, 3, 5, 7	4

values $\phi(d)$ taken over all divisors d of a given positive integer n. For example, when $n = 12$ the divisors d are 1, 2, 3, 4, 6, and 12, and we find that

$$\phi(1) + \phi(2) + \phi(3) + \phi(4) + \phi(6) + \phi(12)$$
$$= 1 + 1 + 2 + 2 + 2 + 4$$
$$= 12.$$

We shall show that the sum is always equal to the given integer n.

Theorem 3.3

For any positive integer n,

$$\sum_{d|n} \phi(d) = n.$$

Proof

Let S denote the set of pairs of integers (d, f) satisfying

$$d \mid n, \qquad 1 \leqslant f \leqslant d, \qquad \gcd(f, d) = 1.$$

Table 3.3.2

d	\multicolumn{12}{c}{f}	$\phi(d)$

d	1	2	3	4	5	6	7	8	9	10	11	12	$\phi(d)$
1	12												1
2	6												1
3	4	8											2
4	3		9										2
6	2				10								2
12	1				5		7				11		4
													12

Table 3.3.2 illustrates S when $n = 12$; the 'mark' indicating that (d, f) is in S is a number whose significance will appear in a moment. In general, the number of 'marks' in row d is just the number of f in the range $1 \leqslant f \leqslant d$ satisfying $\gcd(f, d) = 1$; that is, $\phi(d)$. Hence, counting S by the row method we obtain

$$|S| = \sum_{d \mid n} \phi(d).$$

In order to show that $|S| = n$ we shall construct a bijection β from S to $\mathbb{N}_n$. Given a pair (d, f) in S, we define

$$\beta(d, f) = fn/d.$$

In the table, $\beta(d, f)$ is the 'mark' in row d and column f. Since $d \mid n$ the value of β is an integer, and since $1 \leqslant f \leqslant d$ it lies in $\mathbb{N}_n$.

To show that β is an injection we remark that

$$\beta(d, f) = \beta(d', f') \quad \Rightarrow \quad fn/d = f'n/d' \quad \Rightarrow \quad fd' = f'd.$$

But f and d are coprime, as are f' and d', so we can conclude (as in Section 1.7 (*Example*)) that $d = d'$ and $f = f'$.

To show that β is a surjection, suppose we are given x in $\mathbb{N}_n$. Let g_x denote the gcd of x and n, and let

$$d_x = n/g_x, \qquad f_x = x/g_x.$$

Since g_x is a divisor of x and n both d_x and f_x are integers, and since it is the gcd, d_x and f_x are coprime. Now

$$\beta(d_x, f_x) = f_x n/d_x = x,$$

and so β is a surjection.

Thus β is a bijection and $|S| = n$, as required. □

Exercises 3.3

1 Find the values of $\phi(19)$, $\phi(20)$, $\phi(21)$.

2 Show that if x and n are coprime, so are $n - x$ and n. Deduce that $\phi(n)$ is even for all $n \geqslant 3$.

3 Show that, if p is a prime and m is a positive integer, then an integer x in the range $1 \leqslant x \leqslant p^m$ is *not* coprime to p^m if and only if it is a multiple of p. Deduce that $\phi(p^m) = p^m - p^{m-1}$.

4 Find a counter-example which disproves the conjecture that $\phi(a)\phi(b) = \phi(ab)$ for any positive integers a and b. Try to modify the conjecture so that it cannot be disproved.

3.4 Functions, words, and selections

We shall consider functions (not necessarily bijections) defined on a set of positive integers $\mathbb{N}_m$, and with values in a given set Y. The values of such a function f determine an m-tuple

$$(f(1), f(2), \ldots, f(m))$$

of elements of Y. According to the general definition of a product set (Ex. 3.2.4) this m-tuple belongs to the set $Y \times Y \times \cdots \times Y$ (m factors), which we shall denote by Y^m. Each element of Y^m is an m-tuple $(y_1, y_2, \ldots, y_m)$ and corresponds to a function f from $\mathbb{N}_m$ to Y defined by the equations

$$f(1) = y_1, \qquad f(2) = y_2, \ldots, f(m) = y_m.$$

These remarks lead us to the conclusion that a function from $\mathbb{N}_m$ to Y is logically the same thing as an element of the product set Y^m.

There is another way of looking at this relationship, which is very useful in practice. If we think of the members of Y as the letters of an alphabet, then the sequence $f(1), f(2), \ldots, f(m)$ can be regarded as the m letters of a word. For example, if Y is the simple alphabet $\{a, b, c, d\}$ the words *cab* and *dad* correspond to the functions f and g defined by

$$f(1) = c, \qquad f(2) = a, \qquad f(3) = b,$$
$$g(1) = d, \qquad g(2) = a, \qquad g(3) = d.$$

The function f, the 3-tuple (c, a, b), and the word *cab*, are formally identical, and so we shall define a **word** of **length** m in the **alphabet** Y to be a function from $\mathbb{N}_m$ to Y.

Before we proceed to exploit this idea we shall prove the most general result on counting sets of functions.

Theorem 3.4

Let X and Y be non-empty finite sets, and let F denote the set of functions from X to Y. If $|X| = m$ and $|Y| = n$, then

$$|F| = n^m.$$

Proof

Let $X = \{x_1, x_2, \ldots, x_m\}$. Each member f of the set F is a function from X to Y, and is uniquely determined by the m-tuple of its values $(f(x_1), f(x_2), \ldots, f(x_m))$. This m-tuple belongs to Y^m, and so

$$|F| = |Y^m| = n^m. \qquad \square$$

Equivalently, we may say that the number of words of length m in an alphabet Y of n symbols is n^m. For example, there are 26^3

three-letter words in the usual Roman alphabet—assuming of course that there are no restrictions on spelling.

There is another important way in which we can interpret a function from $\mathbb{N}_m$ to Y, or equivalently, a word of length m in the alphabet Y. Let us consider the task of a compositor in an old-fashioned printing works. In order to set up the word *cab*, he selects a letter c from his stock, then a, then b. We presume that he has an unlimited stock of letters, so that in order to set up *dad*, for example, he can select one d, then an a, then another d. Each word represents an ordered selection of letters from the alphabet, $Y = \{a, b, \ldots, z\}$, with repetition allowed as many times as required.

In general we can say that a function from $\mathbb{N}_m$ to Y is a mathematical model of an *ordered selection with repetition* of m things from the set Y. By Theorem 3.4, the number of such selections is n^m when $|Y| = n$. (In subsequent sections we shall discover how to deal with selections which may be ordered or unordered, and with or without repetition.)

The simple rule for counting functions (or words, or ordered selections with repetition) can be applied to yield some quite general results, as the following *Example* shows.

Example

If X is a set with n members, show that the total number of subsets of X is 2^n.

Solution

Suppose $X = \{x_1, x_2, \ldots, x_n\}$, and let Y be the alphabet $\{0, 1\}$. Any subset S of X corresponds to a word of length n in Y, defined by the function

$$S(i) = \begin{cases} 0 & \text{if} \quad x_i \notin S, \\ 1 & \text{if} \quad x_i \in S. \end{cases}$$

For example, if $n = 7$ and $S = \{x_2, x_4, x_5\}$, the word is 0101100. We usually think of 0 as representing **false**, and 1 as representing **true**; then the word is constructed by checking each member of X in turn and entering **false** if it is not in S, **true** if it is.

Consequently the number of distinct subsets of X is the same as the number of words of length n in the alphabet $\{0, 1\}$, that is, 2^n. $\qquad\square$

Exercises 3.4

1 How many national flags can be constructed from three equal vertical strips, using the colours red, white, blue, and green? (It is

assumed that colours can be repeated, and that one vertical edge of the flag is distinguished as the 'flagpole side'.)

2 Write down all the subsets of the set $\{a, b, c, d\}$ and use the correspondence given in the *Example* to check that your list is complete.

3 Keys are made by cutting incisions of various depths in a number of positions on a blank key. If there are eight possible depths, how many positions are required to make one million different keys? [Hint: to facilitate the calculation use the fact that 2^{10} is slightly greater than 10^3.]

4 Show that there are more than 10^{76} subsets of the set of subsets of a set with eight members.

3.5 Injections as ordered selections without repetition

In many situations we have to make an ordered selection *without* repetition. Whereas a compositor is presumed to have an unlimited supply of letters available, it can happen that there is only one object of each kind. For example, if we are selecting a team for baseball (or cricket) in batting order, then no player can be selected more than once.

The language of functions provides a ready model of this situation. We have seen that an ordered selection of m things from a set Y corresponds to a function f from $\mathbb{N}_m$ to Y, where $f(1)$ is the first member of Y selected, and so on. When repetition is allowed, it is possible that the same object is selected twice, so that $f(r) = f(s)$ for distinct r and s in $\mathbb{N}_m$. If this is prohibited, that is, if f is an *injection*, then we have a model of an ordered selection without repetition.

Theorem 3.5 The number of ordered selections, without repetition, of m things from a set Y of size n is the same as the number of injections from $\mathbb{N}_m$ to Y, and it is

$$n(n-1)(n-2) \cdots (n-m+1).$$

Proof Each injection i from $\mathbb{N}_m$ to Y is uniquely determined by the ordered selection of distinct values $i(1), i(2), \ldots, i(m)$. The first selection $i(1)$ can be any one of the n objects in Y. Since repetition is not allowed, the second selection $i(2)$ must be one of the remaining $n-1$ objects. Similarly, there are $n-2$ possibilities for $i(3)$, and so on. When we come to select $i(m)$, $m-1$ objects have already been selected, and so $i(m)$ must be one of the remaining $n-(m-1)$ objects. Hence the total number is as stated. □

For example, if we have a pool of 16 players the number of ways of selecting a batting order for a baseball team of nine is

$$16 \times 15 \times 14 \times 13 \times 12 \times 11 \times 10 \times 9 \times 8 = 4\ 151\ 347\ 200.$$

Exercises 3.5

1 In how many ways can we select a batting order of 11 from a pool of 14 cricketers?

2 How many four-letter words can be made from an alphabet of 10 symbols if there are no restrictions on spelling except that no letter can be used more than once?

3 Explain briefly how you would make a systematic list of all the ordered selections, without repetition, of three things from the set $\{a, b, c, d, e, f\}$.

4 Let $(n)_m = n(n-1) \cdots (n-m+1)$. By interpreting the result in terms of ordered selections, show that

$$(n)_m \times (n-m)_{r-m} = (n)_r$$

for any positive integers satisfying $n > r > m$.

3.6 Permutations

A **permutation** of a non-empty finite set X is a bijection from X to X. (Frequently we take X to be $\mathbb{N}_n = \{1, 2, \ldots, n\}$.) For example, a typical permutation of $\mathbb{N}_5$ is the function α defined by the equations

$$\alpha(1) = 2, \qquad \alpha(2) = 4, \qquad \alpha(3) = 5, \qquad \alpha(4) = 1, \qquad \alpha(5) = 3.$$

A bijection from a finite set to itself is necessarily an injection, and conversely any such injection is a bijection (Ex. 2.6.3). Thus the number of permutations of an n-set is the same as the number of injections from $\mathbb{N}_n$ to itself, and by Theorem 3.5 this number is

$$n \times (n-1) \times \cdots \times 1 = n!.$$

We shall denote the set of all permutations of $\mathbb{N}_n$ by S_n. For example, S_3 contains the following $3! = 6$ permutations:

$$
\begin{array}{ccc}
1 & 2 & 3 \\
\downarrow & \downarrow & \downarrow \\
1 & 2 & 3
\end{array}
\qquad
\begin{array}{ccc}
1 & 2 & 3 \\
\downarrow & \downarrow & \downarrow \\
1 & 3 & 2
\end{array}
\qquad
\begin{array}{ccc}
1 & 2 & 3 \\
\downarrow & \downarrow & \downarrow \\
2 & 1 & 3
\end{array}
$$

$$
\begin{array}{ccc}
1 & 2 & 3 \\
\downarrow & \downarrow & \downarrow \\
2 & 3 & 1
\end{array}
\qquad
\begin{array}{ccc}
1 & 2 & 3 \\
\downarrow & \downarrow & \downarrow \\
3 & 1 & 2
\end{array}
\qquad
\begin{array}{ccc}
1 & 2 & 3 \\
\downarrow & \downarrow & \downarrow \\
3 & 2 & 1.
\end{array}
$$

In practice, we usually assign some concrete interpretation to an element of S_n. As in the previous section we can use the interpretation as an 'ordered selection without repetition' where, in this case, we select the members of $\{1, 2, \ldots, n\}$ in some order until there are none left. A related interpretation is that a permutation *effects a rearrangement* of $\{1, 2, \ldots, n\}$; for example, the permutation α given above effects the rearrangement of 12345 into 24513, thus:

$$
\begin{array}{ccccc}
1 & 2 & 3 & 4 & 5 \\
\downarrow & \downarrow & \downarrow & \downarrow & \downarrow \\
2 & 4 & 5 & 1 & 3.
\end{array}
$$

In some circumstances it is convenient to regard a permutation and the corresponding rearrangement as the same thing, but this can lead to difficulties when we have to consider successive rearrangements. By and large, it is advisable to remember that

a permutation is a kind of function.

When permutations are treated as functions it is clear how they should be combined. Let us take α to be the permutation of $\mathbb{N}_5$ specified above, and suppose β is the permutation of $\mathbb{N}_5$ given by

$$\beta(1) = 3, \qquad \beta(2) = 5, \qquad \beta(3) = 1, \qquad \beta(4) = 4, \qquad \beta(5) = 2.$$

The composite function $\beta\alpha$ is the permutation defined by $\beta\alpha(i) = \beta(\alpha(i))$ $(1 \leqslant i \leqslant 5)$, that is

$$\beta\alpha(1) = 5, \qquad \beta\alpha(2) = 4, \qquad \beta\alpha(3) = 2, \qquad \beta\alpha(4) = 3, \qquad \beta\alpha(5) = 1.$$

(Recall that, as always, $\beta\alpha$ means 'first α, then β'). In terms of rearrangements we have

$$
\begin{array}{cccccc}
 & 1 & 2 & 3 & 4 & 5 \\
\alpha & \downarrow & \downarrow & \downarrow & \downarrow & \downarrow \\
 & 2 & 4 & 5 & 1 & 3 \\
\beta & \downarrow & \downarrow & \downarrow & \downarrow & \downarrow \\
 & 5 & 4 & 2 & 3 & 1.
\end{array}
$$

There are four features of the composition of permutations which are of paramount importance, and they are listed in the next theorem. In Part III of the course we shall review these properties in a more general context.

Theorem 3.6

The following properties hold in the set S_n of all permutations of $\{1, 2, \ldots, n\}$.

(i) If π and σ are in S_n, so is $\pi\sigma$.

(ii) For any permutations π, σ, τ in S_n,

$$(\pi\sigma)\tau = \pi(\sigma\tau).$$

(iii) The identity function, denoted by id and defined by $\mathrm{id}(r) = r$ for all r in $\mathbb{N}_n$, is a permutation and for any σ in S_n we have

$$\mathrm{id}\,\sigma = \sigma\,\mathrm{id} = \sigma.$$

(iv) For every permutation π in S_n there is an inverse permutation π^{-1} in S_n such that

$$\pi\pi^{-1} = \pi^{-1}\pi = \mathrm{id}.$$

Proof

Statement (i) follows immediately from the fact that the composite of two bijections is a bijection (Theorem 2.2.1), and statement (ii) is a standard property of composition (Ex. 2.1.4). Statement (iii) is plainly true, and statement (iv) follows from the fact that every bijection has an inverse (Theorem 2.2.2). □

It is convenient to have a more compact notation for permutations. Consider once again the permutation α of $\{1, 2, 3, 4, 5\}$, and note in particular that

$$\alpha(1) = 2, \qquad \alpha(2) = 4, \qquad \alpha(4) = 1.$$

Thus α takes 1 to 2, 2 to 4, and 4 back to 1, and for this reason we say that the symbols 1, 2, 4 form a *cycle* (of length 3). Similarly, the symbols 3 and 5 form a cycle of length 2, and we write

$$\alpha = (1\ 2\ 4)(3\ 5).$$

This is the *cycle notation* for α. Any permutation π can be written in cycle notation in the following way:

> begin with any symbol (say 1) and trace the effect of π on it and its successors until we reach 1 again, so that we have a cycle;
> choose a symbol not already dealt with and construct the cycle derived from it;
> repeat the procedure until every symbol has been dealt with.

For example, the permutation β defined above has the cycle notation

$$\beta = (13)(25)(4),$$

where we note especially that the symbol 4 forms a 'degenerate' cycle on its own, since $\beta(4) = 4$. In some circumstances we can omit such cycles of length 1 when writing a permutation in cycle notation, since they correspond to symbols which are not affected by the permutation. However, it is usually helpful *not* to adopt this convention until one is quite familiar with the notation.

Although the representation of a permutation in cycle notation is essentially unique, there are two obvious ways in which we can change the notation without altering the permutation. First, each cycle can begin with any one of its symbols—for example (7 8 2 1 3) and (1 3 7 8 2) describe the same cycle. Secondly, the order of the cycles is unimportant—for example (124)(35) and (35)(124) denote the same permutation. But the important features are the numbers and lengths of the cycles, and the disposition of the symbols within the cycles, and these are uniquely determined. Consequently, the cycle notation tells us many useful things about a permutation.

Example   Cards numbered 1 to 12 are laid out in the manner shown on the left below. They are picked up in row order and re-dealt in the same array, but by columns rather than rows, so that they appear as on the right.

1	2	3	1	5	9
4	5	6	2	6	10
7	8	9	3	7	11
10	11	12	4	8	12

How many times must this procedure be carried out before the cards reappear in their original positions?

Solution   Let π be the permutation which effects the rearrangement; that is $\pi(i) = j$ if card j appears in the position previously occupied by card i. Working out the cycle notation for π we find

$$\pi = (1)(2\ 5\ 6\ 10\ 4)(3\ 9\ 11\ 8\ 7)(12).$$

The degenerate cycles (1) and (12) indicate that cards 1 and 12 never change their positions. The other cycles have length 5, so after the procedure has been carried out five times all the cards will reappear in their original positions. (Try it.) Another way of expressing the result is to say that $\pi^5 = \mathrm{id}$, where π^5 signifies the five-fold repetition of π. $\qquad\qquad\square$

Exercises
3.6

1 Write down the cycle notation for the permutation which effects the rearrangement

$$
\begin{array}{ccccccccc}
1 & 2 & 3 & 4 & 5 & 6 & 7 & 8 & 9 \\
\downarrow & \downarrow & \downarrow & \downarrow & \downarrow & \downarrow & \downarrow & \downarrow & \downarrow \\
3 & 5 & 7 & 8 & 4 & 6 & 1 & 2 & 9.
\end{array}
$$

2 Let σ and τ be the permutations of $\{1, 2, \ldots, 8\}$ whose representations in cycle notation are

$$\sigma = (1\ 2\ 3)(4\ 5\ 6)(7\ 8), \qquad \tau = (1\ 3\ 5\ 7)(2\ 6)(4)(8).$$

Write down the cycle notations for $\sigma\tau$, $\tau\sigma$, σ^2, σ^{-1}, τ^{-1}.

3 Solve the problem posed in the *Example* when there are 20 cards arranged in five rows of four.

4 Show that there are just three members of S_4 which have two cycles of length 2 when written in cycle notation.

5 Let K denote the subset of S_4 which contains the identity permutation id and the three permutations α_1, α_2, α_3 described in the previous exercise. Write out the 'multiplication table' for K, when multiplication is interpreted as composition of permutations.

3.7 Miscellaneous Exercises

1 A committee of nine people must elect a chairman, secretary and treasurer. In how many ways can this be done? (Explain carefully the assumptions you make in your solution.)

2 In the usual set of dominos each domino may be represented by the symbol $[x \mid y]$, where x and y are members of the set $\{0, 1, 2, 3, 4, 5, 6\}$. The numbers x and y may be equal. Explain why the total number of dominos is 28 rather than 49.

3 In how many ways can we select a black square and a white square on a chessboard in such a way that the two squares are not in the same rank or the same file?

4 In how many ways can we place eight Rooks on a chessboard in such a way that no two of them are on the same rank or file?

5   Suppose there are m girls and n boys in a class. What is the number of ways of arranging them in a line so that all the girls are together?

6   If we have nine different subsets of $\mathbb{N}_{12}$, each of which has eight members, and each member of $\mathbb{N}_{12}$ occurs in the same number r of subsets, what is the value of r? Is it possible to fine nine different subsets of $\mathbb{N}_{12}$ each of which has seven members, and such that each member of $\mathbb{N}_{12}$ occurs in the same number of subsets?

7   How many five-digit telephone numbers have a digit which occurs more than once?

8   Calculate the total number of permutations σ of $\mathbb{N}_6$ which satisfy $\sigma^2 = \mathrm{id}$, $\sigma \neq \mathrm{id}$.

9   Let α and β be the permutations of $\mathbb{N}_9$ whose representations in cycle notation are

$$\alpha = (1237)(49)(58)(6), \qquad \beta = (135)(246)(789)$$

Write down the cycle notations for $\alpha\beta$, $\beta\alpha$, α^2, β^2, α^{-1}, β^{-1}.

10   The rooms of the house shown in Fig. 3.1 are to be painted in such a way that rooms with an interconnecting door have different colours. If there are n colours available, how many different colour schemes are possible?

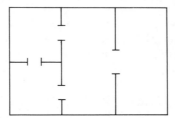

Fig. 3.1
Interior decorating.

11   Let $X_1 = \{0, 1\}$, and for $i \geq 2$ define the set X_i to be the set of subsets of X_{i-1}. Find the least value of i such that $|X_i| > 10^{100}$.

12   Suppose we have a set of 'generalized dominos' in which the numbers range from 0 to n. Let k be any integer in the range $0 \leq k \leq n$. Show that the number of dominos $[x \mid y]$ for which $x + y = n - k$ is equal to the number for which $x + y = n + k$.

13   For each integer i in the range $1 \leq i \leq n - 1$ define τ_i to be the permutation of $\mathbb{N}_n$ which switches i and $i + 1$ and does not affect the other elements of $\mathbb{N}_n$. Explicitly

$$\tau_i = (1)(2) \cdots (i-1)(i \;\; i+1)(i+2) \cdots (n).$$

Show that any permutation of $\mathbb{N}_n$ can be expressed in terms of τ_1, $\tau_2, \ldots, \tau_{n-1}$. (This is the foundation of the old English pastime of change-ringing.)

14   Show that, for any positive integers n and m

$$\phi(n^m) = n^{m-1}\phi(n).$$

15   Calculate $\phi(1000)$ and $\phi(1001)$.

16   A permutation of $\mathbb{N}_n$ which has just one cycle (necessarily of length n) is said to be *cyclic*. Show that there are $(n-1)!$ cyclic permutations of $\mathbb{N}_n$.

17   Let u_n denote the number of words of length n in the alphabet $\{0, 1\}$ which have the property that no two zeros are consecutive. Show that

$$u_1 = 2, \qquad u_2 = 3, \qquad u_n = u_{n-1} + u_{n-2} \qquad (n \geqslant 3).$$

18   A pack of 52 cards is divided into two equal parts and then 'interlaced', so that if the original order was $1, 2, 3, 4, \ldots$, the new order is $1, 27, 2, 28, \ldots$. How many times must this shuffle be repeated before the cards are once again in the original order?

19   Show that in any set of 20 people there is either a set of four mutual friends or a set of four mutual strangers.

20   When s and t are integers satisfying $s \geqslant 2$, $t \geqslant 2$ there is a smallest integer $r(s, t)$ such that whenever $n \geqslant r(s, t)$ any set of n people contains either a set of s mutual friends or a set of t mutual strangers. Show that

$$r(s, 2) = s, \qquad r(2, t) = t, \qquad r(s, t) \leqslant r(s-1, t) + r(s, \ 1).$$

4
Subsets and designs

4.1 Binomial numbers

Many practical questions in Discrete Mathematics take the standard form: in how many ways can a certain number of objects be selected from a given set? The answer to such a question will depend on whether or not *repetition* is allowed, and whether or not the *order* of selection is important. If the order is important, then we must use the models for ordered selections discussed in Sections 3.4 and 3.5; but if it is not then it is appropriate to use different models, as we shall now explain.

The mathematical model of an *unordered selection without repetition* is very simple. When we are given a set X with n members and we select r of them, the result is a subset Y of X with $|Y| = r$. It must be stressed that in this model it is the result of the selection (the subset Y) which is important, rather than the process of selection. Also, there is no possibility of repetition, since each member of X is either in Y or not, and no member can be selected twice.

We shall find it convenient to refer to a set X with n members as an **n-set**, and a subset Y with r members as an **r-subset** of X. Thus the number of unordered selections, without repetition, of r things from a set X of size n is just the number of r-subsets of the n-set X. For example, there are six unordered selections, without repetition, of two things from the set $\{a, b, c, d\}$; they correspond to the subsets

$$\{a, b\}, \quad \{a, c\}, \quad \{a, d\}, \quad \{b, c\}, \quad \{b, d\}, \quad \{c, d\}.$$

In general, the number of r-subsets of an n-set is denoted by the symbol

$$\binom{n}{r}.$$

This is often spoken as 'n choose r', and will be referred to as a **binomial number**. For example, we have just checked that there are six 2-subsets of a 4-set, and so

$$\binom{4}{2} = 6.$$

The following exercises should be solved by using only the definition of the binomial numbers.

Exercises 4.1

1 Show that

$$\binom{n}{r} = 0 \quad \text{if } r > n.$$

2 Find the values of

$$\binom{n}{0}, \binom{n}{1}, \binom{n}{n} \quad \text{for all } n \geqslant 1.$$

3 Prove that

$$\binom{n}{r} = \binom{n}{n-r} \quad \text{for } 0 \leqslant r \leqslant n.$$

The calculation of the binomial numbers in general depends on the following fundamental result.

Theorem 4.1.1

If n and r are positive integers satisfying $1 \leqslant r \leqslant n$ then

$$\binom{n}{r} = \binom{n-1}{r-1} + \binom{n-1}{r}.$$

Proof

Let X be an n-set, and suppose x is a chosen member of X. The set of all r-subsets of X can be split into two disjoint parts U and V in the following way:

$U =$ those r-subsets which contain x;

$V =$ those r-subsets which do not contain x.

An r-subset is in U if (and only if) when x is removed from it we obtain an $(r-1)$-subset of the $(n-1)$-set $X - \{x\}$. Hence

$$|U| = \binom{n-1}{r-1}.$$

On the other hand, an r-subset is in V if (and only if) it is an r-subset of the $(n-1)$-set $X-\{x\}$. Hence

$$|V| = \binom{n-1}{r}.$$

It follows from the addition principle that the total number of r-subsets is equal to $|U|+|V|$, and since this number is $\binom{n}{r}$ we have the result. $\square$

Theorem 4.1.1 provides a recursive method for calculating binomial numbers since, if the numbers $\binom{n-1}{k}$ are known for $0 \le k \le n-1$, then the numbers $\binom{n}{k}$ can be computed. The calculation is often displayed in the form of a triangle, thus:

$$
\begin{array}{ccccccccccccc}
 & & & & & & 1 & & & & & & \\
 & & & & & 1 & & 1 & & & & & \\
 & & & & 1 & & 2 & & 1 & & & & \\
 & & & 1 & & 3 & & 3 & & 1 & & & \\
 & & 1 & & 4 & & 6 & & 4 & & 1 & & \\
 & 1 & & 5 & & 10 & & 10 & & 5 & & 1 & \\
1 & & 6 & & 15 & & 20 & & 15 & & 6 & & 1 \\
1 & 7 & & 21 & & 35 & & 35 & & 21 & & 7 & 1.
\end{array}
$$

This is sometimes called *Pascal's triangle*, after Blaise Pascal (1623–1662), although it was known long before his time. The numbers in the $(n+1)$th row are the binomial numbers $\binom{n}{r}$ for $r = 0, 1, \ldots, n$. The results of Ex. 4.1.2 imply that the border consists entirely of 1's, and Theorem 4.1.1 tells us that each number is the sum of the two numbers immediately above it. Consequently the table can be built up one row at a time. For example, the fourth number in the next row is

$$\binom{8}{3} = \binom{7}{2} + \binom{7}{3} = 21 + 35 = 56.$$

In some circumstances it is convenient to have an explicit formula for the binomial numbers.

Theorem 4.1.2

If n and r are positive integers satisfying $1 \le r \le n$, then
$$\binom{n}{r} = \frac{n(n-1)\cdots(n-r+1)}{r!} = \frac{n!}{r!\,(n-r)!}.$$

Proof

We use the principle of induction. For the induction basis, we remark that the result is true when $n = 1$, since $\binom{1}{1} = 1$ and the formula reduces to $1/1! = 1$ also.

For the induction hypothesis, suppose the result is true when $n = k$. Then, by Theorem 4.1.1 and the induction hypothesis,

$$\binom{k+1}{r} = \binom{k}{r-1} + \binom{k}{r}$$

$$= \frac{k(k-1)\cdots(k-r+2)}{(r-1)!} + \frac{k(k-1)\cdots(k-r+1)}{r!}$$

$$= \frac{k(k-1)\cdots(k-r+2)}{(r-1)!}\left(1 + \frac{k-r+1}{r}\right)$$

$$= \frac{(k+1)k(k-1)\cdots(k-r+2)}{r!}.$$

(If $r = 1$ or $r = k+1$ we have to use the values $\binom{k}{0} = 1$ and $\binom{k}{k+1} = 0$ rather than the formula.) It follows that the result is true when $n = k+1$, and so, by the principle of induction, it is true for all positive integers n. $\quad\square$

There are many useful and interesting identities involving the binomial numbers and, although they can sometimes be proved by means of the formula, it is usually better to rely on the definition itself, or on the recursive technique given in Theorem 4.1.1. We shall give examples of both methods.

Example 1 Show that

$$\binom{n}{0} + \binom{n}{1} + \binom{n}{2} + \cdots + \binom{n}{n-1} + \binom{n}{n} = 2^n.$$

Solution The expression on the left-hand side is the sum of the numbers of r-subsets of an n-set, taken over all possible values of r. Hence the left-hand side is equal to the total number of subsets of an n-set, and according to the *Example* given in Section 3.4, this number is 2^n. $\quad\square$

Example 2 Show that

$$\binom{n}{0} - \binom{n}{1} + \binom{n}{2} - \cdots + (-1)^{n-1}\binom{n}{n-1} + (-1)^n\binom{n}{n} = 0.$$

Solution According to Ex. 4.1.2 and Theorem 4.1.1 the left-hand side is equal to

$$1-\left\{\binom{n-1}{0}+\binom{n-1}{1}\right\}+\left\{\binom{n-1}{1}+\binom{n-1}{2}\right\}-\cdots$$

$$\cdots+(-1)^{n-1}\left\{\binom{n-1}{n-2}+\binom{n-1}{n-1}\right\}+(-1)^n.$$

Each term $\binom{n-1}{k}$ with k in the range $1\leqslant k\leqslant n-2$ occurs with a positive sign and with a negative sign, and so these terms cancel. The remaining terms are

$$1-\binom{n-1}{0}+(-1)^{n-1}\binom{n-1}{n-1}+(-1)^n=1-1+(-1)^{n-1}+(-1)^n,$$

which is 0, as required. □

Exercises 4.1 (continued)

4   Calculate the next three rows of Pascal's triangle, starting from the portion given on p. 64.

5   Evaluate $\binom{16}{4}$ and $\binom{17}{5}$.

6   Show that the number of words of length n in the alphabet $\{0, 1\}$ which contain exactly r zeros is $\binom{n}{r}$.

7   Prove the identity

$$\binom{s-1}{0}+\binom{s}{1}+\cdots+\binom{s+n-2}{n-1}+\binom{s+n-1}{n}=\binom{s+n}{n},$$

where s and n are positive integers. [Hint: if X is an $(s+n)$-set and $Y=\{y_1, y_2, \ldots, y_n\}$ is a specific n-subset of X, what is the number of n-subsets of X for which y_r is the first member of Y not in the subset?]

8   Give an alternative proof of Ex. 7, starting from the formula

$$\binom{s+n}{n}=\binom{s+n-1}{n}+\binom{s+n-1}{n-1},$$

and repeatedly using Theorem 4.1.1 to split up the last term.

4.2   Unordered selections with repetition

The binomial number $\binom{n}{r}$ is defined to be the number of r-subsets of an n-set, or the number of unordered selections *without* repetition of r things from a set of n things. We now turn to unordered selections *with* repetition. When the numbers involved are small, it is easy to list all the possibilities. For example, there are 15 unordered selections of four things from the set $\{a, b, c\}$, with repetition allowed, and they are:

aaaa	*aaab*	*aaac*	*aabb*	*aabc*
aacc	*abbb*	*abbc*	*abcc*	*accc*
bbbb	*bbbc*	*bbcc*	*bccc*	*cccc.*

We shall show that it is possible to give a general formula for the number of unordered selections with repetition, in terms of the binomial numbers. The proof of this fact involves the representation of such selections as words in the alphabet $\{0, 1\}$; for example, the selection *abcc* will be represented by the word 101011. The zeros are markers which separate the kinds of objects, and the ones tell us how many of each object there are, according to the scheme

$$a \quad b \quad c \ c$$
$$1 \ 0 \ 1 \ 0 \ 1 \ 1.$$

Since there are two markers, which can be placed in any of the six positions, the total number of selections in this case is $\binom{6}{2} = 15$, as we found by listing them. We shall now prove the general form of this result.

Theorem 4.2

The number of unordered selections, with repetition, of r objects from a set of n objects is

$$\binom{n+r-1}{r}.$$

Proof

Since the selections are unordered, we may arrange matters so that, within each selection, all the objects of one given kind come first, followed by the objects of another kind, and so on. When this has been done, we can assign to each selection a word of length $n+(r-1)$ in the alphabet $\{0, 1\}$, by the method explained above. That is, if there are k_i objects of the ith kind ($1 \le i \le n$), then the first k_1 letters of the word are 1's, followed by a single 0, followed by k_2 1's, another 0, and so on. The function defined by this rule is a bijection from the set of selections to the set of words of length $n+r-1$ which contain exactly $n-1$ zeros. The zeros can occupy any of the $n+r-1$ positions, so the number of words is

$$\binom{n+r-1}{n-1} = \binom{n+r-1}{r},$$

as required. $\square$

We can now tabulate (Table 4.2.1) our results for the different kinds of selections—ordered and unordered, with and without repetition— of r things from an n-set.

Table 4.2.1

	Ordered	Unordered
Without repetition	$n(n-1) \cdots (n-r+1)$	$\binom{n}{r}$
With repetition	n^r	$\binom{n+r-1}{r}$

Exercises 4.2

1   Write down the values given by the table above when $r = 2$ and $n = 3$, and in each of the four cases make a list of the relevant selections, taking $\{a, b, c\}$ as the 3-set.

2   Show that when three indistinguishable dice are thrown there are 56 possible outcomes. What is the number of outcomes when n indistinguishable dice are thrown?

3   Suppose that the expression $(x+y+z)^n$ is expanded and the terms are collected, according to the usual rules of elementary algebra: for example,

$$(x+y+z)^2 = x^2 + y^2 + z^2 + 2xy + 2yz + 2zx.$$

What is the number of terms in the resulting formula?

4   Show that the number of n-tuples $(x_1, x_2, \ldots, x_n)$ of non-negative integers which satisfy the equation

$$x_1 + x_2 + \cdots + x_n = r$$

is $\binom{n+r-1}{r}$. [Hint: suppose that an unordered selection, with repetition, of r objects from a set of n objects contains x_i copies of the ith object $(1 \leqslant i \leqslant n)$.]

4.3   The binomial theorem

In elementary algebra we learn the formulae

$$(a+b)^2 = a^2 + 2ab + b^2, \qquad (a+b)^3 = a^3 + 3a^2b + 3ab^2 + b^3,$$

and sometimes we are asked to work out the formula for $(a+b)^4$ and higher powers of $a+b$. The general result giving a formula for $(a+b)^n$ is known as the **binomial theorem**.

Theorem 4.3

Let n be a positive integer. The coefficient of the term $a^{n-r}b^r$ in the expansion of $(a+b)^n$ is the binomial number $\binom{n}{r}$. Explicitly, we have

$$(a+b)^n = \binom{n}{0}a^n + \binom{n}{1}a^{n-1}b + \binom{n}{2}a^{n-2}b^2 + \cdots + \binom{n}{n}b^n.$$

Proof

Consider what happens when we multiply n factors

$$(a+b)(a+b)\cdots(a+b).$$

A term in the product is obtained by selecting either a or b from each factor. The number of terms $a^{n-r}b^r$ is just the number of ways of selecting r b's (and consequently $n-r$ a's), and by definition this is the binomial number $\binom{n}{r}$. □

The coefficients in the expansion may therefore be calculated by using the recursion for the binomial numbers (Pascal's triangle) or by using the formula. For example,

$$(a+b)^6 = \binom{6}{0}a^6 + \binom{6}{1}a^5b + \binom{6}{2}a^4b^2 + \binom{6}{3}a^3b^3$$

$$+ \binom{6}{4}a^2b^4 + \binom{6}{5}ab^5 + \binom{6}{6}b^6$$

$$= a^6 + 6a^5b + 15a^4b^2 + 20a^3b^3 + 15a^2b^4 + 6ab^5 + b^6.$$

Of course, we can obtain other useful formula if we replace a and b by suitable expressions. Some typical examples are:

$$(1+x)^4 = 1 + 4x + 6x^2 + 4x^3 + x^4;$$
$$(1-x)^7 = 1 - 7x + 21x^2 - 35x^3 + 35x^4 - 21x^5 + 7x^6 - x^7;$$
$$(x+2y)^5 = x^5 + 10x^4y + 40x^3y^2 + 80x^2y^3 + 80xy^4 + 32y^5;$$
$$(x^2+y)^4 = x^8 + 4x^6y + 6x^4y^2 + 4x^2y^3 + y^4.$$

The expression $a+b$ is known as a *binomial* expression because it has two terms. Since the numbers $\binom{n}{r}$ occur as the coefficients in the expansion of $(a+b)^n$ they are usually referred to as the *binomial coefficients*. However it is clear from the proof of Theorem 4.3 that they occur in this context because they represent the numbers of ways of making certain selections. For this reason we shall continue to use the name *binomial numbers*, which corresponds more closely to the concept underlying the symbols.

In addition to being extremely useful in algebraic manipulations, the binomial theorem can be used to derive identities involving the binomial numbers.

Example   Prove that

$$\binom{n}{0}^2 + \binom{n}{1}^2 + \binom{n}{2}^2 + \cdots + \binom{n}{n}^2 = \binom{2n}{n}.$$

Solution   We use the identity

$$(1+x)^n(1+x)^n = (1+x)^{2n}.$$

According to the binomial theorem the left-hand side is the product of two factors, both equal to

$$1 + \binom{n}{1}x + \cdots + \binom{n}{r}x^r + \cdots + x^n.$$

When the two factors are multiplied, a term in x^n is obtained by taking a term $\binom{n}{r}x^r$ from the first factor and a term $\binom{n}{n-r}x^{n-r}$ from the second factor. Hence the coefficient of x^n in the product is

$$\binom{n}{0}\binom{n}{n} + \binom{n}{1}\binom{n}{n-1} + \binom{n}{2}\binom{n}{n-2} + \cdots + \binom{n}{n}\binom{n}{0}.$$

Since $\binom{n}{n-r} = \binom{n}{r}$, we see that this is the left-hand side of the required identity. But the right-hand side is $\binom{2n}{n}$ which is also the coefficient of x^n in the expansion of $(1+x)^{2n}$, and so we have the equality as stated.   □

A more general form of the binomial theorem, involving negative exponents, will be proved in Chapter 18.

Exercises 4.3

1 Write out the formulae for $(1+x)^8$ and $(1-x)^8$.

2 Calculate the coefficient of

  (i) x^5 in $(1+x)^{11}$;
  (ii) a^2b^8 in $(a+b)^{10}$;
  (iii) a^6b^6 in $(a^2+b^3)^5$;
  (iv) x^3 in $(3+4x)^6$.

3 Use the identity $(1+x)^m(1+x)^n = (1+x)^{m+n}$ to prove that

$$\binom{m+n}{r} = \binom{m}{0}\binom{n}{r} + \binom{m}{1}\binom{n}{r-1} + \cdots + \binom{m}{r}\binom{n}{0}$$

where m, n, and r are positive integers, $m \geqslant r$, and $n \geqslant r$.

4 By making appropriate substitutions in the formulae for $(1+x)^n$

and $(1-x)^n$ give alternative proofs of the results established in *Example 1* and *Example 2* of Section 4.1.

5   Show that if r and s are integers such that $s \mid r$, and p is a prime for which $p \mid r$ but $p \nmid s$, then $p \mid r/s$. [Hint: write $r = st$.] Deduce that

    (i) the binomial number $\binom{p}{i}$ is divisible by p for all values of i in the range $1 \leqslant i \leqslant p-1$;

    (ii) $(a+b)^p - a^p - b^p$ is divisible by p for all integers a and b.

4.4   The sieve principle

The most basic counting principle (Theorem 3.1) asserts that $|A \cup B|$ is the sum of $|A|$ and $|B|$ when A and B are disjoint sets. If A and B are not disjoint, the result of adding $|A|$ and $|B|$ is that the members of $A \cap B$ are counted twice (Fig. 4.1a). So in order to get the correct answer, we must subtract $|A \cap B|$:

$$|A \cup B| = |A| + |B| - |A \cap B|.$$

A similar method can be applied to three sets (Fig. 4.1b). When we add $|A|$, $|B|$, and $|C|$ the members of $A \cap B$, $B \cap C$, and $C \cap A$ are counted twice (if they are not in all three sets). To correct for this we subtract $|A \cap B|$, $|B \cap C|$, and $|C \cap A|$. But now the members of $A \cap B \cap C$, originally counted three times, have been deducted three times. So, in order to get the correct answer, we must *add* $|A \cap B \cap C|$. Thus

$$|A \cup B \cup C| = \alpha_1 - \alpha_2 + \alpha_3,$$

where

$$\alpha_1 = |A| + |B| + |C|, \qquad \alpha_2 = |A \cap B| + |B \cap C| + |C \cap A|,$$
$$\alpha_3 = |A \cap B \cap C|.$$

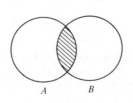

Fig. 4.1
Intersections of two and three sets.

    (a)

    (b)

This result is a simple case of what used to be called, for obvious reasons, the principle of inclusion and exclusion. Nowadays it is usually called the *sieve principle*.

Theorem 4.4

If $A_1, A_2, \ldots, A_n$ are finite sets then

$$|A_1 \cup A_2 \cup \cdots \cup A_n| = \alpha_1 - \alpha_2 + \alpha_3 - \cdots + (-1)^{n-1}\alpha_n,$$

where α_i is the sum of the cardinalities of the intersections of the sets taken i at a time $(1 \leq i \leq n)$.

Proof

We shall show that every member x of the union makes a net contribution of 1 to the right-hand side. Suppose that x belongs to precisely k of the sets $A_1, \ldots, A_n$. Then x will contribute k to the sum $\alpha_1 = |A_1| + \cdots + |A_n|$. In the sum α_2, x contributes 1 to $|A_i \cap A_j|$ when both A_i and A_j are among the k sets which contain x. There are $\binom{k}{2}$ such pairs, and so $\binom{k}{2}$ is the contribution of x to α_2. In general, the contribution of x to α_i $(1 \leq i \leq n)$ is $\binom{k}{i}$. Hence the total contribution of x to the right-hand side is

$$\binom{k}{1} - \binom{k}{2} + \cdots + (-1)^{k-1}\binom{k}{k},$$

since the terms with $i > k$ are zero.

But the identity established in *Example 2*, Section 4.1, implies that this expression is equal to $\binom{k}{0}$, which is 1. $\qquad\square$

A simple corollary of Theorem 4.4 is often most useful in practice. Suppose that $A_1, A_2, \ldots, A_n$ are subsets of a given set X with $|X| = N$. Then the number of members of X which are *not* in any of these subsets is

$$|X \setminus (A_1 \cup A_2 \cup \cdots \cup A_n)| = |X| - |A_1 \cup A_2 \cup \cdots \cup A_n|$$
$$= N - \alpha_1 + \alpha_2 - \cdots + (-1)^n \alpha_n.$$

Example 1

There are 73 students in the first year Humanities class at the University of Folornia. Among them a total of 52 can play the piano, 25 can play the violin, and 10 can play the flute; 17 can play both piano and violin, 12 can play piano and flute, and 7 can play violin and flute; but only Osbert Smugg can play all three instruments. How many of the class cannot play any of them?

Solution

Let P, V, and F denote the sets of students who can play the piano, violin, and flute respectively. Using the information given we have

$$\alpha_1 = |P| + |V| + |F| = 52 + 25 + 10 = 87,$$
$$\alpha_2 = |P \cap V| + |P \cap F| + |V \cap F| = 17 + 12 + 7 = 36,$$
$$\alpha_3 = |P \cap V \cap F| = 1.$$

Hence the number of students who do not belong to any of the sets P, V, F is

$$73 - 87 + 36 - 1 = 21. \qquad \square$$

Example 2 An inefficient secretary has n letters and n addressed envelopes for them. In how many ways can she achieve the feat of putting every letter into the wrong envelope? (This is often called the *derangement problem*: there are several other picturesque formulations.)

Solution We can regard each letter and its correct envelope as being labelled with an integer i in the range $1 \leqslant i \leqslant n$. The act of putting letters into envelopes is described by a permutation π of $\mathbb{N}_n$: $\pi(i) = j$ if letter i goes into envelope j. We require the number of **derangements**, that is, permutations π such that $\pi(i) \neq i$ for all i in $\mathbb{N}_n$.

Let A_i $(1 \leqslant i \leqslant n)$ denote the subset of S_n (the set of all permutations of $\mathbb{N}_n$) containing those π for which $\pi(i) = i$. We say that the members of A_i *fix i*. By the sieve principle, the required number of derangements is

$$d_n = n! - \alpha_1 + \alpha_2 - \cdots + (-1)^n \alpha_n,$$

where α_r is the sum of the cardinalities of the intersections of the A_i taken r at a time. In other words, α_r is the number of permutations which fix r given symbols, taken over all ways of choosing the r symbols. Now there are $\binom{n}{r}$ ways of choosing r symbols, and the number of permutations which fix them is just the number of permutations of the remaining $n - r$ symbols, that is, $(n - r)!$ Hence

$$\alpha_r = \binom{n}{r} \times (n - r)! = \frac{n!}{r!}, \qquad d_n = n! \left(1 - \frac{1}{1!} + \frac{1}{2!} - \cdots + (-1)^n \frac{1}{n!} \right).$$

Exercises 4.4

1 In a class of 67 mathematics students, 47 can read French, 35 can read German and 23 can read both languages. How many can read neither language? If, furthermore, 20 can read Russian, of whom 12 read French also, 11 German also and 5 read all three languages, how many cannot read any of the three languages?

2   Find the number of ways of arranging the letters A, E, M, O, U, Y in a sequence in such a way that the words ME and YOU do not occur.

3   Calculate the number d_4 of derangements of $\{1, 2, 3, 4\}$ and write down the relevant permutations in cycle notation.

4   Use the principle of induction to prove that the formula for d_n satisfies the recursion

$$d_1 = 0, \qquad d_2 = 1, \qquad d_n = (n-1)(d_{n-1} + d_{n-2}) \quad (n \geq 3).$$

5   Show that the number of derangements of $\{1, 2, \ldots, n\}$ in which a given object (say 1) is in a 2-cycle is $(n-1)d_{n-2}$. Hence construct a direct proof of the recursion formula given in the previous exercise.

4.5   Some arithmetical applications

For many hundreds of years mathematicians have studied problems about prime numbers and the factorization of integers. Our brief discussion of these matters in earlier chapters should have convinced the reader that such problems are difficult, because the primes themselves are irregularly distributed, and because there is no straightforward way to find the prime factorization of a given integer.

However, if the prime factorization of an integer is given to us, then it is relatively easy to answer some questions about its arithmetical properties. Suppose, for instance, that we wish to list all the divisors of an integer n, and we know that the prime factorization of n is

$$n = p_1^{e_1} p_2^{e_2} \cdots p_r^{e_r}.$$

Then an integer d is a divisor of n if and only if it has no prime divisors different from those of n, and no prime divides it more often than it divides n. Thus the divisors are precisely the integers which can be written in the form

$$d = p_1^{f_1} p_2^{f_2} \cdots p_r^{f_r},$$

where each f_i $(1 \leq i \leq r)$ satisfies $0 \leq f_i \leq e_i$. For example, given that $60 = 2^2 \times 3 \times 5$ we can quickly list all the divisors of 60: a good way of arranging them is illustrated in Fig. 4.2. (Technically, Fig. 4.2 is a diagram of the *lattice* of divisors of 60, but we shall not need the precise mathematical description of this term.)

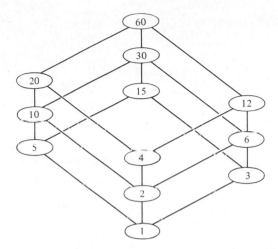

Fig. 4.2
The divisors of 60.

A similar problem is to find the number of integers x in the range $1 \leq x \leq n$ which are coprime to n. In Section 3.3 we denoted this number by $\phi(n)$, the value of Euler's function ϕ at n. We now show that if the prime factorization of n is known then $\phi(n)$ can be calculated by the sieve principle.

Example   What is the value of $\phi(60)$? In other words, how many integers x in the range $1 \leq x \leq 60$ satisfy $\gcd(x, 60) = 1$?

Solution   We know that $60 = 2^2 \times 3 \times 5$, so we have to count the number of integers x in the range $1 \leq x \leq 60$ which are *not* divisible by 2, 3, or 5. Let $A(2)$ denote the subset of $\mathbb{N}_{60}$ containing those integers which *are* divisible by 2, $A(2, 3)$ those which *are* divisible by 2 and 3, and so on. then we have

$$\phi(60) = 60 - |A(2) \cup A(3) \cup A(5)|$$
$$= 60 - (|A(2)| + |A(3)| + |A(5)|)$$
$$+ (|A(2, 3)| + |A(2, 5)| + |A(3, 5)|) - |A(2, 3, 5)|,$$

by the sieve principle. Now $|A(2)|$ is the number of multiples of 2 in $\mathbb{N}_{60}$, which is $60/2 = 30$. Similarly, $|A(2, 3)|$ is the number of multiples of 2×3, which is $60/(2 \times 3) = 10$, and so on. Hence

$$\phi(60) = 60 - (30 + 20 + 12) + (10 + 6 + 4) - 2 = 16. \qquad \square$$

The same method can be used to give an explicit formula for $\phi(n)$ in the general case.

Theorem 4.5.1

Let $n \geqslant 2$ be an integer whose prime factorization is $n = p_1^{e_1} p_2^{e_2} \cdots p_r^{e_r}$. Then

$$\phi(n) = n\left(1 - \frac{1}{p_1}\right)\left(1 - \frac{1}{p_2}\right) \cdots \left(1 - \frac{1}{p_r}\right).$$

Proof

Let A_j denote the subset of $\mathbb{N}_n$ containing the multiples of p_j $(1 \leqslant j \leqslant r)$. Then

$$\phi(n) = n - |A_1 \cup A_2 \cup \cdots \cup A_r|$$
$$= n - \alpha_1 + \alpha_2 - \cdots + (-1)^r \alpha_r,$$

where α_i is the sum of the cardinalities of the intersections taken i at a time. Now a typical intersection like

$$A_{j_1} \cap A_{j_2} \cap \cdots \cap A_{j_i}$$

contains the multiples of $P = p_{j_1} \times p_{j_2} \times \cdots \times p_{j_i}$ in $\mathbb{N}_n$, and these are just the integers

$$P, 2P, 3P, \ldots, \left(\frac{n}{P}\right)P.$$

Hence the cardinality of the typical intersection is n/P, and α_i is the sum of all terms like

$$\frac{n}{P} = n\left(\frac{1}{p_{j_1}}\right)\left(\frac{1}{p_{j_2}}\right) \cdots \left(\frac{1}{p_{j_i}}\right).$$

It follows that

$$\phi(n) = n - n\left(\frac{1}{p_1} + \frac{1}{p_2} + \cdots + \frac{1}{p_r}\right) + n\left(\frac{1}{p_1 p_2} + \frac{1}{p_1 p_3} + \cdots\right) + \cdots$$
$$\cdots + (-1)^r n\left(\frac{1}{p_1 p_2 \cdots p_r}\right)$$
$$= n\left(1 - \frac{1}{p_1}\right)\left(1 - \frac{1}{p_2}\right) \cdots \left(1 - \frac{1}{p_r}\right). \quad \square$$

It is easy to use the formula—provided, of course, that we know the prime factorization of n. For example,

$$\phi(60) = 60(1 - \tfrac{1}{2})(1 - \tfrac{1}{3})(1 - \tfrac{1}{5}) = 60 \times \tfrac{1}{2} \times \tfrac{2}{3} \times \tfrac{4}{5} = 16.$$

The sieve principle and the resulting formula for $\phi(n)$ also have some important theoretical consequences, which we shall now discuss.

Suppose we reverse the last step of the proof, and multiply out the factors in the formula for $\phi(n)$. In the case $n = 60$ we get

$$\phi(60) = \frac{60}{1} - \left(\frac{60}{2} + \frac{60}{3} + \frac{60}{5}\right) + \left(\frac{60}{6} + \frac{60}{10} + \frac{60}{15}\right) - \frac{60}{30}.$$

There is a term $60/d$ for each divisor d of 60 which is the product of distinct primes, and its coefficient is $+1$ or -1 according as the number of primes is even or odd. The divisors 4, 12, and 20, which have repeated prime factors (2^2 in this case) do not contribute, but for the sake of uniformity we may say that they contribute a term with coefficient 0. In general, we can express $\phi(n)$ as a sum of terms $\mu(d) \times (n/d)$, one for each divisor d of n; that is

$$\phi(n) = \sum_{d \mid n} \mu(d) \frac{n}{d},$$

where the coefficients $\mu(d)$ are given by

$$\mu(d) = \begin{cases} 1 & \text{if} \quad d = 1, \\ (-1)^k & \text{if} \quad d \text{ is a product of } k \text{ distinct primes} \\ 0 & \text{if} \quad d \text{ has a repeated prime factor.} \end{cases}$$

The function μ is known as the **Möbius function**, after A. F. Möbius (1790–1868); it will play a vital part in some of the algebraic theorems in Part III of this book. For the moment we shall show that it has some rather unexpected properties.

We begin by showing that for any integer $n \geq 2$ the sum of $\mu(d)$ taken over all divisors d of n is zero; that is

$$\sum_{d \mid n} \mu(d) = 0.$$

To prove this, suppose $n = p_1^{e_1} p_2^{e_2} \ldots p_r^{e_r}$. Each divisor d has the form $p_1^{f_1} p_2^{f_2} \ldots p_r^{f_r}$ with $0 \leq f_i \leq e_i$, and $\mu(d)$ is zero unless each f_i is 0 or 1. Thus each divisor d with $\mu(d) \neq 0$ corresponds to the subset of $\{p_1, p_2, \ldots, p_r\}$ containing those p_i with $f_i = 1$. The number of such subsets of size k is $\binom{r}{k}$, and $\mu(d)$ is $(-1)^k$, so we have

$$\sum_{d \mid n} \mu(d) = 1 - \binom{r}{1} + \binom{r}{2} - \cdots + (-1)^r \binom{r}{r} = 0.$$

Once again, we have used the fundamental fact about the alternating sum of binomial numbers (*Example 2*, Section 4.1).

Now we are ready to establish the characteristic property of the Möbius function; it is usually known as the **Möbius inversion formula**.

Theorem 4.5.2

Let g be a function defined on $\mathbb{N}$ and suppose that f is the function obtained from g by the rule

$$f(n) = \sum_{d|n} g(d).$$

Then g can be obtained from f by the rule

$$g(n) = \sum_{d|n} \mu(d) f\left(\frac{n}{d}\right).$$

Proof

Substituting for $f(n/d)$ in the right-hand side of the second equation we get

$$\sum_{d|n} \mu(d) f\left(\frac{n}{d}\right) = \sum_{d|n} \mu(d) \sum_{c|n/d} g(c)$$

$$= \sum_{(c,d)\in S} \mu(d) g(c).$$

The double sum is taken over the set S of all pairs (c, d) for which $d|n$ and $c|n/d$. But this is the same as the set of pairs (c, d) for which $c|n$ and $d|n/c$ (Section 1.6, *Example*), and so we may rearrange the sum as follows:

$$\sum_{c|n} g(c) \left(\sum_{d|n/c} \mu(d) \right).$$

The sum in brackets is zero when $n/c \geq 2$, by the result obtained above. Hence the only term remaining is the one for which $n = c$, which reduces to

$$g(n) \sum_{d|1} \mu(d) = g(n)\mu(1) = g(n),$$

as required. □

Exercises 4.5

1 Calculate $\phi(n)$ and $\mu(n)$ for each n in the range $95 \leq n \leq 100$.

2 Show that if the prime factorization of n is $p_1^{e_1} p_2^{e_2} \ldots p_r^{e_r}$ then the number of divisors of n is

$$(e_1 + 1)(e_2 + 1) \ldots (e_r + 1).$$

3 Use Theorem 4.5.1 to show that if $\gcd(m, n) = 1$ then $\phi(mn) = \phi(m)\phi(n)$.

4 Show that if $1 \leqslant x \leqslant n$ then

$$\gcd(x, n) = \gcd(n - x, n).$$

Hence prove that the sum of all the integers x which satisfy $1 \leqslant x \leqslant n$ and $\gcd(x, n) = 1$ is $\frac{1}{2}n\phi(n)$.

5 Suppose that the function f is obtained from g by the rule

$$f(n) = \sum_{d \mid n} \mu(d) g\left(\frac{n}{d}\right).$$

Show that g can be obtained from f by the rule

$$g(n) = \sum_{d \mid n} f(d).$$

(This is the converse of the Möbius inversion formula.) Hence use the formula

$$\phi(n) = \sum_{d \mid n} \mu(n) \frac{n}{d}$$

to give another proof of Theorem 3.3; that is,

$$\sum_{d \mid n} \phi(d) = n.$$

4.6 Designs

Suppose that a manufacturer has developed a new product, and wishes to evaluate a certain number (say v) of varieties of it by asking some typical consumers to test them. It may be impracticable for each consumer to test all the varieties, in which case it would be reasonable to impose the following conditions.

(i) Each consumer should test the same number (k) of varieties.
(ii) Each variety should be tested by the same number (r) of consumers.

For example, when $v = 8$, $k = 4$, and $r = 3$ a possible scheme would be to ask six people to test the varieties

    1234,   5678,   1357,   2468,   1247,   3658,

respectively (where the varieties are represented by the numbers 1, 2, 3, 4, 5, 6, 7, 8).

In general, let X be any set of size v. We say that a set **B** of

k-subsets of X is a **design**, with parameters (v, k, r), if each member x of X belongs to exactly r of the subsets in **B**. It is usual to refer to a subset B which belongs to **B** as a **block** of the design. In the example given above the parameters are $(8, 4, 3)$ and the blocks are $\{1, 2, 3, 4\}$, $\{5, 6, 7, 8\}$, and so on.

It is clear that there must be some restrictions on the parameters of a design; for instance, there is no design with parameters $(8, 3, 5)$. (You proved this in Ex. 3.2.3.) We shall find that the most obvious necessary conditions for the existence of a design with parameters (v, k, r) are also sufficient.

Suppose that **C** is any set of k-subsets of a v-set X, not necessarily a design. We may draw up a table (along the lines of McBrain's table, Section 3.2) in which the rows correspond to the members x of X, the columns correspond to the subsets C in **C**, and there is a mark in row x and column C whenever x belongs to the subset C. In the following example (Table 4.6.1) X is the set $\{1, 2, 3, 4, 5, 6\}$ and there are four subsets C_1, C_2, C_3, C_4. In general, the marks indicate the pairs (x, C) which are in the set

$$S = \{(x, C) \mid x \in C\}.$$

The 'row total' $r(x)$ gives the number of times that x occurs as a member of a subset C: this is sometimes referred to as the number of **replications** of x. The 'column total' is k in all cases, since we stipulate that each C is a k-subset of X. Hence the two methods of counting the set S lead to the equation

$$\sum_{x \in X} r(x) = |\mathbf{C}| \times k.$$

Table 4.6.1

x	C_1	C_2	C_3	C_4	$r(x)$
1	✓	✓		✓	3
2	✓		✓		2
3		✓			1
4	✓		✓	✓	3
5			✓		1
6		✓		✓	2
k	3	3	3	3	

In the case when we have a design **B** the replication numbers $r(x)$ are all equal to a constant, r. So the left-hand side of the equation reduces to vr, and we obtain

$$vr = bk,$$

where $b = |\mathbf{B}|$ is the number of blocks. It follows that k must be a divisor of vr. Furthermore, since the total number of k-subsets of X is $\binom{v}{k}$, the number b of blocks cannot exceed this number, that is

$$b = \frac{vr}{k} \le \binom{v}{k}.$$

Remarkably these simple conditions on v, k, and r are also sufficient for the existence of a design.

Theorem 4.6

There is a design with parameters (v, k, r) if and only if

$$k \mid vr \quad \text{and} \quad \frac{vr}{k} \le \binom{v}{k}.$$

Proof

The necessity of the conditions has been established above. Conversely, suppose we are given positive integers v, k, and r satisfying the conditions, so that $b = vr/k$ is a positive integer not exceeding $\binom{v}{k}$. Let $\mathbf{C}$ be any collection of b distinct k-subsets of a v-set X. The replication numbers $r(x)$ for $\mathbf{C}$ satisfy the equation

$$\sum_{x \in X} r(x) = bk = vr,$$

and if each $r(x)$ is equal to r then $\mathbf{C}$ is already a design. If not, there must be two objects x_1 and x_2 such that $r(x_1) > r > r(x_2)$. (See Fig. 4.3.)

Let q_{12} denote the number of sets C which contain x_1 but not x_2, let $q_{\bar{1}2}$ denote the number which contain x_2 but not x_1, and let q_{12}

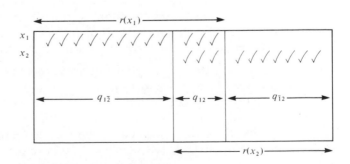

Fig. 4.3
Illustrating the proof of Theorem 4.6.

denote the number which contain x_1 and x_2. Then

$$q_{1\bar{2}} = r(x_1) - q_{12}, \qquad q_{\bar{1}2} = r(x_2) - q_{12}$$

and so

$$q_{1\bar{2}} - q_{\bar{1}2} = r(x_1) - r(x_2) > 0.$$

For each of the $q_{1\bar{2}}$ sets C containing x_1 but not x_2 let C^* be the set obtained from C by removing x_1 and inserting x_2. Each C^* contains x_2 but not x_1, and since $q_{1\bar{2}} > q_{\bar{1}2}$ there must be at least one C^* which is not in the original collection $\mathbf{C}$. Let C_0^* be one such set. If we remove C_0 from $\mathbf{C}$ and replace it by C_0^* we obtain a new collection $\mathbf{C}^*$ of distinct k-subsets of X. The replication numbers $r^*(x)$ for $\mathbf{C}^*$ are the same as for $\mathbf{C}$, except that

$$r^*(x_1) = r(x_1) - 1, \qquad r^*(x_2) = r(x_2) + 1.$$

If $\mathbf{C}^*$ is a design we are finished; if not we can repeat the process. At each step we get closer to a design, since two of the replication numbers are forced to differ less from the required constant r. Hence after a finite number of steps we shall find that all the replication numbers are equal to r, and a design is obtained. $\square$

We remark that, using the explicit formula for the binomial numbers, the second condition can be rewritten as

$$r \leq \binom{v-1}{k-1}.$$

In this form we can see that the condition is necessary by noting that each time an object x occurs in a block it is accompanied by $k-1$ of the remaining $v-1$ objects. Consequently, x cannot occur more than $\binom{v-1}{k-1}$ times.

Exercises 4.6

1   Undeterred by the apparent failure of his scheme to force every student to take exactly four of the seven mathematics courses (Section 3.2), Professor McBrain is working on a plan to ensure that, in addition, each course has the same number of takers.

(i) If there are v students, and k take each course, what is the relationship between v and k?

(ii) If the actual number of students is 53, how many must Professor McBrain expel in order to make his plan feasible?

(iii) Write down explicit designs for McBrain's plan in the cases $v = 7$ and $v = 14$. (Remember that in a design no block can be repeated—that is, no two courses can have the same set of takers.)

2   For the following values of (v, k, r) either construct a design with those parameters or explain why such a design cannot exist.

> (i) $(v, k, r) = (6, 3, 1)$;   (ii) $(v, k, r) = (5, 2, 1)$;
>
> (iii) $(v, k, r) = (7, 3, 3)$;   (iv) $(v, k, r) = (9, 6, 4)$.

3   What is the value of r in the design whose blocks are *all* the k-subsets of a v-set?

4   Let **B** be the set of blocks of a design with parameters (v, k, r), and let **B'** denote the set of complements $\bar{B}$ of the blocks B in **B**. Show that **B'** is also a design, and write down its parameters.

4.7 *t*-designs

The condition that each object or variety belongs to the same number of blocks can be strengthened in various ways. We could require that each *pair* of objects is contained in the same number of blocks, so that in an experiment each pair is compared in the same number of tests. More generally, we can formulate a similar condition for the objects taken t at a time, where t is any positive integer.

Definition If X is a set with cardinality v, then a set **B** of k-subsets of X is said to be a **t-design** with parameters (v, k, r_t) if, for each t-subset T of X, the number of blocks which contain T is a constant r_t.

According to the new definition, what we originally referred to as a design is a *1-design*. The example we gave of a 1-design with $v = 8$, $k = 4$, and $r_1 = 3$, was

$$1234, \quad 5678, \quad 1357, \quad 2468, \quad 1247, \quad 3568.$$

This is not a 2-design since some 2-subsets, such as $\{1, 2\}$, occur twice whereas others, such as $\{1, 5\}$, occur only once and others, such as $\{1, 6\}$, do not occur at all. Also, it is not a t-design for any larger value of t. However, there is a 3-design with $v = 8$ and $k = 4$:

1 2 3 5	4 6 7 8
1 3 4 6	2 5 7 8
1 4 5 7	2 3 6 8
1 5 6 8	2 3 4 7
1 2 6 7	3 4 5 8
1 3 7 8	2 4 5 6
1 2 4 8	3 5 6 7.

With a lot of checking it can be verified that each 3-subset of $\{1, 2, \ldots, 8\}$ occurs as a subset of just one of the 14 blocks. For instance $\{2, 3, 8\}$ occurs in the block 2368, and in no other block. Therefore we have a 3-design with parameters $(v, k, r_3) = (8, 4, 1)$. Clearly it is not a 4-design, since some 4-subsets (such as $\{1, 2, 3, 6\}$) do not occur while others (such as $\{1, 2, 3, 5\}$) do occur. It is, however, both a 2-design and a 1-design, and the next theorem shows that this is no accident.

Theorem 4.7.1   If **B** is a t-design then it is also an s-design for $s = 1, 2, \ldots, t-1$.

Proof   It is sufficient to show that **B** is a $(t-1)$-design, since this result with t replaced by $t-1$ will then establish that it is a $(t-2)$-design, and so on.

Let X be the set of objects and suppose that S is any $(t-1)$-subset of X. We shall count the pairs (x, B), where x is an object and b is a block such that

$$x \notin S \quad \text{and} \quad \{x\} \cup S \subseteq B.$$

Since x is not in the $(t-1)$-subset S, there are $v - (t-1)$ possibilities for x; for each such x, the t-subset $\{x\} \cup S$ is contained in r_t blocks, since B is a t-design. Hence the number of pairs is $(v - (t-1)) \times r_t$.

On the other hand, suppose that r_S is the number of blocks B containing S; for each such B any one of the $k - (t-1)$ members of $B - S$ is a possible x. Hence the number of pairs is $(k - (t-1)) \times r_S$. Equating the two expressions for the number of pairs we have

$$(v - (t-1))r_t = (k - (t-1))r_S.$$

This equation shows that r_S depends only on $t, k, v,$ and r_t, and so it is a constant, r_{t-1}, for each $(t-1)$-subset of X. Hence B is a $(t-1)$-design.   □

In the proof of Theorem 4.7.1 we obtain a useful formula for r_{t-1} in terms of r_t:

$$r_{t-1} = r_t \times \frac{v-t+1}{k-t+1}.$$

For example, in our 3-design with $v = 8$, $k = 4$, and $r_3 = 1$, we can calculate r_2 and r_1 as follows.

$$r_2 = r_3 \times \frac{v-2}{k-2} = 3, \qquad r_1 = r_2 \times \frac{v-1}{k-1} = 7.$$

We conclude that each pair occurs 3 times and each object occurs 7 times, as can be verified directly.

It is also worth remarking that the argument used in the proof of Theorem 4.7.1 holds even when $t = 1$ and S is the empty set. In that case the number r_S (or r_0) of blocks containing S is just the total number of blocks, $b = |\mathbf{B}|$. In our example,

$$b = r_0 = r_1 \times \frac{v}{k} = 14.$$

Such calculations not only enable us to compute the numbers r_s $(0 \leqslant s \leqslant t-1)$, but also provide necessary conditions for the existence of a t-design.

Theorem 4.7.2 (i) If $\mathbf{B}$ is a t-design with parameters (v, k, r_t) then, when $\mathbf{B}$ is considered as an s-design with $1 \leqslant s \leqslant t-1$, the parameter r_s is given by

$$r_s = r_t \frac{(v-s)(v-s-1) \cdots (v-t+1)}{(k-s)(k-s-1) \cdots (k-t+1)}.$$

(ii) If there is a t-design with parameters (v, k, r_t) then for each s in the range $0 \leqslant s \leqslant t-1$ we must have

$$(k-s)(k-s-1) \cdots (k-t+1) \mid r_t(v-s)(v-s-1) \cdots (v-t+1).$$

Proof (i) This formula is obtained by repeated application of the expression for r_{t-1} in terms of r_t.

(ii) Since the numbers r_s $(0 \leqslant s \leqslant t-1)$ must be integers, the denominator must divide the numerator. $\qquad \square$

The divisibility conditions (ii) exclude many sets of parameters. For example, when $v = 56$, $k = 11$, and $r_2 = 1$ the conditions are

$$(s = 0): \quad 11 \times 10 \mid 56 \times 55,$$
$$(s = 1): \quad 10 \mid 55.$$

Since the $s = 1$ condition is violated there is no such 2-design.

It is very important to note that, when $t \geqslant 2$, the divisibility conditions are necessary for the existence of a t-design, but *not* sufficient. (The proof of this result is part of the more advanced mathematical theory of t-designs.) This situation is in contrast to the case $t = 1$ when, according to Theorem 4.6, the divisibility condition is both necessary and sufficient.

Exercises 4.7

1   Given that there is a 5-design with parameters $v = 12$, $k = 6$, and $r_5 = 1$, find the values of r_4, r_3, r_2, r_1, and b.

2   Is it possible that designs of the following kind exist?
(i) A 3-design with $v = 15$, $k = 6$, and $r_3 = 2$.
(ii) A 4-design with $v = 11$, $k = 5$, and $r_4 = 1$.

3   A 2-design with $k = 3$ and $r_2 = 1$ is usually called a *Steiner triple system* (STS). Determine for which values of v in the range $3 \leq v \leq 12$ an STS with v varieties can exist, and construct an STS in these cases.

4   Show that an STS with v varieties can exist only if v is a positive integer of the form $6n + 1$ or $6n + 3$.

5   In 1850 the Rev. T. P. Kirkman proposed the following problem. 'Fifteen young ladies in a school walk out three abreast for seven days in succession: it is required to arrange them daily so that no two shall walk twice abreast.'
    Solve Kirkman's problem and explain how your solution is related to the existence of a special kind of STS with fifteen varieties.

4.8 Miscellaneous Exercises

1   Write out the formulae for $(x + y)^9$ and $(x - y)^9$.

2   Calculate the coefficient of
(i) x^6 in $(1 + x)^{12}$,
(ii) $a^3 b^7$ in $(a + b)^{10}$,
(iii) $a^4 b^6$ in $(a^2 + b)^8$.

3   Prove that
$$\binom{n}{r}\binom{r}{k} = \binom{n}{k}\binom{n-k}{r-k}.$$

4   When all possible diagonals connecting a set of n points on a circle are drawn, no three of them are concurrent. How many internal points of intersection are there?

5   Show that the number of ways of distributing n identical balls into m labelled boxes, some of which may be empty, is
$$\binom{n+m-1}{n}.$$

6 Show that when $n \geqslant m$

$$\binom{m}{m} + \binom{m+1}{m} + \cdots + \binom{n}{m} = \binom{n+1}{m+1}.$$

7 Let X be an n-set. Show that
  (i) there is a set of $\binom{n-1}{k-1}$ k-subsets of X each pair of which has a non-empty intersection;
  (ii) there is a set of $\binom{n}{n*}$ subsets of X with the property that no one of them is contained in another, where n^* is equal to $\frac{1}{2}n$ if n is even and $\frac{1}{2}(n-1)$ if n is odd.

8 If $\mathbf{u}$ and $\mathbf{v}$ are words of length n in the alphabet $\{0, 1\}$ let $\mathbf{u} + \mathbf{v}$ denote the word obtained by adding corresponding digits of $\mathbf{u}$ and $\mathbf{v}$ according to the rules $0+0=0$, $0+1=1$, $1+0=1$, $1+1=0$. Let X denote the set of all such words, with the exception of $00\ldots0$. Show that the set of all 3-subsets of X which have the form

$$\{\mathbf{u}, \mathbf{v}, \mathbf{u}+\mathbf{v}\}$$

is a 2-design with parameters $(2^n - 1, 3, 1)$. (In other words, it is a Steiner triple system with $2^n - 1$ varieties).

9 Show that the parameters

$$v = q^3 + 1, \qquad k = q + 1, \qquad r_2 = 1$$

satisfy the divisibility conditions for a 2-design. What are the values of r_1 and r_0?

10 Suppose $\mathbf{B}$ is the set of blocks of a t-design on a set X, with parameters (v, k, r), and choose x in X. Let $\mathbf{B'}$ be the set of blocks obtained by deleting all the blocks which do *not* contain x, and removing x from those which do. Show that $\mathbf{B'}$ is a $(t-1)$-design on $X-\{x\}$, with parameters $(v-1, k-1, r)$.

11 Show that a 2-design with parameters $(v, 4, 1)$ can exist only if v is a positive integer of the form $12n+1$ or $12n+4$.

12 Construct a 3-design with parameters $(10, 4, 1)$.

13 How many integers x in the range $1 \leqslant x \leqslant 1000$ are not divisible by 2, 3, or 5?

14 Professor McBrain has taught the same course for the last twelve years and tells three jokes each year. He has never told the same set of three jokes twice (the order of the jokes is unimportant). How many jokes must he know?

15 A number of men enter a disreputable establishment, and each one leaves a coat and an umbrella at the door. When a message is received saying that the establishment is about to be raided by the police, the men leave hurriedly, and each man takes a wrong coat and a wrong umbrella. If there are n men, show that the number of ways in which this can happen is

$$n!\left(n! - \frac{(n-1)!}{1!} + \frac{(n-2)!}{2!} - \cdots + (-1)^n \frac{1}{n!} \right).$$

16 A function of f defined on the set of positive integers is said to be **multiplicative** if

$$f(nm) = f(n)f(m) \quad \text{whenever} \quad \gcd(n, m) = 1.$$

Show that if f is multiplicative then so is the function g defined by

$$g(n) = \sum_{d \mid n} f(d).$$

17 Write down formulae for

$$\text{(i)} \quad \sum_{d \mid n} \mu(d)\phi(d), \qquad \text{(ii)} \quad \sum_{d \mid n} \frac{\mu(d)}{\phi(d)}.$$

18 Let $\sigma_k(n)$ denote the sum of the kth powers d^k taken over all divisors d of n. Show that $\sigma_k(n)$ is multiplicative (Ex. 16) and find a formula for it.

19 Let $S_r(n)$ be the sum of the rth powers of the first n positive integers. Show that for $r \geq 1$

$$(n+1)^{r+1} - (n+1) = \sum_{i=1}^{r} \binom{r+1}{i} S_{r-i+1}(n).$$

Deduce that there is a formula for $S_r(n)$ which is a polynomial of degree $r+1$ in n.

20 Give an alternative proof of the binomial theorem based on the principle of induction and Theorem 4.1.1.

5
Partition, classification, and distribution

5.1 Partitions of a set

In this chapter we shall study three kinds of counting problems, associated with the partition of a set into subsets, the classification of a set of objects, and the distribution of a set of objects into a set of boxes. It will become clear that these three ideas are all variations on the same theme.

We begin with some useful notation. Let I be a non-empty set, finite or infinite, and suppose that for each i in I we are given a set X_i. Then we say that we have a **family** of sets, written

$$\mathcal{X} = \{X_i \mid i \in I\},$$

and we refer to I as the **index set**. Notice that we do not insist that the sets are all different, although in the application which we have in mind the sets are not only different, but disjoint.

Definition A **partition** of a set X is a family $\{X_i \mid i \in I\}$ of non-empty subsets of X such that

(i) X is the union of the sets X_i $(i \in I)$,
(ii) each pair X_i, X_j $(i \neq j)$ is disjoint.

The subsets X_i are called the **parts** of the partition.

Another way of stating the definition is to say that every member of X must belong to one, and only one, part. For example Fig. 5.1 shows a partition of $\mathbb{N}_{16}$ with five parts X_1, X_2, X_3, X_4, X_5, where

$$X_1 = \{1, 5, 9\}, \qquad X_2 = \{2, 3, 4, 6, 7\}, \qquad X_3 = \{8\},$$
$$X_4 = \{10, 11, 13, 14\}, \qquad X_5 = \{12, 15, 16\}.$$

In general, when X is a finite set, a partition of X must have a finite number k of parts, so we can use $\mathbb{N}_k$ as the index set and list the parts as $X_1, X_2, \ldots, X_k$. But it must be stressed that the order of the parts is irrelevant.

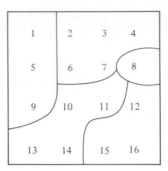

Fig. 5.1
A partition of $\mathbb{N}_{16}$.

Theorem Let $S(n, k)$ denote the number of partitions of an n-set X into k parts, where $1 \leqslant k \leqslant n$. Then

$$S(n, 1) = 1, \qquad S(n, n) = 1,$$
$$S(n, k) = S(n-1, k-1) + kS(n-1, k) \quad (2 \leqslant k \leqslant n-1).$$

Proof Clearly there is only one partition with one part, X itself, and there is only one partition with n parts, the singleton subsets $\{x\}$.

Let z be any given member of X. Every partition of X has the property that either (i) the singleton set $\{z\}$ is a part or (ii) the part containing z also has other members. When the part $\{z\}$ is removed from a partition of type (i) we obtain a partition of the $(n-1)$-set $X \backslash \{z\}$ into $k-1$ parts, and there are $S(n-1, k-1)$ of those. Conversely, if we are given such a partition we can restore the part $\{z\}$, so that the correspondence is a bijection.

Now suppose we have a partition $\mathcal{X}$ of type (ii), with parts $X_1, X_2, \ldots, X_k$. This situation determines a pair of objects $(i, \mathcal{X}_0)$, such that z is in X_i and $\mathcal{X}_0$ is the partition of the $(n-1)$-set $X \backslash \{z\}$ with parts $X_1, X_2, \ldots, X_i \backslash \{z\}, \ldots, X_k$ (Fig. 5.2). There are k possible values of i and $S(n-1, k)$ possible partitions $\mathcal{X}_0$, so we have $kS(n-1, k)$ such pairs. Furthermore, if we are given such a pair we

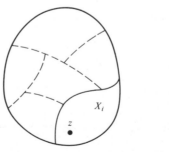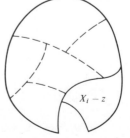

Fig. 5.2
The partitions $\mathcal{X}$ and $\mathcal{X}_0$.

can restore z to the part X_i, and recover $\mathcal{X}$. Hence this correspondence is also a bijection. Since every partition is of type (i) or type (ii) we have the result. □

The numbers $S(n, k)$ are sometimes called **Stirling numbers** (of the second kind). As a consequence of Theorem 5.1 they may be tabulated (Table 5.1.1) in much the same way as the binomial numbers are arranged in Pascal's triangle.

Table 5.1.1

					1				
				1		1			
			1		3		1		
		1		7		6		1	
	1		15		25		10		1
1		31		90		65		15		1
1	63		301		350		140		21	1

The table can be built up one row at a time, using the recursion formula obtained in the theorem. Each entry is calculated from the two immediately above it; for example, the fourth entry in the bottom row is

$$S(6, 4) = S(5, 3) + 4S(5, 4) - 90 + (4 \times 65) = 350.$$

Exercises 5.1

1 Compute one more row of the table.

2 Give direct proofs of the identities

$$S(n, 2) = 2^{n-1} - 1, \qquad S(n, n-1) = \binom{n}{2}.$$

3 Suppose we are given a partition of the n-set X into k parts, and we delete the part containing a given element x. We get a partition of a subset Y of X into $k-1$ parts, where $r = |Y|$ is in the range $0 \leqslant r \leqslant n-1$. Use this idea to show that

$$S(n, k) = \sum_{r=0}^{n-1} \binom{n-1}{r} S(r, k-1).$$

5.2 Classification and equivalence relations

There is another way of looking at a partition of a set of objects. When a child sorts a pile of bricks into subsets according to their colours, we could say that she is constructing a partition of the set of bricks. Alternatively, we could say that she is *classifying* the bricks, using the fact that bricks with the same colour are related, and must be put into the same class.

More formally, suppose that $\{X_i \mid i \in I\}$ is a partition of the set X. Then we write

$$x \, R \, x' \quad \text{or} \quad x \text{ is related to } x',$$

whenever x and x' are in the same part X_i. The relation R defined in this way has three properties which are trivial consequences of the definition. If x, y, z are any members of X, not necessarily different, then we have the following statements, which are known by the names indicated:

$$x \, R \, x \qquad\qquad\qquad \text{(the } \textit{reflexive} \text{ property)}$$
$$x \, R \, y \;\Rightarrow\; y \, R \, x \qquad\quad \text{(the } \textit{symmetric} \text{ property)}$$
$$x \, R \, y \text{ and } y \, R \, z \;\Rightarrow\; x \, R \, z \quad \text{(the } \textit{transitive} \text{ property).}$$

A relation R on a set X is said to be an **equivalence relation** if it is reflexive, symmetric, and transitive.

Exercises 5.2

1 The symbols in the left-hand column of Table 5.2.1 denote relations between integers. Answer yes or no to each question about each relation.

Table 5.2.1

$x \, R \, y$ means	Reflexive?	Symmetric?	Transitive?	Equivalence relation?
$x \leqslant y$				
$x \mid y$				
$3 \mid (x - y)$				
$x + y = 7$				

Just as a partition gives rise to an equivalence relation, so conversely an equivalence relation R defined on a set X determines a partition of X. In order to see how this comes about, we define the **equivalence class** of x, as follows:

$$C_x = \{x' \in X \mid x' \, R \, x\}.$$

C_x is the subset of X containing all those x' which are related to x in the given equivalence relation R. It is important to notice that, in general, the subset C_x will have several different names. In fact, if x and y satisfy $x \, R \, y$, then the equivalence classes C_x and C_y are the same. For suppose $x \, R \, y$, and let z be any member of C_x. Then

	$z \, R \, x$	(definition of C_x)
and	$x \, R \, y$	(given)
so	$z \, R \, y$	(transitive property).

In other words, z is in C_y, and we have proved $C_x \subseteq C_y$. A similar argument shows that $C_y \subseteq C_x$, so $C_x = C_y$ as we claimed.

Thus every equivalence class has several aliases; indeed it uses the name C_x for every x which belongs to it. Once this rather confusing point has been understood the proof of the next theorem is almost self-evident.

Theorem 5.2   If R is an equivalence relation on a set X then the distinct equivalence classes with respect to R form a partition of X.

Proof   Since (by the reflexive property) each element x belongs to its class C_x, the classes are non-empty and their union is the whole set X.

It remains to show that two distinct classes are disjoint; or equivalently, that overlapping classes are identical. Suppose then that z belongs to the classes C_x and C_y. We have

	$z \, R \, x$	(definition of C_x)
so	$x \, R \, z$	(symmetric property)
also	$z \, R \, y$	(definition of C_y)
so	$x \, R \, y$	(transitive property).

Now, by the discussion preceding the theorem, $x \, R \, y$ means that $C_x = C_y$, as required.   □

We have established that a partition of X and an equivalence relation on X are essentially the same thing. The idea of a partition is easy to grasp, whereas an equivalence relation is, at first sight, a more elusive concept. But in mathematics it is often convenient to begin by defining a relation on a set and then prove that it has the three properties that characterize an equivalence relation. The resulting partition, or set of distinct equivalence classes, is sometimes surprisingly familiar. This is the case in the following example, where we follow the standard practice of using a sign resembling equality, such as $\sim$, to denote an equivalence relation.

Example   Let X denote the set of ordered pairs of integers (a, b) with $b \neq 0$. Show that the relation $\sim$ on X defined by

$$(a, b) \sim (c, d) \quad \Leftrightarrow \quad ad = bc$$

is an equivalence relation.

Solution   We check the three properties in turn.

Reflexive property: Since $ab = ba$, we have $(a, b) \sim (a, b)$.

Symmetric property: Using the definition of $\sim$ we have the implications

$$(a, b) \sim (c, d) \quad \Rightarrow \quad ad = bc \quad \Rightarrow \quad cb = da \quad \Rightarrow \quad (c, d) \sim (a, b).$$

Transitive property: Suppose $(a, b) \sim (c, d)$ and $(c, d) \sim (e, f)$, so that $ad = bc$ and $cf = de$. Then

$$afd = adf = bcf = bde = bed,$$

and since $d \neq 0$, $af = be$. So $(a, b) \sim (e, f)$ as required. $\square$

In this example the equivalence classes are indeed familiar objects, for we may identify the class of (a, b) with the fraction a/b. The class of $(1, 2)$, for example, contains the pairs $(2, 4)$, $(3, 6)$, $(4, 8)$, and so on, just as the fraction $\frac{1}{2}$ is equivalent to $\frac{2}{4}$, $\frac{3}{6}$, $\frac{4}{8}$, and so on. This provides a means of *constructing* the fractions, using only the properties of the integers stated in Chapter 1. The fact that a fraction is really an equivalence class is not a problem, since we have all been trained to treat $\frac{1}{2}$ and $\frac{2}{4}$ as being equivalent.

The kind of construction mentioned above is very common in mathematics; indeed, the next chapter is devoted to just such a construction.

Exercises 5.2 (continued)   2   Let $X = \{1, 2, 5, 6, 7, 9, 11\}$ and define $x \sim x'$ to mean that $x - x'$ is divisible by 5. Check that $\sim$ is an equivalence relation and describe the partition of X into equivalence classes.

3 Suppose that four chairs labelled 1, 2, 3, 4 are arranged round a circular table with equal spaces between them. A *seating plan* for four people A, B, C, D can be described by an array such as

$$1 \quad 2 \quad 3 \quad 4$$
$$C \quad A \quad B \quad D,$$

which means that C occupies chair 1, A occupies chair 2, and so on. Two seating plans are *related* (in the relation R) if one of them can be obtained from the other by moving everyone the same number of places to the right.

(i) Show that the total number of seating plans is 24.
(ii) Show that R is an equivalence relation.
(iii) Find the number of equivalence classes and give a representative of each one.

4 Find the number of seating plans and the number of equivalence classes when there are n people and n chairs.

5 Define a relation $\approx$ on $\mathbb{Z}$ by the rule that $n \approx n'$ means $n\,n' > 0$. Show that this relation is symmetric and transitive but not reflexive.

6 What is wrong with the following attempt to prove that if a relation $\sim$ is both symmetric and transitive, then it must be reflexive?
If $a \sim b$ then $b \sim a$ (symmetric property).
But $a \sim b$ and $b \sim a$ imply $a \sim a$ (transitive property).
Hence $a \sim a$ for all a.

5.3 Distributions and the multinomial numbers

Yet another way of looking at a partition $\{X_i \mid i \in I\}$ of a set X is to consider the associated function p from X to I, defined by

$$p(x) = i \quad \text{whenever } x \in X_i.$$

Since each x belongs to precisely one part X_i, p is well-defined by this rule. In concrete terms, instead of partitioning the set of objects, we are making a *distribution* of the objects into boxes labelled by the members of the index set. An example is illustrated in Fig. 5.3.

By definition, each part of a partition is a non-empty set, and it follows that p is a surjection: each box receives at least one object.

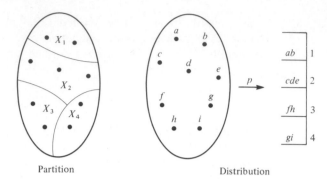

Fig. 5.3
A partition and the corresponding distribution.

Partition · · · · · · · · · · · · Distribution

(It may be recalled that we previously used the objects-into-boxes picture of a function in our discussion of the pigeonhole principle (Section 2.4), but there we were concerned with injections rather than surjections.)

The connection between partitions and distributions also works in the other direction. If we are given a surjection p from a set X to a set Y then the subsets of X defined for each y in Y by

$$X_y = \{x \in X \mid p(x) = y\}$$

form a partition of X. This relationship is the basis for a simple method of counting surjections.

Theorem 5.3.1

Let J denote the set of surjections from an n-set X to a k-set Y. Then

$$|J| = k! \, S(n, k).$$

Proof

Each surjection p from X to Y induces a partition of X into k parts as above. Conversely, if we are given a partition of X into k parts, there are $k!$ surjections which induce it, since the k parts can be assigned to the k members of Y in any bijective way. Hence the number of surjections is $k!$ times the number of partitions $S(n, k)$, as claimed.   □

Many practical problems can be analysed in terms of the objects-into-boxes picture of a surjection. For instance, suppose we wish to count the number of ways of dealing cards for a game of Bridge. In this case we have 52 objects (the cards) and four boxes (the players), and we have to assign 13 cards to each player. One player (say N) is designated as the first bidder, and the others (E, S, W) must bid in turn, so it matters which player gets which cards. In other words, we do not require the number of partitions of a 52-set X into four 13-sets, but rather the number of surjections of a 52-set X onto the 4-set $Y = \{N, E, S, W\}$ with the property that each member of Y receives 13 members of X.

More generally we may ask for the number of surjections from an n-set to a k-set $\{y_1, y_2, \ldots, y_k\}$, with the property that n_1 objects go into the first box y_1, n_2 go into y_2 and so on. This number is denoted by

$$\binom{n}{n_1, n_2, \ldots, n_k}$$

and is called a **multinomial number**. Of course, we must have $n_1 + n_2 + \cdots + n_k = n$.

The multinomial numbers are a generalization of the binomial numbers. When $k = 2$ any multinomial number is equal to a binomial number; specifically

$$\binom{n}{n_1, n_2} = \binom{n}{n_1},$$

since both numbers count the number of ways of selecting the n_1 objects to go into the first box, and the remaining $n_2 = n - n_1$ objects must go into the second box. It is clear that we can obtain recursion formulae for the multinomial numbers similar to the basic recursion formula for the binomial numbers (as in Ex. 5.3.4), but since these formulae involve complicated multiple recursions we shall proceed directly to an explicit formula.

Theorem 5.3.2   Given any positive integers $n, n_1, \ldots, n_k$ satisfying $n_1 + n_2 + \cdots + n_k = n$, we have

$$\binom{n}{n_1, n_2, \ldots, n_k} = \frac{n!}{n_1! \, n_2! \cdots n_k!}.$$

Proof   Let $X = \{x_1, x_2, \ldots, x_n\}$, $Y = \{y_1, y_2, \ldots, y_k\}$. Any permutation π of $\mathbb{N}_n$ effects a rearrangement of X into the order

$$x_{\pi(1)}, x_{\pi(2)}, \ldots, x_{\pi(n)}.$$

Define a surjection of X onto Y by sending the first n_1 members of this list to y_1, the next n_2 to y_2, and so on. This is a surjection of the kind counted by the multinomial number.

However, we get the same surjection if we rearrange the first n_1 objects among themselves in any way, and rearrange the next n_2 among themselves, and so on. It follows that, of the $n!$ permutations

π, there are $n_1! \, n_2! \cdots n_k!$ which induce any given surjection. Hence the total number of surjections is as stated. $\qquad\square$

Example   How many 11-letter words can be made from the letters of the word ABRACADABRA?

Solution   Each word has 11 letters $x_1 x_2 x_3 \ldots x_{11}$, and five of the letters are A's, two of them are B's, two of them are R's, one is a C, and one is a D. Thus each word corresponds to a surjection from the set of 11 objects $\{x_1, x_2, x_3, \ldots, x_{11}\}$ to the set of five boxes $\{A, B, R, C, D\}$, such that five objects go into box A, two go into box B, two go into box R, one goes into box C, and one goes into box D. So the required number is the multinomial number

$$\binom{11}{5, 2, 2, 1, 1},$$

and by Theorem 5.3.2 its value is

$$\frac{11!}{5! \, 2! \, 2! \, 1! \, 1!} = 11 \times 9 \times 7 \times 6 \times 5 \times 4 = 83 \; 160. \qquad\square$$

It is convenient to extend the definition of the multinomial numbers to cover the case when one or more of the integers n_i ($1 \le i \le k$) is zero. Thus if $n_1, n_2, \ldots, n_k$ are *non-negative* integers we define the symbol

$$\binom{n}{n_1, n_2, \ldots, n_k}$$

to be the number of functions from an n-set to a set of k boxes, with the property that n_1 objects go into the first box, n_2 objects go into the second box, and so on. In this case the functions are not necessarily surjections, since any of the n_i may be zero. However, with the convention $0! = 1$, it is clear that the formula

$$\binom{n}{n_1, n_2, \ldots, n_k} = \frac{n!}{n_1! \, n_2! \cdots n_k!}$$

remains true.

Since the multinomial numbers are a generalization of the binomial numbers, it is not surprising that there is a corresponding generalization of the binomial theorem. It is known as the **multinomial theorem**.

Theorem 5.3.3

For any positive integers n and k

$$(x_1 + x_2 + \cdots + x_k)^n = \sum \binom{n}{n_1, n_2, \ldots, n_k} x_1^{n_1} x_2^{n_2} \cdots x_k^{n_k},$$

where the sum is taken over all k-tuples of non-negative integers $(n_1, n_2, \ldots, n_k)$ such that $n_1 + n_2 + \cdots + n_k = n$.

Proof

When the n factors $x_1 + x_2 + \cdots + x_k$ are multiplied, a typical product term $x_1^{n_1} x_2^{n_2} \cdots x_k^{n_k}$ arises by choosing the x_1 term from n_1 of the factors, the x_2 term from n_2 of the factors, and so on. In other words, the typical term corresponds to a function from the set of n factors to the set $\{x_1, x_2, \ldots, x_k\}$, with the property that n_1 of the factors go to x_1, n_2 of them go to x_2, and so on. By the definition of the multinomial numbers, there are

$$\binom{n}{n_1, n_2, \ldots, n_k}$$

functions of this kind, and so this is the number of terms $x_1^{n_1} x_2^{n_2} \cdots x_k^{n_k}$ in the product. □

Because of their occurrence as coefficients of the terms in the expansion of $(x_1 + x_2 + \cdots + x_k)^n$ the multinomial numbers are often referred to as multinomial *coefficients*. However the proof of the multinomial theorem shows that they occur in this way because they represent the numbers of functions of a certian kind, and for this reason we prefer to use a name which emphasizes their definition as basic counting numbers.

Exercises 5.3

1 Formulate and solve the following problem as a question about surjections of an 11-set onto a 4-set: 'How many 11-letter words can be made from the letters of the word MISSISSIPPI?'

2 Evaluate the multinomial numbers

$$\binom{10}{4, 3, 2, 1} \quad \text{and} \quad \binom{9}{5, 2, 2}.$$

3 Show that the number of different positions possible after four moves in a game of noughts-and-crosses (tic-tac-toe) is 756.

4   Show that if $a + b + c = n$, then

$$\binom{n}{a, b, c} = \binom{n-1}{a-1, b, c} + \binom{n-1}{a, b-1, c} + \binom{n-1}{a, b, c-1}.$$

Write down the analogous formula for a general multinomial number.

5   Calculate the coefficient of

(i) $x^5 y^3 z^2$ in $(x + y + z)^{10}$,
(ii) $x^3 y z^4 t$ in $(x + y + z + t)^9$.

6   Let p be a prime. Show that the multinomial number

$$\binom{p}{n_1, n_2, \ldots, n_k}$$

is divisible by p unless one of the n_i $(1 \leqslant i \leqslant k)$ is equal to p.

5.4 Partitions of a positive integer

If we are given a partition of an n-set X, such as

$$X = X_1 \cup X_2 \cup \cdots \cup X_k,$$

then there is a corresponding equation

$$n = n_1 + n_2 + \cdots + n_k,$$

where n_i is the size of X_i $(1 \leqslant i \leqslant k)$. We refer to this equation as a **partition of the integer n** into k parts. It must be stressed that the parts are non-zero (since the sets X_i are non-empty) and that the order of the parts is unimportant. Thus the partitions of the integer 6 are

6,	5 + 1,	4 + 2,
4 + 1 + 1,	3 + 3,	3 + 2 + 1,
3 + 1 + 1 + 1,	2 + 2 + 2,	2 + 2 + 1 + 1.
2 + 1 + 1 + 1 + 1,	1 + 1 + 1 + 1 + 1 + 1.	

The standard notation for partitions of a positive integer n involves counting the number of parts of each size. If there are α_i parts of size i, then the partition is written as

$$[1^{\alpha_1} 2^{\alpha_2} \cdots n^{\alpha_n}].$$

With this notation, and making some trivial abbreviations, the partitions of 6 are:

[6],	[3²],
[15],	[2³],
[1²4],	[24]
[1³3],	[123]
[1⁴2],	[1²2²].
[1⁶],	

It is rather unfortunate that this conventional notation for partitions uses a multiplicative symbol for an additive decomposition. It is worth remembering that when we say, for example, that [2³] is a partition of 6, we mean that 6 is the *sum* of three twos.

The problem of counting partitions of n is a very interesting one, but it requires techniques which we have not yet dealt with in this book. We shall return to the problem in Chapter 19.

Exercises 5.4

1   Write down the partitions of 7 in standard notation.

2   Let $p_k(n)$ denote the number of partitions of n into k parts. Prove that

$$p_k(n) = p_k(n-k) + p_{k-1}(n-k) + \cdots + p_1(n-k).$$

3   Use the formula given in Ex. 2 to construct a table of the numbers $p_k(n)$ for $1 \leqslant k \leqslant n \leqslant 7$, and hence check your answer to Ex. 1.

5.5   Classification of permutations

In Section 3.6 we showed how any permutation can be written in terms of disjoint cycles. For example, the permutation of $\mathbb{N}_9$ indicated in Fig. 5.4 is written as (13624)(587)(9); clearly its cycles are the parts of a partition of $\mathbb{N}_9$, as shown on the right. This observation may be used to provide a formal justification for the cycle notation, by means of equivalence relations. The details are indicated in Ex. 5.7.18.

Fig. 5.4
A permutation of $\mathbb{N}_9$ and the corresponding partition.

In this section we shall study the classification of permutations according to their cycle structure. Recall that S_n denotes the set of all permutations of $\mathbb{N}_n$. Associated with each permutation π in S_n is the partition of $\mathbb{N}_n$ whose parts are the cycles of π, and this in turn yields a corresponding partition of the integer n. We shall refer to the latter as the **type** of π. In other words, if π has α_i cycles of length i $(1 \leq i \leq n)$, then the type of π is the partition $[1^{\alpha_1} 2^{\alpha_2} \cdots n^{\alpha_n}]$ of n. The permutation depicted in Fig. 5.4 has type $[135]$.

It is fairly easy to count the number of permutations of a given type, provided we remember the conventions of the cycle notation. Suppose, for instance, we wish to count the number of members of S_{14} which have type $[2^2 3^2 4]$. We have to put the symbols $1, 2, \ldots, 14$ into the cycle pattern

$$(\cdot\,\cdot)(\cdot\,\cdot)(\cdot\,\cdot\,\cdot)(\cdot\,\cdot\,\cdot)(\cdot\,\cdot\,\cdot\,\cdot),$$

and there are $14!$ ways of doing this. However, a given permutation π arises from this procedure in many different ways. With regard to each cycle, any member of that cycle can be put in the first position and the order of the rest is then determined by π. So there are two ways of getting each 2-cycle, three ways of getting each 3-cycle, and four ways of getting the 4-cycle. Hence the internal arrangement of the cycles can be made in $2^2 \times 3^2 \times 4$ ways for each given π. In general, for a permutation of type $[1^{\alpha_1} 2^{\alpha_2} \cdots n^{\alpha_n}]$ there are

$$1^{\alpha_1} 2^{\alpha_2} \cdots n^{\alpha_n}$$

ways of making the internal arrangements.

Also, the order of cycles of equal length is arbitrary. In the example, there are $2!$ ways of ordering the two 2-cycles and $2!$ ways of ordering the two 3-cycles. In general, the relevant number is

$$\alpha_1! \, \alpha_2! \cdots \alpha_n!.$$

Thus the number of permutations of type $[2^2 \, 3^2 \, 4]$ is

$$\frac{14!}{2^2 \times 3^2 \times 4 \times 2! \times 2!};$$

and the number of type $[1^{\alpha_1}2^{\alpha_2}\cdots n^{\alpha_n}]$ is

$$\frac{n!}{1^{\alpha_1}2^{\alpha_2}\cdots n^{\alpha_n}\alpha_1!\,\alpha_2!\cdots\alpha_n!}.$$

In simple cases it is often easier to use commonsense methods rather than this cumbersome formula. For example, the numbers in the classification of S_5 shown in Table 5.5.1 can be obtained in a variety of simple ways.

Table 5.5.1

Type	Example	Number
$[1^5]$	id	1
$[1^3 2]$	(12)(3)(4)(5)	10
$[1^2 3]$	(123)(4)(5)	20
$[12^2]$	(12)(34)(5)	15
$[14]$	(1234)(5)	30
$[23]$	(123)(45)	20
$[5]$	(12345)	24
		120

The classification of permutations by type is doubly useful, because there is an alternative description of the classes which has useful consequences in the algebraic theory of permutations (Chapters 14 and 20). Let α and β be permutations in S_n. If there is a permutation σ in S_n such that

$$\sigma\alpha\sigma^{-1} = \beta,$$

then we say that α and β are **conjugate**.

Theorem 5.5

Two permutations are conjugate if and only if they have the same type.

Proof

Suppose α and β conjugate, so that $\sigma\alpha\sigma^{-1} = \beta$. If $\alpha(x_1) = x_2$, put $y_1 = \sigma(x_1)$, $y_2 = \sigma(x_2)$; if $\alpha(x_2) = x_3$ put $y_3 = \sigma(x_3)$; and so on (Fig. 5.5). Then we have

$$\beta(y_1) = \sigma\alpha\sigma^{-1}(\sigma(x_1)) = \sigma\alpha(x_1) = \sigma(x_2) = y_2.$$

Similarly $\beta(y_2) = y_3$, $\beta(y_3) = y_4$, and so on. Consequently, for each cycle $(x_1 x_2 \cdots x_r)$ of α there is a corresponding cycle $(y_1 y_2 \ldots y_r)$ of β, and it follows that α and β have the same type.

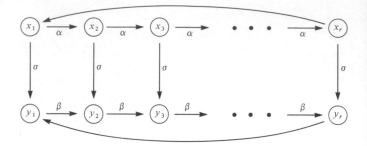

Fig. 5.5
Conjugate permutations.

Conversely, suppose α and β have the same type. Since they have the same number of cycles of each length, we can set up a bijective correspondence between their cycles, in which a typical cycle $(x_1 x_2 \ldots x_r)$ of α will correspond to a cycle $(z_1 z_2 \ldots z_r)$ of β. Let us define σ by the rule $\sigma(x_i) = z_i$ $(1 \leqslant i \leqslant r)$, and use similar rules for the other cycles. Then $\sigma \alpha \sigma^{-1} = \beta$, since

$$\sigma \alpha \sigma^{-1}(z_1) = \sigma \alpha(x_1) = \sigma(x_2) = z_2 = \beta(z_1).$$

and so on. Hence α and β are conjugate. $\qquad \square$

Exercises 5.5

1 Write out the classification of the elements of S_6 in the same form as the table for S_5 given on p. 103).

2 Let $\alpha = (13624)(587)(9)$, $\beta = (15862)(394)(7)$. Write down a permutation σ in S_9 such that $\sigma \alpha \sigma^{-1} = \beta$.

3 Prove that if π and τ are any members of S_n then $\pi\tau$ and $\tau\pi$ have the same type. [Hint: Use Theorem 5.5.]

4 Prove directly from the definition (without using Theorem 5.5) that conjugacy is an equivalence relation on S_n.

5 Use the classification of S_6 obtained in Ex. 1 to find the number of *derangements* in S_6 (as defined in Section 4.4).

6 Find the number of permutations σ which have the property specified in Ex. 2.

5.6   Even and odd permutations

The classification of S_5 obtained in the previous section has a remarkable feature, which should be evident from the following tabulation of the classes.

Type	Number	Type	Number
$[1^5]$	1	$[13]$	10
$[1^2 3]$	20	$[14]$	30
$[1\,2^2]$	15	$[23]$	20
$[5]$	24		
	60		60

Clearly we have a partition of S_5 into two parts of equal size with 60 permutations in each. In this section we shall prove that S_n can be split into two equal parts for every integer $n \geqslant 2$, and we shall see that there is a very simple way of deciding which part contains a given permutation.

The key observation is that every rearrangement can be achieved by successively switching certain pairs of objects. For example, in order to obtain 35142 from 12345 we need just two switches:

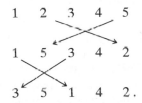

In terms of permutations, the permutation which effects the transformation of 12345 into 35142 is (13) (25) (4), and clearly the two 2-cycles correspond exactly to the switches carried out above. More generally it may not be so obvious that a given permutation can be expressed in terms of 2-cycles in this way, and our first task is to show that this is indeed so.

The technical name for a permutation which interchanges two objects and leaves the rest unaltered is a **transposition**. Thus an element of S_n is a transposition if it has type $[1^{n-2}2]$: it has $n-2$ 1-cycles and just one 2-cycle. Now any cycle, such as $(x_1 x_2 \ldots x_{r-1} x_r)$, effects the rearrangement taking

$$x_1 x_2 \cdots x_{r-1} x_r \quad \text{to} \quad x_2 x_3 \cdots x_r x_1,$$

and this can be achieved by successive transpositions in the following way:

$$
\begin{array}{cccccc}
x_1 & x_2 & x_3 & \ldots & x_{r-1} & x_r \\
\end{array}
$$

$$
\begin{array}{cccccc}
x_2 & x_1 & x_3 & \ldots & x_{r-1} & x_r \\
\end{array}
$$

$$
\begin{array}{cccccc}
x_2 & x_3 & x_1 & \ldots & x_{r-1} & x_r \\
\end{array}
$$

$$
\begin{array}{cccccc}
x_2 & x_3 & x_4 & \ldots & x_1 & x_r \\
\end{array}
$$

$$
\begin{array}{cccccc}
x_2 & x_3 & x_4 & \ldots & x_r & x_1 \, .
\end{array}
$$

So we can write

$$(x_1 x_2 \ldots x_{r-1} x_r) = (x_1 x_r) \ldots (x_1 x_3)(x_1 x_2).$$

Here each transposition is represented by the corresponding 2-cycle, and the 1-cycles are suppressed since they do not have any effect. Also it must be remembered that the transposition written last is the first one to be carried out, in accordance with the rule for combining permutations. Since any permutation can be decomposed into cycles it can thereby be decomposed into transpositions, using the rule given above. For example

$$(136)(2457) = (16)(13)(27)(25)(24).$$

Two things must be stressed about the decomposition into transpositions. First, the transpositions are not disjoint, so that one object may be moved several times. Secondly, the decomposition is far from unique, for not only can the order of the transpositions be altered, but we can actually use an entirely different set of transpositions, such as

$$(136)(2457) = (15)(35)(36)(57)(14)(27)(12).$$

Our aim is to show that, although there are many different ways of decomposing a given permutation, there is one feature common to all of them.

Let $c(\pi)$ denote the total number of cycles of a permutation π, so that if π has type $[1^{\alpha_1} 2^{\alpha_2} \ldots n^{\alpha_n}]$ then $c(\pi) = \alpha_1 + \alpha_2 + \cdots + \alpha_n$. Suppose we combine π with a transposition τ, giving a new permutation $\tau\pi$. What is the relation between $c(\tau\pi)$ and $c(\pi)$?

Suppose τ switches a and b, so that $\tau(a) = b$, $\tau(b) = a$, and $\tau(k) = k$ when $k \neq a, b$. If a and b are in the same cycle of π we can

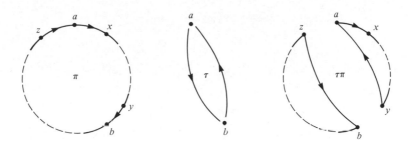

Fig. 5.6
Combining τ with a cycle
of π (first case).

write

$$\pi = (ax \ldots yb \ldots z) \ldots \text{and some other cycles.}$$

The composite permutation $\tau\pi$ (first π, then τ) can be calculated quite simply; as illustrated in Fig. 5.6 it is

$$\tau\pi = (ax \ldots y)(b \ldots z) \ldots \text{and the same other cycles.}$$

In this case $c(\tau\pi) = c(\pi) + 1$. On the other hand, if a and b are in different cycles of π, so that

$$\pi = (ax \ldots y)(b \ldots z) \ldots \text{and some other cycles,}$$

then a similar calculation shows that

$$\tau\pi = (ax \ldots yb \ldots z) \ldots \text{and the same other cycles.}$$

In this case $c(\tau\pi) = c(\pi) - 1$. In both cases the result of following π with a transposition is to alter the number of cycles by one, and this simple fact leads to the next theorem.

Theorem 5.6.1

Suppose the permutation π in S_n can be written as the composite of r transpositions, and also as the composite of r' transpositions. Then either r and r' are both even or r and r' are both odd.

Proof

Let $\pi = \tau_r\tau_{r-1} \cdots \tau_2\tau_1$, where τ_i $(1 \leq i \leq r)$ is a transposition. Since τ_1 has one 2-cycle and $n-2$ 1-cycles we have

$$c(\tau_1) = 1 + (n-2) = n - 1.$$

When $\tau_2, \tau_3, \ldots, \tau_r$ are combined in turn with τ_1 the final result is π. At each stage, the number of cycles is altered by one: suppose it increases by one g times and decreases by one h times. Then the final number of cycles is

$$(n-1)+g-h=c(\pi).$$

But $g+h$ is the total number of stages, $r-1$. So we have

$$r=1+g+h=1+g+(n-1+g-c(\pi))$$
$$=n-c(\pi)+2g.$$

By the same argument, if π is the composite of r' transpositions, there is an integer g' such that $r'=n-c(\pi)+2g'$. Hence

$$r-r'=2(g-g'),$$

and since the right-hand side is even, we have the result. $\qquad\square$

As a consequence of the theorem we can say that a permutation is **even** or **odd**, according as the number of transpositions in any decomposition of it is even or odd. We can also define the **sign** of a permutation π, written sgn π, to be $+1$ if π is even, and -1 if π is odd. Thus

$$\text{sgn } \pi = (-1)^r,$$

where r is the number of transpositions in a decomposition of π. In particular sgn id $=(-1)^0=+1$. If π can be decomposed into r transpositions and σ can be decomposed into s transpositions, then clearly the composite $\pi\sigma$ can be decomposed into $r+s$ transpositions and so

$$\text{sgn } \pi\sigma = (-1)^{r+s} = (-1)^r(-1)^s = \text{sgn } \pi \text{ sgn } \sigma.$$

This implies, for example, that sgn $\pi^{-1}=$ sgn π, since $\pi^{-1}\pi$ is the identity and

$$\text{sgn } \pi^{-1} \text{ sgn } \pi = \text{sgn } \pi^{-1}\pi = \text{sgn id} = +1.$$

Now we can establish the general form of the partition of S_n observed at the beginning of the Section in the case $n=5$.

Theorem 5.6.2 For any integer $n\geqslant 2$ exactly half of the permutations in S_n are even and exactly half of them are odd.

Proof Let $\pi_1, \pi_2, \ldots, \pi_k$ be a list of the even permutations in S_n. (Certainly there are such permutations, since id is one of them.) Let τ be any transposition in S_n, say $\tau=(12)(3)(4)\ldots(n)$.

The permutations $\tau\pi_1, \tau\pi_2, \ldots, \tau\pi_k$ are all distinct, since if $\tau\pi_i=\tau\pi_j$ we should have

$$\pi_i = (\tau^{-1}\tau)\pi_i = \tau^{-1}(\tau\pi_i) = \tau^{-1}(\tau\pi_j) = (\tau^{-1}\tau)\pi_j = \pi_j,$$

using the fundamental rules given in Theorem 3.6. Furthermore, these permutations are all odd, since

$$\text{sgn } \tau\pi_i = \text{sgn } \tau \text{ sgn } \pi_i = (-1)\times(+1) = -1.$$

Finally, we show that *any* odd permutation ρ is one of the $\tau\pi_i$ $(1 \leqslant i \leqslant n)$. Since

$$\text{sgn } \tau^{-1}\rho = \text{sgn } \tau^{-1} \text{ sgn } \rho = (-1)\times(-1) = +1,$$

it follows that $\tau^{-1}\rho$ is one of the even permutations π_i. Thus

$$\rho = (\tau\tau^{-1})\rho = \tau(\tau^{-1}\rho) = \tau\pi_i,$$

as claimed. Hence there are just as many odd permutations as even permutations, and the result follows. □

The partition of S_n into two equal parts has many interesting consequences. The following is an example from 'recreational' mathematics.

Example Eight blocks labelled A, E, I, O, U, Y, R, T, are placed in a square frame as shown in the first diagram (Fig. 5.7). A legal move consists of sliding one block into the currently empty space. Prove that it is impossible to obtain the arrangement shown in the second diagram by a sequence of legal moves.

Fig. 5.7
Can it be done?

Solution Let us denote the space by □ so that the initial arrangement is AEIOUYRT□ and the final arrangement is YOUAREIT□. Moving a letter X into the space corresponds to the 2-cycle $(X\square)$. If we are given any final arrangement with □ in its original place, then in order to achieve it □ must have been moved upwards the same number of times as downwards, and leftwards the same number of times as rightwards. Consequently, the total number of moves required is even and, since each move is a transposition, the final arrangement must be effected by an *even* permutation. However, the permutation which effects the rearrangement of

$$AEIOUYRT\square$$

into

$$YOUAREIT\square$$

is (AYEO)(IUR)(T)($\square$). This is an *odd* permutation, since the 4-cycle is equivalent to three transpositions and the 3-cycle is equivalent to two transpositions. It follows that it is impossible to achieve the required result by a sequence of legal moves. $\square$

Exercises 5.6

1   Express the following members of S_8 in terms of 2-cycles and find the sign of each one.

$$\alpha = (1357)(2468),$$

$$\beta = (127)(356)(48),$$

$$\gamma = (135)(678)(2)(4).$$

2   Without using Theorem 5.5 prove that

$$\text{sgn } \pi\sigma\pi^{-1} = \text{sgn } \sigma \qquad (\sigma, \pi \in S_n).$$

3   Show that if π has type $[1^{\alpha_1}2^{\alpha_2}\ldots n^{\alpha_n}]$ then

$$\text{sgn } \pi = (-1)^{\alpha_2+\alpha_4+\alpha_6+\cdots}.$$

4   Which of the following positions can be obtained by legal moves starting from the initial arrangement specified in the *Example*?

(i)   A R E   (ii)   Y E A
      Y O U   T O
      I T   U R I

5   Show that any cycle of odd length can be written as the composite of (not necessarily disjoint) 3-cycles. Deduce that a permutation is even if and only if it can be expressed as a composite of 3-cycles.

5.7   Miscellaneous Exercises

1   How many 14-letter words can be made from the letters of the word CLASSIFICATION?

2   Calculate the coefficient of

(i) $x^3y^2z^4$ in $(x+y+z)^9$,
(ii) xy^3zt^2u in $(x+y+z+t+u)^8$.

3 Calculate $p(8)$, the total number of partitions of 8, and verify that the number which have distinct parts is equal to the number whose parts are all odd. Can you explain this equality?

4 Let α, β be the elements of S_8 represented in cycle notation as

$$\alpha = (123)(456)(78), \qquad \beta = (1357)(26)(4)(8).$$

Find sgn α and sgn β, and express α and β in terms of transpositions, using the least number of transpositions possible in each case.

5 Show that

$$S(n, 3) = \tfrac{1}{2}(3^{n-1}+1) - 2^{n-1}.$$

6 Let $\sim$ denote the relation on $\mathbb{Z}$ defined by

$$a \sim b \quad \Leftrightarrow \quad a - b \text{ is divisible by } 11.$$

Show that $\sim$ is an equivalence relation on $\mathbb{Z}$. What is the number of equivalence classes?

7 Define a relation $\approx$ on the set $\mathbb{N} \times \mathbb{N}$ of ordered pairs of positive integers by the rule

$$(a, b) \approx (c, d) \quad \Leftrightarrow \quad a + d = b + c.$$

Show that $\approx$ is an equivalence relation on $\mathbb{N} \times \mathbb{N}$. Explain how this result could be used to construct $\mathbb{Z}$, starting from $\mathbb{N}$.

8 Prove that

$$\sum_{k=1}^{m} S(m, k)n(n-1) \cdots (n-k+1) = n^m.$$

9 The United Nations General Assembly has decreed that the national flag of every country must consist of m vertical bands, each coloured with one of n colours in such a way that adjacent bands are coloured differently. Show that $n! \, S(m-1, n-1)$ flags can be constructed in this way. It may be assumed that one edge of each flag is distinguished by being attached to a flagpole, so that ABC and CBA represent different flags.

10 Let q_n denote the total number of partitions of a set with n elements. Show that

$$q_n = \sum_{k=1}^{n} \binom{n-1}{k-1} q_{n-k} = \sum_{k=0}^{n-1} \binom{n-1}{k} q_k.$$

11 Use the sieve principle to show that the number of surjections from an n-set to a k-set is

$$\sum_{i=0}^{n} (-1)^i \binom{n}{i} (n-i)^k.$$

12 Show that

$$S(n, k) = \frac{1}{k!} \sum \binom{n}{n_1, n_2, \ldots, n_k}$$

where the sum is taken over all k-tuples $(n_1, n_2, \ldots, n_k)$ of *positive* integers satisfying $n_1 + n_2 + \cdots + n_k = n$.

13   Show that if the sum on the right-hand side of the previous equation is taken over all *non-negative* integers satisfying $n_1 + n_2 + \cdots + n_k = n$, then the result is k^n.

14   In how many ways can mn objects be distributed among m boxes so that there are n objects in each box?

15   By using the multinomial numbers show that, for any positive integer n,

   (i) 2^n divides $(2n)!$ and the quotient is even;
   (ii) $(n!)^{n+1}$ divides $(n^2!)!$.

16   Show that the number of permutations in S_6 of type $[1^4 2]$ is the same as the number of type $[2^3]$. If α is of the first type, find the number of permutations β of the second type which satisfy $\alpha\beta = \beta\alpha$.

17   List the types of permutations in S_7 which are derangements, and hence check the value of d_7.

18   Let π be a permutation of the finite set X.
   (i) Show that for each x in X there is a non-negative integer k such that $\pi^k(x) = x$, where π^k denotes the k-fold interation of π, and $\pi^0 = \mathrm{id}$.
   (ii) Define a relation $\sim$ on X by the rule that

$$x \sim x' \quad \text{means} \quad x' = \pi^r(x) \text{ for some } r \geqslant 0.$$

   Show that $\sim$ is an equivalence relation on X and explain how this result justifies the cycle notation for permutations.

19   The 'Fifteen Puzzle' consists of square blocks labelled $1, 2, \ldots, 15$ arranged in a 4×4 frame, with one space. As in the *Example* of Section 5.6, a legal move consists of sliding one block into the space, thereby creating a new space. Classify the following positions, putting two positions in the same class if and only if one of them can be obtained from the other by a sequence of legal moves.

1	2	3	4
5	6	7	8
9	10	11	12
13	14	15	

1	2	3	4
5	6	7	8
9	10	11	12
13	15	14	

1	8	9	
2	7	10	15
3	6	11	14
4	5	12	13

1	2	3	4
12	13	14	5
11		15	6
10	9	8	7

20   Find the number of ten-digit positive integers (in base 10 notation) which have the property that the number of distinct odd digits is half the number of distinct even digits.

6
Modular arithmetic

6.1 Congruences

One of the most familiar partitions is the partition of $\mathbb{Z}$ into the even integers and the odd integers. According to the general theory discussed in Section 5.2, this partition corresponds to an equivalence relation on $\mathbb{Z}$ which (in this case) we may define by saying that x_1 is related to x_2 whenever $x_1 - x_2$ is divisible by 2. It is customary to use the notation

$$x_1 \equiv x_2 \pmod{2}$$

for this relation, and to say that x_1 is *congruent* to x_2 modulo 2. Thus, x_1 and x_2 are in the same part of the partition if and only if x_1 is congruent to x_2 modulo 2.

Clearly any positive integer m can be used instead of 2.

Definition Let x_1 and x_2 be integers, and m a positive integer. We say that x_1 is **congruent** to x_2 **modulo** m, and write

$$x_1 \equiv x_2 \pmod{m}$$

whenever $x_1 - x_2$ is divisible by m.

It is easy to check that congruence modulo m is an equivalence relation. It is reflexive, since $x - x$ is zero and is divisible by m for any x. It is symmetric, since if $x_1 - x_2 = km$, then $x_2 - x_1 = (-k)m$. It is transitive, since if $x_1 - x_2 = km$ and $x_2 - x_3 = lm$, then $x_1 - x_3 = (k + l)m$.

The usefulness of congruence relations stems mainly from the fact that they are compatible with the arithmetical operations. Specifically, we have the following theorem.

Theorem 6.1 Let m be a positive integer and x_1, x_2, y_1, y_2 integers such that

$$x_1 \equiv x_2 \pmod{m}, \qquad y_1 \equiv y_2 \pmod{m}.$$

Then

(i) $x_1 + y_1 \equiv x_2 + y_2 \pmod{m}$, (ii) $x_1 y_1 \equiv x_2 y_2 \pmod{m}$.

Proof   (i) We are given that $x_1 - x_2 = mx$, $y_1 - y_2 = my$ for some x and y in $\mathbb{N}$. It follows that

$$(x_1 + y_1) - (x_2 + y_2) = (x_1 - x_2) + (y_1 - y_2)$$
$$= mx + my$$
$$= m(x + y),$$

and so the left-hand side is divisible by m, as required.

(ii) Here we have

$$x_1 y_1 - x_2 y_2 = (x_1 - x_2)y_1 + x_2(y_1 - y_2)$$
$$= mxy_1 + x_2 my$$
$$= m(xy_1 + x_2 y),$$

and again the left-hand side is divisible by m, as required.   □

Example   Let $(x_n x_{n-1} \ldots x_0)_{10}$ be the representation of the positive integer x in base 10. Show that

$$x \equiv x_0 + x_1 + \cdots + x_n \pmod 9$$

and use this result to check the calculation

$$54\,321 \times 98\,765 = 5\,363\,013\,565.$$

Solution   Using the definition of the base 10 representation, we have

$$x - (x_0 + x_1 + \cdots + x_n) = (x_0 + 10x_1 + \cdots 10^n x_n) - (x_0 + x_1 + \cdots + x_n)$$
$$= (10^1 - 1)x_1 + \cdots + (10^n - 1)x_n.$$

Now, for each $r \geqslant 1$,

$$10^r - 1 = (99 \ldots 9)_{10} \quad (r \text{ nines})$$
$$= 9 \times (11 \ldots 1)_{10}.$$

Thus 9 divides $x - (x_0 + x_1 + \cdots + x_n)$, as required.

For convenience, let us write $\theta(x)$ instead of $x_0 + x_1 + \cdots + x_n$. We have shown that $\theta(x) \equiv x \pmod 9$. By part (ii) of Theorem 6.1

$$\theta(x)\theta(y) \equiv xy \pmod 9,$$

and so if $xy = z$ we must have $\theta(x)\theta(y) \equiv \theta(z) \pmod 9$. In the given calculation

$$\theta(54\,321) = 15, \qquad \theta(98\,765) = 35, \qquad \theta(5\,363\,013\,565) = 37,$$

and

$$\theta(15) = 6, \qquad \theta(35) = 8, \qquad \theta(37) = 10.$$

Since 6×8 is not congruent to $10 \pmod 9$ it follows that 15×35 is not congruent to $37 \pmod 9$ and $54\,321 \times 98\,765$ is not congruent to $5\,363\,013\,565 \pmod 9$. If the calculation were correct, the expressions would be equal, and consequently congruent mod 9. Hence the calculation is wrong.

This procedure is known as 'casting out nines'. □

**Exercises
5.1**

1   Without doing any 'long multiplication' show that

(i) $1\,234\,567 \times 90\,123 \equiv 1 \pmod{10}$,
(ii) $2468 \times 13\,579 \equiv -3 \pmod{25}$.

2   Use the method of casting out nines to show that two of the following equations are false. What can be said about the other equation?

(i) $5783 \times 40\,162 = 233\,256\,846$,
(ii) $9787 \times 1258 \quad = \quad 12\,342\,046$,
(iii) $8901 \times 5743 \quad = \quad 52\,018\,443$.

3   Suppose we are given $m \geqslant 2$ and an integer x. The remainder r when x is divided by m satisfies

$$x \equiv r \pmod{m}, \qquad 0 \leqslant r \leqslant m - 1,$$

and is sometimes called the *least non-negative residue* of $x \pmod m$. Find the least non-negative residue of $3^{15} \pmod{17}$ and $15^{81} \pmod{13}$.

4   Let $(x_n x_{n-1} \dots x_0)_{10}$ be the base 10 representation of the positive integer x. Show that

$$x \equiv x_0 - x_1 + x_2 - \dots + (-1)^n x_n \pmod{11},$$

and use this result to test whether $1\,213\,141\,516\,171\,819$ is divisible by 11.

6.2 $\mathbb{Z}_m$ and its arithmetic

In this section we shall introduce a more compact method of dealing with congruence properties of integers.

For any integer x and positive integer m we shall use the notation $[x]_m$ for the equivalence class of x with respect to congruence modulo m. In other words, $[x]_m$ consists of all the integers x' for which $x' - x$ is a multiple of m. For example,

$$[5]_3 = \{\ldots, -4, -1, 2, 5, 8, 11, \ldots\},$$
$$[6]_7 = \{\ldots, -8, -1, 6, 13, 20, \ldots\}.$$

As usual, each equivalence class has many aliases, depending on which representative is used. For example,

$$\ldots = [-8]_7 = [-1]_7 = [6]_7 = [13]_7 = [20]_7 = \ldots .$$

The general theory of equivalence relations ensures that, for each m, the set $\mathbb{Z}$ is partitioned into disjoint equivalence classes by the relation of congruence modulo m. When $m = 3$ we have

$$\mathbb{Z} = X_0 \cup X_1 \cup X_2,$$

where

$$X_0 = [0]_3 = \{\ldots, -3, 0, 3, 6, \ldots\},$$
$$X_1 = [1]_3 = \{\ldots, -2, 1, 4, 7, \ldots\},$$
$$X_2 = [2]_3 = \{\ldots, -1, 2, 5, 8, \ldots\}.$$

In this example, and for any given m, there are m distinct equivalence classes $[0]_m, [1]_m, \ldots, [m-1]_m$. This follows from the fact that any x in $\mathbb{Z}$ can be expressed uniquely in the form $qm + r$ with $0 \leqslant r \leqslant m - 1$ (Theorem 1.5), so that x is in $[r]_m$ for just one such r.

Definition   The set of **integers modulo m**, written $\mathbb{Z}_m$, is the set of distinct equivalence classes under the relation of congruence modulo m in $\mathbb{Z}$.

So $\mathbb{Z}_m$ is the set $\{[0]_m, [1]_m, \ldots, [m-1]_m\}$. It must be emphasized that the members of $\mathbb{Z}_m$ are defined as *subsets* of $\mathbb{Z}$. However, it is often convenient to think of them as the integers $0, 1, 2, \ldots, m-1$ with a modified arithmetical structure, and we can make this idea precise in the following way.

Let us define new operations of 'addition' and 'multiplication' for the members of $\mathbb{Z}_m$, written $\oplus$ and $\otimes$, by the rules

$$[x]_m \oplus [y]_m = [x+y]_m, \qquad [x]_m \otimes [y]_m = [xy]_m.$$

Since x and y are integers, the expressions $x + y$ and xy on the right-hand side are defined, and they have all the properties listed in the first section of this book. The new operations inherit their properties from those properties of the familiar operations. But before we study them we must deal with a difficulty concerning the definitions.

The difficulty arises because each equivalence class has many names. Suppose, for example, that $[x]_m$ and $[x']_m$ denote the same class, and $[y]_m$ and $[y']_m$ denote the same class. Then, in order that the definition of $\oplus$ be reasonable, we must ensure that $[x]_m \oplus [y]_m$ and $[x']_m \oplus [y']_m$ denote the same class. The fact that this is so is a simple consequence of Theorem 6.1. For we are given that $x \equiv x' \pmod{m}$ and $y \equiv y' \pmod{m}$, and so $x + x' \equiv y + y' \pmod{m}$; consequently $[x + x']_m = [y + y']_m$ as required. A similar proof holds for multiplication.

We can now list the arithmetical properties of $\mathbb{Z}_m$. They are numbered **M1 – M6**, to correspond with the axioms for $\mathbb{Z}$ listed in Section 1.1.

Theorem 6.2

The operations $\oplus$ and $\otimes$ satisfy the following rules, where a, b, c denote any members of $\mathbb{Z}_m$, and $0 = [0]_m$, $1 = [1]_m$.

M1. $a \oplus b$ and $a \otimes b$ are in $\mathbb{Z}_m$.
M2. $a \oplus b = b \oplus a$, $\quad a \otimes b = b \otimes a$.
M3. $(a \oplus b) \oplus c = a \oplus (b \oplus c)$, $\quad (a \otimes b) \otimes c = a \otimes (b \otimes c)$.
M4. $a \oplus 0 = a$, $\quad a \otimes 1 = a$.
M5. $a \otimes (b \oplus c) = (a \otimes b) \oplus (a \otimes c)$.
M6. For each a in $\mathbb{Z}_m$ there is a unique element $-a$ in $\mathbb{Z}_m$ such that $a \oplus (-a) = 0$.

Proof

M1 is a direct consequence of the definitions of $\oplus$ and $\otimes$. For the first part of **M2**, suppose $a = [x]_m$ and $b = [y]_m$; then

$$
\begin{aligned}
a \oplus b = [x]_m \oplus [y]_m &= [x + y]_m & \text{(definition of } \oplus) \\
&= [y + x]_m & \text{(Axiom I2 for } \mathbb{Z}) \\
&= [y]_m \oplus [x]_m & \text{(definition of } \oplus) \\
&= b \oplus a.
\end{aligned}
$$

Similar proofs hold for the second part of **M2** and for **M3**, **M4**, **M5**. For **M6**, if $a = [x]_m$ we put $-a = [-x]_m$, and check:

$$
\begin{aligned}
a \oplus (-a) &= [x]_m \oplus [-x]_m \\
&= [x + (-x)]_m \\
&= [0]_m \\
&= 0,
\end{aligned}
$$

as required. $\qquad\qquad\qquad\qquad\qquad\qquad\qquad\qquad\qquad\qquad\qquad\square$

In practice we usually dispense with the cumbersome $[x]_m$ notation for the members of $\mathbb{Z}_m$, and use the integers $0, 1, \ldots, m-1$ to denote the classes $[0]_m, [1]_m, \ldots, [m-1]_m$. The specific value of m under consideration must either be clear from the context, or stated explicitly. Also, we use the usual notations for addition and multiplication, rather than $\oplus$ and $\otimes$. Thus, we write

$$7 + 5 = 3 \text{ (in } \mathbb{Z}_9) \quad \text{instead of} \quad [7]_9 \oplus [5]_9 = [3]_9,$$

and with precisely the same meaning as $7 + 5 \equiv 3 \pmod{9}$. The properties stated in Theorem 6.2 justify most of the usual arithmetical manipulations in $\mathbb{Z}_m$, just as in $\mathbb{Z}$.

However, there are some important differences between $\mathbb{Z}_m$ and $\mathbb{Z}$. Recall that Axiom **I7** for $\mathbb{Z}$ asserts that if $ab = ac$ and $a \neq 0$ then $b = c$. This does *not* hold in $\mathbb{Z}_m$: for example, in $\mathbb{Z}_6$ we have

$$3 \times 1 = 3 \times 5 \text{ and } 3 \neq 0, \text{ but } 1 \neq 5.$$

So we must be careful about 'cancellation' in $\mathbb{Z}_m$, a topic we shall discuss in more detail in the next section.

Finally, we remark that there is no order relation in $\mathbb{Z}_m$ resembling the relation $\leqslant$ in $\mathbb{Z}$. Our intuitive picture of $\mathbb{Z}$ as a regularly-spaced set of points on a line, extending indefinitely in either direction, represents the properties of that relation. In $\mathbb{Z}_m$ we have instead a kind of *cyclic* order, represented by a regularly-spaced set of points on a circle (Fig. 6.1). For this reason, arithmetic in $\mathbb{Z}_m$, or as we shall say, *modular arithmetic*, is often taught in schools as 'clock arithmetic'.

Fig. 6.1
Pictures of $\mathbb{Z}$ and $\mathbb{Z}_5$.

Exercises 6.2

1 Complete Tables 6.2.1(a, b), the addition and multiplication tables for $\mathbb{Z}_6$.

Table 6.2.1(a)

$\oplus$	0	1	2	3	4	5
0	0	1	2	3	4	5
1	1	2	3	4	5	0
2						
3						
4						
5						

Table 6.2.1(b)

$\otimes$	0	1	2	3	4	5
0	0	0	0	0	0	0
1	0	1	2	3	4	5
2						
3						
4						
5						

2 Deduce from Axiom **I7** for $\mathbb{Z}$ that if x and y are integers such that $xy = 0$ and $x \neq 0$, then $y = 0$. Show by counter-examples that this axiom does hold in $\mathbb{Z}_6$, $\mathbb{Z}_8$, and $\mathbb{Z}_{15}$. Is there a counter-example in $\mathbb{Z}_7$?

3 Solve the simultaneous equations

$$x + 2y = 4, \qquad 4x + 3y = 4$$

in $\mathbb{Z}_7$. Is there a solution in $\mathbb{Z}_5$?

4 Solve the quadratic equation

$$x^2 + 3x + 4 = 0$$

in $\mathbb{Z}_{11}$.

6.3 Invertible elements of $\mathbb{Z}_m$

In Chapter 1 we stressed the fact that the symbol s/r need not represent an integer even though r and s are integers. In other words, if we are given integers r and s then it may be impossible to solve the equation $rx = s$ with x in $\mathbb{Z}$. In this section we shall investigate the corresponding problem in $\mathbb{Z}_m$.

Definition An element r in $\mathbb{Z}_m$ is said to be **invertible** if there is some x in $\mathbb{Z}_m$ such that $rx = 1$ in $\mathbb{Z}_m$. In that case, x is called the **inverse** of r, and we write $x = r^{-1}$.

Since $rx = xr$ in $\mathbb{Z}_m$ we have $xr = 1$ also, and $r = x^{-1}$.

Exercises 6.3

1 Find the invertible elements of $\mathbb{Z}_6$, $\mathbb{Z}_7$, and $\mathbb{Z}_8$.

2 Show that 0 is not invertible in any $\mathbb{Z}_m$, but 1 is always invertible.

3 Show that if x and y are invertible in $\mathbb{Z}_m$ then xy and x^{-1} are invertible in $\mathbb{Z}_m$.

Theorem 6.3.1 The element r in $\mathbb{Z}_m$ is invertible if and only if r and m are coprime in $\mathbb{Z}$. In particular, when p is a prime every element of $\mathbb{Z}_p$ except 0 is invertible.

Proof   Suppose r is invertible, so that $rx = 1$ in $\mathbb{Z}_m$. It follows that, in $\mathbb{Z}$, we have $rx - 1 = km$ for some integer k, or

$$rx - km = 1.$$

Now any common divisor of r and m must divide $rx - km$, which is 1, so gcd $(r, m) = 1$.

Conversely, suppose gcd $(r, m) = 1$. Then by Theorem 1.7 there are integers x and y such that $rx + my = 1$. Rearranging this equation we obtain $rx \equiv 1 \pmod{m}$, or $rx = 1$ (in $\mathbb{Z}_m$), as required.   □

Recall that the value $\phi(m)$ of Euler's function (Section 3.3) is the number of integers in the range $1 \leq r \leq m$ which are coprime to m. It follows from Theorem 6.3.1 that the number of invertible elements of $\mathbb{Z}_m$ is $\phi(m)$.

The next theorem is one of the classics of elementary number theory, and it has many useful consequences. We prepare for it with a simple observation concerning the set U_m of invertible elements of $\mathbb{Z}_m$. If y is in U_m define yU_m to be the set obtained when we multiply each member of U_m by y; that is

$$yU_m = \{z \in \mathbb{Z}_m \mid z = yx \text{ for some } x \text{ in } U_m\}.$$

We shall show that $yU_m = U_m$. For example, taking $m = 9$ and $y = 5$ we get

$$U_9 = \{1, 2, 4, 5, 7, 8\}, \qquad 5U_9 = \{5, 1, 2, 7, 8, 4\}.$$

First, we have $yU_m \subseteq U_m$ since if $z = yx$ and both y and x are in U_m then z is also in U_m (Ex. 6.3.3). On the other hand, $U_m \subseteq yU_m$, since given x in U_m we can write

$$x = y(y^{-1}x),$$

which is clearly an element of yU_m (again by Ex. 6.3.3). Thus $yU_m = U_m$, as claimed.

Theorem 6.3.2   If y is invertible in $\mathbb{Z}_m$ then

$$y^{\phi(m)} = 1 \quad \text{in } \mathbb{Z}_m.$$

Proof   Let u denote the product of all the members of U_m; say $u = x_1 x_2 \cdots x_k$, where (as noted above) $k = \phi(m)$. Since $yU_m = U_m$ the elements $yx_1, yx_2, \ldots, yx_k$ are just a rearrangement of $x_1, x_2, \ldots, x_k$. It follows that

$$u = x_1 x_2 \cdots x_k = (yx_1)(yx_2) \cdots (yx_k)$$

$$= y^k u.$$

But u itself is invertible (its inverse is $x_k^{-1} \cdots x_2^{-1} x_1^{-1}$), so multiplying by u^{-1} we get $y^k = 1$ as required. $\qquad\square$

Theorem 6.3.2 can also be stated as a theorem about integers, in the following way:

$$\text{if} \quad \gcd(y, m) = 1 \quad \text{then} \quad y^{\phi(m)} \equiv 1 \pmod{m}.$$

This is known as **Euler's Theorem.** The particular case when m is replaced by a prime p is **Fermat's Theorem:**

$$\text{if} \quad p \nmid y \quad \text{then} \quad y^{p-1} \equiv 1 \pmod{p}.$$

Example

Show that for any positive integer n and prime p

$$n^p \equiv n \pmod{p}.$$

Deduce that, in base 10, the last digits of n and n^5 are always the same.

Solution

Suppose $p \nmid n$; then by Fermat's theorem $n^{p-1} \equiv 1 \pmod{p}$ and so $n^p \equiv n \pmod{p}$. On the other hand, if $p \mid n$ then both n and n^p are congruent to 0 modulo p.

Using this result with $p = 5$ we have $n^5 - n \equiv 0 \pmod{5}$. Also $n^5 - n = n(n-1)(n^3 + n^2 + n + 1)$ and, since one of the first two factors is even, $n^5 - n \equiv 0 \pmod{2}$. Thus $n^5 - n$ is divisible by 5 and by 2, and so it is divisible by 10, which is equivalent to the result stated.

$\qquad\square$

Exercises 6.3 (continued)

4 Find the inverses of

    (i) 2 in $\mathbb{Z}_{11}$,   (ii) 7 in $\mathbb{Z}_{15}$,

    (iii) 7 in $\mathbb{Z}_{16}$,   (iv) 5 in $\mathbb{Z}_{13}$.

5 Use Fermat's Theorem to calculate the remainder when 3^{47} is divided by 23.

6 Suppose that a and b are integers and p is a prime. Use Fermat's Theorem to show that

$$(a+b)^p \equiv a^p + b^p \pmod{p}.$$

7 Show that the equation $x = x^{-1}$ in $\mathbb{Z}_p$ implies that $x^2 - 1 = 0$, and deduce that 1 and -1 are the only elements of $\mathbb{Z}_p$ which are equal to their own inverse.

8   By considering the product of all the non-zero elements of $\mathbb{Z}_p$ show that

$$(p-1)! \equiv -1 \pmod{p}.$$

6.4   Cyclic constructions for designs

In this section and the next one we shall study some constructions based on the cyclic properties of modular arithmetic.

If S is a subset of $\mathbb{Z}_m$ and i is any member of $\mathbb{Z}_m$, then we denote by $S+i$ the subset of $\mathbb{Z}_m$ obtained by adding i to each member of S. For example, when $m = 12$ and $S = \{0, 1, 3, 11\}$, we have

$$S+1 = \{1, 2, 4, 0\}, \qquad S+2 = \{2, 3, 5, 1\},$$

and so on. We shall investigate the possibility of using the subsets $S+i$ $(i \in \mathbb{Z}_m)$ as the blocks of a design.

The first point to notice is that, although there are m possible values of i, the subsets $S+i$ $(i \in \mathbb{Z}_m)$ are not necessarily all distinct. For the subset $T = \{0, 3, 6, 9\}$ of $\mathbb{Z}_{12}$ we have

$$T+0 = T+3 = T+6 = T+ 9 = \{0, 3, 6, 9\},$$
$$T+1 = T+4 = T+7 = T+10 = \{1, 4, 7, 10\},$$
$$T+2 = T+5 = T+8 = T+11 = \{2, 5, 8, 11\},$$

and so we get just three distinct subsets. However, the subset $S = \{0, 1, 3, 11\}$ of $\mathbb{Z}_{12}$ does give twelve distinct subsets:

0	1	3	11	6	7	9	5
1	2	4	0	7	8	10	6
2	3	5	1	8	9	11	7
3	4	6	2	9	10	0	8
4	5	7	3	10	11	1	9
5	6	8	4	11	0	2	10.

In general, if K is a subset of $\mathbb{Z}_m$ and the subsets $K+i$ $(i \in \mathbb{Z}_m)$ are all distinct, then these subsets are the blocks of a 1-design with parameters

$$v = m, \qquad k = |K|, \qquad r = k.$$

In order to prove this, we merely remark that the element α of $\mathbb{Z}_m$ occurs in $K+i$ if and only if

$$\alpha = x+i \quad \text{for some } x \in K.$$

This is equivalent to

$$0 = x + (i - \alpha),$$

which means that 0 is in the block $K + (i - \alpha)$. Thus α is in the blocks

$$K + i_1, K + i_2, \ldots, K + i_r$$

if and only if 0 is in the blocks

$$K + (i_1 - \alpha), K + (i_2 - \alpha), \ldots,, K + (i_r - \alpha).$$

It follows that 0 and α belong to the same number (r) of blocks. Since this holds for any α in $\mathbb{Z}_m$ we have a 1-design. Now the number of objects (v) is equal to m and the number of blocks (b) is also equal to m, so the equation $bk = vr$ shows that the replication number r and the block size k are equal, and $r = k = |K|$, as claimed.

Taking $K = \{0, 1, 3, 11\}$ in $\mathbb{Z}_{12}$ as above, we get a 1-design with parameters $(12, 4, 4)$. In this case we do not get a 2-design since, for example, the pair $\{0, 1\}$ occurs twice but $\{0, 6\}$ does not occur at all. However, we can obtain 2-designs by this method if we insist on a suitable property of the *differences* in the basic subset K.

Definition The subset K of $\mathbb{Z}_m$ is a **difference set** if the difference $x - y$ $(x, y \in K, x \neq y)$ takes each non-zero value in $\mathbb{Z}_m$ the same number of times.

The subset $\{0, 2, 3, 4, 8\}$ of $\mathbb{Z}_{11}$ is a difference set, as may be verified by constructing the table of differences (Table 6.4.1). In this example, each non-zero value occurs twice as a difference. In general, when K is a k-subset of $\mathbb{Z}_m$ there are $k(k-1)$ differences and $m-1$ non-zero values, so each difference occurs $k(k-1)/(m-1)$ times. It turns out that this number is also the r_2 parameter of the corresponding 2-design.

Table 6.4.1

	0	2	3	4	8
0	—	9	8	7	3
2	2	—	10	9	5
3	3	1	—	10	6
4	4	2	1	—	7
8	8	6	5	4	—

Theorem 6.4

If K is a difference set in $\mathbb{Z}_m$ then the sets $K + i$ ($i \in \mathbb{Z}_m$) are the blocks of a 2-design with parameters

$$v = m, \qquad k = |K|, \qquad r_2 = k(k-1)/(m-1).$$

Proof

Suppose we are given α and β in $\mathbb{Z}_m$. Since K is a difference set the equation

$$x - y = \alpha - \beta$$

has $k(k-1)/(m-1)$ solutions with x and y in K. For each solution (x, y) let $i = \alpha - x$. Then

$$\alpha = x + i, \qquad \beta = \alpha - (x - y) = y + i,$$

so α and β both belong to $K + i$. Thus each 2-subset $\{\alpha, \beta\}$ of $\mathbb{Z}_m$ is contained in $r_2 = k(k-1)/(m-1)$ blocks $K + i$, and we have a 2-design with the parameters as stated. $\qquad\qquad\square$

According to Theorem 4.6.1 the design is also a 1-design, and the parameters r_1 and $r_0 = b$ are given by

$$r_1 = \frac{v-1}{k-1} \times r_2 = \frac{m-1}{k-1} \times \frac{k(k-1)}{m-1} = k, \qquad r_0 = \frac{v}{k} \times r_1 = m.$$

In particular, there are m blocks, so the assumption that K is a difference set automatically ensures that the blocks $K + i$ ($i \in \mathbb{Z}_m$) are all distinct.

It is frequently impossible to find a difference set with given values of m and k, even when we specify the obvious necessary condition that $m - 1$ divides $k(k-1)$ so that r_2 is an integer. However, there are some special methods of finding difference sets, and these are very useful in the practical construction of designs.

Example

Let K denote the subset $\mathbb{Z}_{23}$ containing all the non-zero elements which can be written as squares in $\mathbb{Z}_{23}$. Show that K is a difference set and find the parameters of the associated 2-design.

Solution

In $\mathbb{Z}_{23}$ we can calculate squares as follows:

$$1^2 = 1, \quad 2^2 = 4, \quad 3^2 = 9, \quad 4^2 = 16, \quad 5^2 = 2, \quad 6^2 = 13,$$
$$7^2 = 3, \quad 8^2 = 18, \quad 9^2 = 12, \quad 10^2 = 8, \quad 11^2 = 6, \quad \dots.$$

We need not compute any more, since $12^2 = (-11)^2 = 6$, $13^2 = (-10)^2 = 8$, and so on. Thus we have found all the squares, and

$$K = \{1, 2, 3, 4, 6, 8, 9, 12, 13, 16, 18\},$$

so that $k = |K| = 11$. Calculating differences for K we obtain Table 6.4.2.

Table 6.4.2

	1	2	3	4	6	8	9	12	13	16	18
1	—	22	21	20	18	16	15	12	11	8	6
2	1	—	22	21	19	17	16	13	12	9	7
3	2	1	—	22	20	18	17	14	13	10	8
4	3	2	1	—	21	19	18	15	14	11	9
6	5	4	3	2	—	21	20	17	16	13	11
8	7	6	5	4	2	—	22	19	18	15	13
9	8	7	6	5	3	1	—	20	19	16	14
12	11	10	9	8	6	4	3	—	22	19	17
13	12	11	10	9	7	5	4	1	—	20	18
16	15	14	13	12	10	8	7	4	3	—	21
18	17	16	15	14	12	10	9	6	5	2	—

By inspection, each non-zero element of $\mathbb{Z}_{23}$ occurs the same number of times— specifically, five times, in agreement with the formula

$$\frac{k(k-1)}{m-1} = \frac{11 \times 10}{22} = 5.$$

Hence the parameters of the associated 2-design are $v = 23$, $k = 11$, and $r_2 = 5$.   □

Exercises 6.4

1   Which of the following are difference sets?

   (i) $\{2, 3, 5, 11\}$ in $\mathbb{Z}_{13}$,   (ii) $\{0, 1, 3, 5\}$ in $\mathbb{Z}_{13}$,
   (iii) $\{3, 6, 7, 12, 14\}$ in $\mathbb{Z}_{21}$.

2   Repeat the *Example* given above with 23 replaced by 11, and with 23 replaced by 31. Make a conjecture about the parameters of the associated design when m is any prime of the form $4n + 3$.

3   Show that if K is a difference set in $\mathbb{Z}_m$ then so is $K + i$ for each i in $\mathbb{Z}_m$. Deduce that we can always assume, if we wish, that any difference set contains both 0 and 1.

4   Show that a Steiner triple system (Ex. 4.7.3) with seven objects can be constructed by the method given in the *Example*. Prove that

this is essentially the only way of constructing an STS with seven objects. [Hint: the three blocks containing a given object must contain all seven objects between them.]

6.5 Latin squares

Suppose we have to plan an agricultural experiment in which five new kinds of fertilizer are to be tested in a square field. We could divide the field into 25 small squares and apply the fertilizers according to the scheme

$$
\begin{array}{ccccc}
A & B & C & D & E \\
B & C & A & E & D \\
C & D & E & A & B \\
D & E & B & C & A \\
E & A & D & B & C.
\end{array}
$$

Here we have arranged that each fertilizer appears just once in each row and column, in the hope that certain effects (such as the direction of the prevailing wind) will thereby be minimized.

Definition A **latin square** of order n is an $n \times n$ array in which each one of n symbols occurs once in each row and once in each column.

When the rows and columns of the square L are labelled in some way, the symbol in row i and column j will be denoted by $L(i, j)$. In the following discussion the labels for the rows and columns will be elements of $\mathbb{Z}_m$, and the symbols will also be elements of $\mathbb{Z}_m$.

Theorem 6.5.1 For each $m \geq 2$ the $m \times m$ array defined by

$$
L(i, j) = i + j \quad (i, j \in \mathbb{Z}_m)
$$

is a latin square.

Proof Suppose the symbols in positions (i, j) and (i, j') are the same. Then we have

$$
i + j = L(i, j) = L(i, j') = i + j'.
$$

Now $\mathbb{Z}_n$ contains an element $-i$, and adding it to both sides we get $j = j'$. Hence each symbol occurs at most once in row i, and since there are m symbols and m positions each symbol occurs exactly once. A similar argument holds for the columns. Thus L is a latin square. $\qquad\square$

The theorem shows that there is always at least one latin square of any given order. Of course, the latin square constructed in the theorem is just the 'addition table' for $\mathbb{Z}_m$ (see Exercise 6.2.1), and the fact that we can find one latin square in this way is not particularly exciting. Indeed, it is very easy to construct individual latin squares by trial and error.

A rather more difficult problem is to find pairs of latin squares of the same order which are, in a sense, as different as possible. We say that two latin squares L_1 and L_2 of the same order are **orthogonal** if, for each ordered pair of symbols (k, k'), there is just one position (i, j) for which

$$L_1(i, j) = k, \qquad L_2(i, j) = k'.$$

For example, in the latin squares

A	B	C	D		A	B	C	D
---	---	---	---		---	---	---	---
B	A	D	C		C	D	A	B
C	D	A	B		D	C	B	A
D	C	B	A		B	A	D	C

each of the 16 pairs $(A, A), (A, B), \ldots, (D, D)$ occurs in one of the 16 positions.

In 1781 Euler proposed the following problem. Given 36 officers of six different ranks from six different regiments, can they be arranged in a square in such a way that each row and each column contains one officer of each rank and one officer from each regiment? In modern terminology, Euler was seeking an orthogonal pair of latin squares of order six. He was unable to solve the problem, and it is now known that no such pair exists.

Although Euler's problem indicates that orthogonal latin squares are not easy to find, there is a very useful construction based on the arithmetical properties of $\mathbb{Z}_p$ when p is prime.

Theorem 6.5.2

Let p be a prime and t a non-zero element of $\mathbb{Z}_p$. Then the rule

$$L_t(i, j) = ti + j \quad (i, j \in \mathbb{Z}_p)$$

defines a latin square. Furthermore, when $t \neq u$ the latin squares L_t and L_u are orthogonal.

Proof

In order to show that L_t is a latin square we can use the same method as in the previous theorem. For instance, if $L_t(i, j) = L_t(i', j)$, then $ti + j = ti' + j$, and since t is invertible in $\mathbb{Z}_p$ it follows that $i = i'$. A similar argument proves that if $L_t(i, j) = L_t(i, j')$ then $j = j'$.

In order to show that when $t \neq u$ the squares L_t and L_u are orthogonal, suppose there are two different positions (i_1, j_1) and (i_2, j_2) such that L_t and L_u have the symbols k and k' respectively in both positions. That is,

$$ti_1 + j_1 = k, \qquad ui_1 + j_1 = k',$$
$$ti_2 + j_2 = k, \qquad ui_2 + j_2 = k'.$$

It follows that

$$t(i_1 - i_2) = j_1 - j_2, \qquad u(i_1 - i_2) = j_1 - j_2.$$

If $i_1 - i_2 = 0$ then these equations imply that $j_1 - j_2 = 0$, and so the positions (i_1, j_1) and (i_2, j_2) are the same, contrary to the hypothesis. Hence $i_1 - i_2 \neq 0$ and $i_1 - i_2$ has an inverse in $\mathbb{Z}_p$. In this case we can solve the equations for t and u, obtaining

$$t = u = (i_1 - i_2)^{-1}(j_1 - j_2).$$

In other words if we insist that $t \neq u$, then the symbols k and k' can occur together in only one position, and L_t and L_u are orthogonal. $\square$

We usually say that the theorem yields a set of $p - 1$ **mutually orthogonal** latin squares of order p, for each prime p. For example, when $p = 3$ we get the squares

$$L_1 = \begin{matrix} 0 & 1 & 2 \\ 1 & 2 & 0 \\ 2 & 0 & 1 \end{matrix} \qquad L_2 = \begin{matrix} 0 & 1 & 2 \\ 2 & 0 & 1 \\ 1 & 2 & 0 \end{matrix}.$$

Exercises 6.5

1 An ordinary pack (deck) of cards contains four Jacks, four Queens, four Kings, and four Aces, one of each denomination from each of the four suits—Hearts, Clubs, Diamonds, and Spades. Explain how to arrange these 16 cards in a 4×4 square in such a way that each row contains one card from each suit and one card from each denomination. Interpret your result in terms of latin squares.

2 Use the construction given in Theorem 6.5.2 to obtain four mutually orthogonal latin squares of order 5.

3 Why is it not possible to use the method of Theorem 6.5.2 to find three mutually orthogonal latin squares of order 4? Use the method of trial and error to find three such squares.

4 Suppose that the first row of an $n \times n$ array is

$$x_1 \quad x_2 \quad x_3 \quad \cdots \quad x_{n-1} \quad x_n,$$

and suppose also that each successive row is obtained from the previous one by a cyclic shift of r places, so that the second row is

$$x_{r+1} \quad x_{r+2} \quad x_{r+3} \quad \cdots \quad x_{r-1} \quad x_r,$$

and so on. If n is given, for which values of r does this construction yield a latin square?

6.6 Miscellaneous Exercises

1 Determine all possible solutions of the congruences

(i) $5x \equiv 1 \pmod{11}$, (ii) $5x \equiv 7 \pmod{15}$.

2 Without doing any long division show that 192 837 465 564 738 291 is divisible by 11.

3 Prove that the congruence $ax \equiv b \pmod{m}$ has a solution x if and only if b is a multiple of gcd (a, m).

4 Solve the equations

(i) $5x = 12$ in $\mathbb{Z}_{13}$, (ii) $x^2 - x \cdot 1 - 0$ in $\mathbb{Z}_{11}$.

5 What is the last digit in the base 10 representation of 7^{93}?

6 Use the fact that $1001 = 7 \times 11 \times 13$ to construct a test for divisibility by 7, 11, or 13, similar to 'casting out nines' and the test for divisibility by 11 given in Ex. 6.1.4.

7 Find the inverses of

(i) 6 in $\mathbb{Z}_{11}$, (ii) 6 in $\mathbb{Z}_{17}$,

(iii) 3 in $\mathbb{Z}_{10}$, (iv) 5 in $\mathbb{Z}_{12}$.

8 Suppose that gcd $(r, m) = 1$. Use the proof of Theorem 6.3.1 to formulate a practical method (based on the Euclidean algorithm) for finding the inverse of r in $\mathbb{Z}_m$.

9 Show that if gcd $(m_1, m_2) = 1$ then there is a solution x of the congruences

$$x \equiv a_1 \pmod{m_1}, \qquad x \equiv a_2 \pmod{m_2}.$$

Show that any two solutions are congruent modulo $m_1 m_2$.

10   State and prove the generalization of the previous exercise to n simultaneous congruences. (This is known as the **Chinese Remainder Theorem.**)

11   It follows from Fermat's Theorem that for any prime $p > 2$ we have $2^{p-1} \equiv 1 \pmod{p}$. Determine the least prime p for which

$$2^{p-1} \equiv 1 \pmod{p^2}.$$

12   Let $F = \mathbb{Z}_3 \times \mathbb{Z}_3$; in other words f is the set of ordered pairs (x, y) with x and y in $\mathbb{Z}_3$. Define an operation $\otimes$ on F by the rule

$$(x_1, y_1) \otimes (x_2, y_2) = (x_1 x_2 - y_1 y_2, x_1 y_2 + x_2 y_1),$$

where the symbols on the right have their usual meanings in $\mathbb{Z}_3$. Show that every element of F except $(0, 0)$ has an inverse with respect to the $\otimes$ operation.

13   Determine for which values of k in the range $2 \leqslant k \leqslant 6$ there is a difference set of size k in $\mathbb{Z}_7$.

14   Let S be the set of non-zero squares in $\mathbb{Z}_{13}$ and T the set of non-zero non-squares. Verify that the result of applying the cyclic construction to S *and* T is a 2-design with parameters $(13, 6, 5)$.

15   Show that if S is a difference set in $\mathbb{Z}_m$ then so is $\mathbb{Z}_m - S$. Write down the parameters of the latter difference set in terms of the parameters of the former.

16   Determine the value of x for which the set $\{1, 5, 24, 25, 27, x\}$ is a difference set in $\mathbb{Z}_{31}$.

17   Let **B** be a 2-design with parameters (v, k, r_2) obtained from a difference set in $\mathbb{Z}_v$. Prove that any two distinct blocks of **B** have exactly r_2 common members.

18   Construct a pair of orthogonal latin squares of order 8.

19   Suppose (without loss of generality) that the first row of a latin square of order n is $1, 2, 3, \ldots, n$. How many ways of filling in the second row are there?

20   Let S be the set of blocks of a Steiner triple system on the set $\{1, 2, \ldots, n\}$ (Ex. 4.7.3). Show that the following rule defines a latin square:

$$L(i, j) = \begin{cases} i & \text{if } i = j; \\ k & \text{if } i \neq j; \end{cases}$$

where $\{i, j, k\}$ is the unique block in S containing i and j.

Part II
Graphs and algorithms

In this part of the book we shall discuss problems which can be solved using step-by-step methods. Many such problems can be described in terms of a 'graph' or 'network', and consequently we shall devote some time to studying the properties of these structures.

The prerequisites for reading this part are essentially covered in Part I. Almost all the calculations involve only the set $\mathbb{Z}$ of integers, but in Chapter 12 it will be convenient to assume the standard arithmetical properties of the sets of real and complex numbers, denoted by $\mathbb{R}$ and $\mathbb{C}$ respectively. Some very simple facts about matrices are used in a few places, but no knowledge of Computer Science or any specific programming language is assumed.

7
Algorithms and their efficiency

7.1 What is an algorithm?

In order to give a completely satisfactory definition of what we mean by an algorithm we should have to delve quite deeply into the realms of mathematical logic. But the idea is so simple, and so natural, that an informal approach is sufficient for our purposes. Originally, the word 'algorithm' was applied to procedures of elementary arithmetic, such as long multiplication and long division. A child who knows the procedure for long multiplication, and who is given two positive integers in base 10 notation, can work out their product, also in base 10 notation. The child does not have to understand the logical foundation of the procedure; all that is required is the ability to carry out the correct steps in the correct order.

Roughly speaking then, we can say that an algorithm is a sequence of instructions. Each individual instruction must be carried out, in its proper place, by the person or machine for whom the algorithm is intended. Consequently, an algorithm should always be considered in the context of certain assumptions. For example, the basic algorithms of arithmetic (as taught to young children) operate with integers in base 10 notation. They depend upon the fact that the sum and product of two single-digit integers is an integer with at most two digits, and so it is possible for children to learn the tables for single-digit operations 'by heart'.

In order to study these algorithms more closely it is convenient to make the following definitions. For each integer m in the range $0 \leq m \leq 99$ let $t(m)$ and $u(m)$ denote the 'tens-digit' and the 'units-digit' of m, respectively. For example, $t(73) = 7$, $u(73) = 3$. Formally, if

$$m = 10t + u \quad (0 \leq t \leq 9, \, 0 \leq u \leq 9),$$

then we have $t(m) = t$ and $u(m) = u$. Using this notation it is possible give a careful description of some of the elementary algorithms.

Example   Let a and b be two integers which have n digits in their base 10 representations, and let $s = a + b$. The algorithm for calculating the base 10 representation of s is usually set out as in the following example.

$$
\begin{array}{rl}
7\ 8\ 1\ 5\ 3 & a \text{ (in base 10)} \\
+\ \ \ 3\ 7\ 4\ 2\ 9 & b \text{ (in base 10)} \\
\hline
1\ 1\ 0\ 0\ 1 & \text{(carrying digits)} \\
\hline
1\ 1\ 5\ 5\ 8\ 2 & s \text{ (in base 10)}
\end{array}
$$

Write down the formulae for calculating the digits of s in the general case, using the t and u notation introduced above.

Solution   In the general case the working looks like this:

$$
\begin{array}{rl}
a_{n-1}\, a_{n-2} \cdots a_2\, a_1\, a_0 & a \\
+\ \ \ b_{n-1}\, b_{n-2} \cdots b_2\, b_1\, b_0 & b \\
\hline
c_n \ \ \ c_{n-1} \ \ \cdots \ \ c_2\ c_1 & \text{(carrying digits)} \\
\hline
s_n\, s_{n-1}\, s_{n-2} \cdots s_2\, s_1\, s_0 & s.
\end{array}
$$

(Note that s_n may be zero.) At the first step, we add a_0 and b_0, take s_0 to be the units-digit of the answer, and 'carry' c_1, the tens-digit of the answer. That is,

$$s_0 = u(a_0 + b_0), \qquad c_1 = t(a_0 + b_0).$$

At the second step we calculate $a_1 + b_1$ and add c_1; the units-digit of the answer is s_1 and the tens-digit is c_2. We continue in this way, calculating for each i in the range $1 \leqslant i \leqslant n - 1$ the digits

$$s_i = u(a_i + b_i + c_i), \qquad c_{i+1} = t(a_i + b_i + c_i).$$

Finally we put $s_n = c_n$. $\square$

The *Example* illustrates the point that an algorithm must be considered in a specific context. In this case, the context is a person who knows the tables for adding and multiplying small integers. The knowledge could be available in his or her memory, or in a book of tables. When such calculations are done by a machine, the memory is part of the 'hardware', and the book is part of the 'software'.

Exercises 7.1   1   The following layout describes the usual algorithm for multiplying an n-digit integer x by a single-digit integer y, in base 10 notation.

$$\begin{array}{lll}
x_{n-1}\,x_{n-2}\cdots x_2 \quad x_1 \quad x_0 & & x \\
\times \qquad\qquad\qquad\qquad y & & y \\
\hline
c_n \quad c_{n-1} \quad \cdots \quad c_2 \quad c_1 & & \text{(carrying digits)} \\
\\
p_n \; p_{n-1}\,p_{n-2}\cdots p_2 \quad p_1 \quad p_0 & & xy
\end{array}$$

Write down the formulae which determine the digits of xy and the carrying digits.

2   Explain carefully how the algorithm described in Ex. 1 can be extended to yield the usual 'long multiplication' algorithm for multiplying an n-digit integer by an m-digit integer ($n \geqslant m \geqslant 1$).

3   Suppose we have a machine which can perform the following tasks:

      (i) add 1 to an integer;
     (ii) subtract 1 from an integer;
    (iii) decide if an integer is zero.

Let m be an integer and n a non-negative integer. Write an algorithm to instruct the machine to calculate $m + n$. (You may assume that the machine can store the results of its intermediate calculations, and can refer to them when necessary.)

7.2 Programs

The traditional way of describing an algorithm is to use ordinary language, with mathematical symbols where appropriate. Nowadays, the development of computers means that we often use a kind of shorthand, midway between ordinary language and a formal computer language like Algol or Fortran. This shorthand can be constructed from a very few simple elements, which we shall now describe.

It is helpful to think in terms of a computer which has a number of locations, each having a *name* like x, y, z, a, b, c, and so on. Each location may be used to store a *value* which, for our purposes, will usually be an integer. Thus, at a given instant, the locations and their values might be as follows:

    Name: x y z a b c $\cdots$
    Value: 17 5 45 67 8 -6 $\cdots$.

The basic unit of our shorthand is the *assignment statement*, which is written using standard mathematical notation and one new symbol, the backward arrow ←. A typical assignment statement has the form

$$x \leftarrow q$$

and it is interpreted as meaning that q is placed in the location x, while the value previously in x is discarded to make way for q. The object q may be specified in several ways, as in the following examples.

$$x \leftarrow y \qquad x \leftarrow 107 \qquad x \leftarrow x + 1$$

In the first example the current value of location y is placed in x (and the same value remains in y). In the second example the value 107 is placed in x. In the third example the value of x is set equal to its previous value plus one.

The simplest kind of *program* is a sequence of assignment statements, separated by semicolons for clarity. For example, here is a program for computing the binomial number $\binom{n}{2}$ when n is a given positive integer.

$$x \leftarrow n; \quad y \leftarrow x - 1; \quad x \leftarrow xy; \quad b \leftarrow x/2$$

It is assumed that the machine for which the program is written can be told that n is a positive integer, and that it can perform the usual arithmetical operations on integers. The program puts the given value n into location x, subtracts 1 and puts the result in y, multiplies the value of x by y, and finally divides the new value of x by 2 and puts the result in location b.

In order to write more useful and interesting programs we need to assume that our machine can decide if an assertion is true or false. Frequently, the assertion in question will be something like '$x > 0$', or '$y = z$', or '$X = \varnothing$'. Given this capability, we can introduce two very useful commands.

The command

while A do S

means that the sequence of statements S is executed repeatedly so long as A remains true. The assertion A may change each time S is executed, and if A eventually becomes false the repetition will cease. For example, here is another way to compute $\binom{n}{2}$, using the fact that it is the sum of the integers $1, 2, \ldots, n-1$.

$$x \leftarrow n; \quad y \leftarrow 0; \quad z \leftarrow 0;$$
$$\textbf{while} \cdot y < x - 1 \ \textbf{do} \ y \leftarrow y + 1; \quad z \leftarrow z + y;$$
$$b \leftarrow z$$

In this program y takes the values $0, 1, 2, \ldots$ and z takes the values $0, 0+1, 0+1+2, \ldots$. The final execution of the **while-do** statement begins with $y = n - 2$; in it the value of y is increased by 1 and z is increased by $n - 1$. Since y is now equal to $x - 1$ the repetition ceases and b is assigned the final value of z, which is $0 + 1 + 2 + \cdots + (n - 1)$.

It must be remembered that if A always remains true, the command **while** A **do** S will result in S being repeated indefinitely and the program will have no outcome. A simple example is the program

$$x \leftarrow 0; \quad \textbf{while } x \geq 0 \textbf{ do } x \leftarrow x + 1; \quad c \leftarrow x$$

which never assigns a value to c. This is an example of a *program* which could be executed by a machine, but which has no conclusion and so cannot properly be described as an *algorithm*.

The command

$$\textbf{if } A \textbf{ then } S \textbf{ else } T$$

means that if A is true then the sequence of statements S must be executed, whereas if A is false the sequence of statements T must be executed. For example, here is yet another program for computing $\binom{n}{2}$.

$x \leftarrow n;$
if x is even **then** $y \leftarrow x - 1; x \leftarrow x/2$ **else** $y \leftarrow (x - 1)/2;$
$b \leftarrow xy$

In this program we use the formula $\binom{n}{2} = \frac{1}{2}n(n - 1)$, and the fact that either n or $n - 1$ is divisible by 2. Frequently the 'else' part of an **if–then–else** command is null, and in such a case we drop the word **else** altogether. An example is given in Ex. 7.2.2.

Using the **if–then–else** and **while–do** constructions it is possible to build up any program we may need. It is convenient to write out programs using indentation to indicate the segments governed by the conditional commands. This is not only an aid to comprehension, but it enables us to dispense with the semicolons at the end of each line. (Of course, such liberties are not allowed when we write programs in a formal computer language.) Here then is a typical program, in which we assume that we have a machine which can add and subtract integers, and we tell it how to find the product mn of two given positive integers m and n.

$$x \leftarrow m; \quad y \leftarrow n; \quad z \leftarrow 0$$
$$\textbf{while } y \neq 0 \textbf{ do}$$
$$z \leftarrow z + x$$
$$y \leftarrow y - 1$$
$$p \leftarrow z$$

We can follow the operation of the program by tabulating (Table 7.2.1) the values in each location at the end of each execution of the **while–do** segment.

Table 7.2.1

z	y	x
0	n	m
m	$n-1$	m
$2m$	$n-2$	m
.	.	.
.	.	.
.	.	.
nm	0	m

At this point $y = 0$ and so the **while–do** segment is omitted. The current value of z (that is, nm) is placed in location p and the program ends.

Often a **while–do** command is used to make one variable y take a regular sequence of values. A typical program segment of this kind is

$$y \leftarrow r; \quad z \leftarrow s$$
$$\textbf{while} \quad y \leqslant z \quad \textbf{do}$$
$$S$$
$$y \leftarrow y+1$$

which results in S being executed for each integer value of y in the range $r \leqslant y \leqslant s$. It is convenient to use the notation

$$\textbf{for} \quad y = r \textbf{ to } s \quad \textbf{do} \quad S$$

for this entire segment, but it must be stressed that it is merely an abbreviation, not a new kind of instruction. Any program we may care to write can be reduced to a sequence of the three basic instructions:

> (i) the assignment statement,
> (ii) the **while–do** command,
> (iii) the **if–then–else** command.

This fact is the basis for the discipline of *structured programming.*

Exercises 7.2

1 Suppose we have a machine which, when given integers m and n with $m \geq n \geq 0$, can calculate the remainder $r(m, n)$ on dividing m by n. Tabulate the values of z, y, and x after each execution of the **while–do** segment of the following program, and hence describe the purpose of the program.

$$x \leftarrow 2406; \quad y \leftarrow 654; \quad z \leftarrow 0$$
$$\textbf{while} \quad y \neq 0 \quad \textbf{do}$$
$$z \leftarrow y$$
$$y \leftarrow r(x, y)$$
$$x \leftarrow z$$
$$d \leftarrow x$$

2 Suppose we are given n integers $x_1, x_2, \ldots, x_n$. Describe in ordinary language the value of b calculated by the following program, and explain how the calculation is done.

$$y \leftarrow x_1$$
$$\textbf{for} \quad j = 2 \textbf{ to } n \quad \textbf{do}$$
$$\textbf{if} \quad x_j < y \quad \textbf{then} \quad y \leftarrow x_j$$
$$b \leftarrow y$$

3 Write a program which, given n integers $x_1, x_2, \ldots, x_n$, will assign to z the value 1 when any two of them are equal, and assign to z the value 0 otherwise.

4 Modify the program for computing $p = mn$ given above so that it will work for any integer n (positive, negative, or zero).

7.3 Flowcharts

Flowcharts are often helpful in planning complex programs. The reader is almost certainly familiar with the general idea, since flowcharts are used nowadays in many everyday situations where a sequence of instructions has to be carried out. In this book we shall use only flowcharts which can be precisely described by the framework of programming commands introduced in Section 7.2.

The basic elements of our flowcharts are illustrated in Figs. 7.1, 7.2, 7.3, which are largely self-explanatory.

Fig. 7.1
$S_1; S_2$

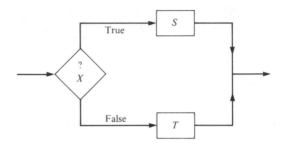

Fig. 7.2
if X **then** S **else** T

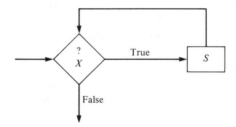

Fig. 7.3
while X **do** S

A few details are worth mentioning. First, the shape of a box—rectangle, rhombus, or circle—indicates whether it represents a statement, a question or a terminus. Secondly, it is often convenient to put several statements, or even a compound statement written in plain English, in one box. Thirdly, our flowcharts will have just one 'begin' and one 'end', in contrast to our informal programs (which

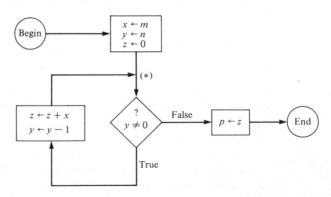

Fig. 7.4
Flowchart for the $p = mn$
program.

have none), and programs written in high-level computer languages (which have many).

Using these conventions, we can construct a flowchart (Fig. 7.4) for the multiplication program given in the previous section. One point is marked (∗) for future reference.

Exercises 7.3

1 Construct flowcharts for the programs given in Ex. 7.2.1 and Ex. 7.2.2.

2 Construct a flowchart for the segment of a program corresponding to the command: **for** $i = r$ **to** s **do** S.

3 Suppose we have a machine which can recognize the integers 0, 1, and 7. Draw a flowchart for a program which, when given a word $w_1 w_2 w_3 w_4 w_5 w_6 w_7$ of length 7 in the alphabet $\{0, 1\}$, will compute the number of 1's in the word. Write out the program in the usual way.

7.4 Verification of programs

In general, the question of what exactly a program does is both deep and difficult. One difficulty is that the program may be, quite simply, wrong. Even when it is formally correct, we may not be able to decide what it does. Consider, for example, the following program, which purports to assign a value to b when n is a given positive integer.

$$x \leftarrow n$$
while $x \neq 1$ **do**
$\quad$ **if** x is even
$\quad\quad$ **then** $x \leftarrow x/2$
$\quad\quad$ **else** $x \leftarrow 3x + 1$
$b \leftarrow x$

When $n = 15$ the successive values of x computed by this program are

15, 46, 23, 70, 35, 106, 53, 160, 80, 40, 20, 10, 5, 16, 8, 4, 2, 1,

and so the program stops with $b = 1$. In fact, it is suspected that the program assigns the value $b = 1$ for any input n, but no one has succeeded in proving that the commands in the **while–do** segment will always result in x eventually taking the value 1. It is conceivable that for some initial values of n the variable x will either return to a previous value and repeat itself thereafter, or go on taking new values indefinitely. So we cannot yet say what the program does. However, if our suspicions are correct, it is just a rather complicated algorithm for computing the values of the function $b : \mathbb{N} \to \mathbb{N}$ defined by $b(n) = 1$ for all $n \in \mathbb{N}$.

In practice, the problem of deciding what a program does is rather different. Usually we have a program which has been constructed in order to perform a specific task, and we have to convince ourselves that it really does work. This is basically a question of mathematical proof in just the same way as we are obliged to formulate proofs of our theorems. As one might expect, the principle of induction is our major weapon in this enterprise.

Example   Prove that if $n \geqslant 0$ the following program will eventually end with $s = n + m$.

$$x \leftarrow n; \quad y \leftarrow m$$
$$(*) \qquad \textbf{while} \quad x \neq 0 \quad \textbf{do}$$
$$x \leftarrow x - 1$$
$$y \leftarrow y + 1$$
$$s \leftarrow y$$

Solution   Let $P(k)$ denote the assertion: if we reach the line $(*)$ with the value $x = k$ and any value of y, then we eventually end with $s = k + y$. The truth of $P(n)$ for all $n \geqslant 0$ can be established as follows.

$P(0)$ is true, since if we arrive at $(*)$ with $x = 0$, the **while–do** loop is omitted, s is assigned the value y, and we end with $s = y$ $(= 0 + y)$.

Suppose $P(k)$ is true. If we arrive at $(*)$ with values $x = k + 1$ and y, then $k + 1 \neq 0$ and so the loop is executed. New values $x' = x - 1 = k$ and $y' = y + 1$ are assigned, and we return to $(*)$ with these values. By the assumption $P(k)$, we shall eventually end with

$$s = x' + y' = k + y + 1 = (k + 1) + y.$$

Thus $P(k + 1)$ is true. By the principle of induction, $P(n)$ is true for all $n \geqslant 0$.

But we arrive at $(*)$ for the first time with $x = n$ and $y = m$. Hence the truth of $P(n)$ implies the truth of the required result.   □

It is obvious that the kind of argument used above is rarely needed in the workaday routine of computer programming. We usually assume that when a program executes a loop, it really does what we expect it to do. Indeed, the argument above is just a formal statement, based on the principle of induction, of our intuitive picture of the positive integers. However, it must be remembered that ultimately the correctness of any program depends upon a mathematical proof, and that the heart of such a proof is the induction principle.

Exercises 7.4

1 Prove that if $n \geq 0$ the program illustrated in Fig. 7.4 will eventually end with $p = mn$. [Hint: the induction hypothesis $P(k)$ is that if the program reaches $(*)$ with $y = k$ and any values of x and z, then it will eventually end with $p = z + xk$.]

2 What happens to the $p = mn$ program if $n < 0$?

3 In the program for computing $s = m + n$ in the *Example* the quantity $x + y$ remains constant throughout, and is called a *loop invariant*. Find a loop invariant for the multiplication program in Fig. 7.4 which takes the constant value mn throughout.

4 Tabulate the values of a, b, and c during the course of the following program when (i) $n = 12$ and $m = 31$, (ii) $n = 77$ and $m = 15$. On the basis of your results make a conjecture as to the purpose of the program.

$$a \leftarrow 0; \quad b \leftarrow n, \quad c \leftarrow m$$
$$\textbf{while} \quad b \neq 0 \quad \textbf{do}$$
$$\textbf{while} \quad b \text{ is even} \quad \textbf{do}$$
$$b \leftarrow b/2; \quad c \leftarrow 2c$$
$$a \leftarrow a + c; \quad b \leftarrow b - 1$$
$$z \leftarrow a$$

5 Construct a formal proof of your conjecture about the program in Ex. 4, using the method outlined in this section.

7.5 Analysis of algorithms

Roughly speaking, the efficiency of an algorithm is determined by the relationship between the *effort* required to solve any specific case of a problem and the *size* of that case. In this section we shall attempt to clarify this vague idea.

When we are discussing algorithms it is customary to use the term **instance** to denote a specific case of a problem. Thus if the problem is the multiplication of two positive integers a typical instance is the pair $(7631, 205)$. We might reasonably decide to use the number of digits in the larger integer as a measure of the size of an instance, so that the size of the instance $(7631, 205)$ is 4.

In order to measure the effort required we usually count the number of significant operations which the algorithm performs. For example, if we use the long multiplication algorithm then, for an instance of size n we need at most n^2 single-digit multiplications (and some shifts and additions). So we might say that the efficiency of this algorithm is n^2. On the other hand the traditional algorithm for calculating the *sum* of two positive integers requires at most n single-digit additions (and some carrying), so we might say that its efficiency is n.

In general, there are three steps in the analysis of an algorithm:

(A) describe the algorithm precisely;
(B) define the size of an instance, n;
(C) calculate $f(n)$, the number of operations required.

This sort of analysis requires subjective judgements at several points. What is the 'best' measure of the size of an instance? Which are the most significant of the operations involved, as opposed to those which require relatively little effort? In order to illustrate the methods which can be used, we shall discuss in detail the algorithm (described in Section 1.5) for finding the binary representation of a positive integer given in decimal form. We follow the three headings given above.

Binary representation algorithm

$y \leftarrow N; \quad i \leftarrow 0$
while $y \neq 0$ **do**
    if y is even
        then $r_i \leftarrow 0$
        else $r_i \leftarrow 1; \quad y \leftarrow y - 1$
    $y \leftarrow y/2$
    $i \leftarrow i + 1$
$k \leftarrow i - 1$
$N_2 \leftarrow r_k r_{k-1} \cdots r_1 r_0$

(A) The algorithm is based on repeated division by 2. For a given positive integer N we compute

$$N = 2q_0 + r_0$$
$$q_0 = 2q_1 + r_1$$
$$\ldots$$
$$q_{k-1} = 2q_k + r_k.$$

Each remainder r_i is 0 or 1, and we stop when $q_k = 0$. The required binary representation is the sequence $N_2 = r_k r_{k-1} \ldots r_1 r_0$. The algorithm is described precisely by the program at the foot of p. 144.

(B) There are two possible measures of the size of an instance. We could use the value of N itself, or we could use the number n of decimal digits in N. If

$$N = (x_{n-1} x_{n-2} \ldots x_1 x_0)_{10}$$

then N lies between 10^{n-1} and $10^n - 1$, so that $n - 1$ is the **integer part** of $\log_{10} N$, written $\lfloor \log_{10} N \rfloor$. (For any real number x, the symbol $\lfloor x \rfloor$ denotes the greatest integer m for which $m \leq x$.) Hence the number n of decimal digits in N is given by

$$n = \lfloor \log_{10} N \rfloor + 1.$$

(C) The most significant operation is the division by 2, which is performed $k + 1$ times in the algorithm. This is also the number of digits in the binary representation N_2 so, by the same argument as above,

$$k + 1 = \lfloor \log_2 N \rfloor + 1.$$

Now the elementary theory of logarithms tells us that

$$\log_2 N = \frac{\log_{10} N}{\log_{10} 2} = (3.3219 \ldots) \times \log_{10} N.$$

Hence the number of operations, $k + 1$, is approximately:

$\frac{10}{3} \log_{10} N$ (as a function of N),

$\frac{10}{3} n$ (as a function of n).

Note that we are content with an approximate formula for the efficiency. This is because we are concerned with predicting the behaviour of the algorithm for large instances, rather than with the details of its performance (which, in any event, will depend upon external factors).

In this particular example all instances with n decimal digits require approximately the same number of operations. For some problems the effort required may vary considerably for different instances of the same size. In such circumstances it is customary to be pessimistic and consider the **worst–case** for an instance of a given size.

1 Suppose we wish to find the least member of a finite sequence of integers $x_1, x_2, \ldots, x_n$, using the algorithm indicated in Ex. 7.2.2.

(i) Suggest a good measure of the size of an instance.
(ii) How many comparisons are needed?
(iii) What is the worst-case, with respect to the number of data transfers (assignment statements) required?

2 The rule for multiplying two complex numbers is

$$(a + ib)(c + id) = x + iy$$

where $x = ac - bd$, $y = ad + bc$. This reduces the problem to one involving four multiplications of real numbers, and two additions and subtractions. Show that it is possible to reorganize the rule in such a way that only the three products $a \times c$, $b \times d$, and $(a + b) \times (c + d)$ are needed. How many additions and subtractions are required in this method?

3 Table 7.5.1 gives the number of steps required to find gcd (a, b) by the Euclidean algorithm when $a \geqslant b \geqslant 1$ and $a \leqslant 5$.

(i) Extend the table to the range $a \leqslant 13$.
(ii) For each n in the range $2 \leqslant n \leqslant 5$ find the smallest pair (a_n, b_n) for which n steps are required.
(iii) Suggest a rule for finding the pair (a_n, b_n) in general, and try to prove it.

Table 7.5.1

b	$a = 1$	$a = 2$	$a = 3$	$a = 4$	$a = 5$
1	1	1	1	1	1
2		1	2	1	2
3			1	2	3
4				1	2
5					1

7.6 Estimating the efficiency of an algorithm

On the basis of the discussion in the previous section we now have a fairly clear idea of the meaning of the word 'efficiency' in the context of algorithms. From now on we shall say that the **efficiency** of a given algorithm is the number $f(n)$ of operations (of a specified kind) required by the algorithm for an instance of size n, in the worst-case.

It is instructive to look at some numerical data concerning the relationship between the efficiency of an algorithm and the time required to execute it. Let us suppose that we have a machine which carries out one million operations per second, and an algorithm whose efficiency is $f(n)$. Table 7.6.1 gives the time required for some typical functions f and some typical values of n.

Table 7.6.1

$f(n)$	$n = 20$	$n = 40$	$n = 60$
n	0.00002 sec	0.00004 sec	0.00006 sec
n^2	0.0004 sec	0.0016 sec	0.0036 sec
n^3	0.008 sec	0.064 sec	0.216 sec
2^n	1.0 sec	12.7 days	366 centuries

An example of an algorithm needing 2^n operations would be one in which we are required to examine all the 2^n subsets of a set of size n. Clearly, such an algorithm is not very practical, and we should be well-advised to seek an alternative algorithm requiring only a number of operations like n^2 or n^3. An algorithm which involves n^c operations (where $c \geq 1$ is a constant) is said to be a 'polynomial-time' algorithm, whereas one which requires c^n operations ($c > 1$) is said to be an 'exponential-time' algorithm.

We now introduce some mathematical notation which is very useful in estimating the efficiency of an algorithm. When we describe how the number of operations $f(n)$ depends on the size n, we can ignore relatively small contributions. Also, we can overlook the special features which sometimes occur when n itself is fairly small. What we require is an estimate of the order of magnitude of $f(n)$, valid for all n except possibly a finite number of special values. The following definition is appropriate for such purposes.

Definition Let f be a function from $\mathbb{N}$ to $\mathbb{N}$. We say that

$$f(n) \quad \text{is} \quad O(g(n))$$

if there is a positive constant k such that $f(n) \le kg(n)$ for all n in $\mathbb{N}$ (with possibly a finite number of exceptions). The symbol $O(g(n))$ is pronounced 'big-oh of $g(n)$'.

Suppose we have calculated that the number of operations involved in a certain algorithm is $3n^3 + 20n^2 + 5n + 11$. Since $n \le n^3$ and $n^2 \le n^3$ we have the inequality

$$3n^3 + 20n^2 + 5n + 11 \le (3 + 20 + 5 + 11)n^3,$$

where the right-hand side is a constant multiple of n^3. Thus we may say that the efficiency of the algorithm is $O(n^3)$ in this case. Roughly speaking, the O symbol picks out the most important term in an expression and ignores constant factors. In this way we obtain estimates such as

$$n^2 + 17n + 3 \quad \text{is} \quad O(n^2),$$
$$2^n + 3n^5 + 12n^4 \quad \text{is} \quad O(2^n).$$

There are a couple of points about the O notation which can cause difficulty. First, we must avoid writing *equations* like $f(n) = O(g(n))$, since this can easily lead to the (false) conclusion that $g(n)$ is $O(f(n))$. Secondly, when we say that $f(n)$ is $O(g(n))$ it is sometimes inferred that $g(n)$ is the 'best' estimate of $f(n)$, although this is not strictly part of the definition. For example, it is perfectly correct to say that the expression $n^2 + 17n + 3$ is $O(n^3)$, rather than $O(n^2)$. For this reason we sometimes say that $n^2 + 17n + 3$ is *at least* $O(n^2)$, where it is implied that no better estimate is possible.

Exercises 7.6

1 Find a function $g(n)$ (of the form A^B) such that $f(n)$ is $O(g(n))$, in each of the cases:

(i) $f(n) = \binom{n}{2}$;

(ii) $f(n) = \dfrac{5n^3 + 6}{n + 2}$;

(iii) $f(n) = \dfrac{(n^2 + 7)3^n}{2^n}$;

(iv) $f(n) = n!$.

2 Show that for any positive constants $C_0, C_1, \ldots, C_k$, the expression

$$C_0 + C_1 n + C_2 n^2 + \cdots + C_k n^k$$

is $O(n^k)$. Prove that the expression is *not* $O(n^{k-1})$.

3   Suppose that we are given an algorithm which, for a (worst-case) instance of size n, involves n stages, and that the ith stage ($1 \le i \le n$) requires i^2 operations. Show that the efficiency of the algorithm is $O(n^3)$.

4   When we say that an expression $f(n)$ is $O(\log n)$, why is it unnecessary to specify the base of the logarithm?

7.7   Comparison of algorithms

For any given problem, there will usually be several different algorithms for solving that problem. Sometimes the most straightforward algorithm is adequate for all practical purposes, but often we must seek better algorithms in order to solve larger instances of the problem. In this context, an algorithm for which the efficiency is $O(\log n)$ is better than $O(n^2)$, and so on. For example, the number of single digit multiplications involved in the long multiplication algorithm is $O(n^2)$, but there is a better algorithm for which the efficiency is $O(n^{1.59})$. (See Ex. 7.9.5 and Ex. 12.7.9.) In this case the more efficient algorithm is also more complicated, and it is unnecessary to use it for routine calculations. But if we have to multiply a great many very large integers the increase in efficiency might outweigh the complications.

We shall illustrate these ideas by considering the problem of computing the nth power of a fixed number m. The size of an instance will be taken as n, and the significant operations will be the multiplications.

The simplest method (let us call it *Algorithm A*) for computing m^n is to calculate $m^2, m^3, \ldots, m^n$ in turn, by multiplying each term by m. Since this involves $n - 1$ multiplications, and $n - 1$ is $O(n)$, the efficiency of *Algorithm A* is $O(n)$.

But it is easy to see that better algorithms are possible. For example, in order to compute m^{23} we could first calculate m^2, m^4, m^8, m^{16} by multiplying each term by itself, and then compute

$$m \times m^2 = m^3, \qquad m^3 \times m^4 = m^7, \qquad m^7 \times m^{16} = m^{23}.$$

This procedure involves seven multiplications, instead of the 22 required by *Algorithm A*. The general rule behind this example can be deduced from Fig. 7.5.

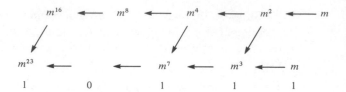

Fig. 7.5
Calculation of m^{23}.

Proceeding along the top line from right to left we carry out the successive multiplications which yield m^2, m^4, m^8, and so on, stopping at m^{16} in this case. The bottom line is governed by the fact that 10111 is the binary representation of 23. Since the last binary digit is 1, we begin on the right with $m^1 = m$. (If the last binary digit were 0 we should begin with $m^0 = 1$.) As we proceed along the bottom line from right to left we multiply the current value by m^{2^i} if and only if the ith binary digit is 1. A flowchart for this procedure, which we shall call *Algorithm B*, is shown in Fig. 7.6. When $n = 23$ the variable x assumes the values m, m^2, m^4, m^8, m^{16}, and the variable z assumes the values 1, m, m^3, m^7, m^{23}. The operations on y amount to finding the binary expansion of n, using the program given in Section 7.5.

The general validity of *Algorithm B* can be established by the inductive method described in Section 7.4 (see Ex. 7.7.2). If the number of digits in the binary representation of n is $L(n)$ then the number of multiplications required is at most $2(L(n) - 1)$. Since $L(n)$ is approximately $\log_2 n$ we may say that the efficiency of *Algorithm B* is $O(\log n)$, and thus it is a significant improvement on *Algorithm A*.

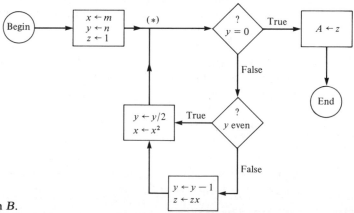

Fig. 7.6
Flowchart for Algorithm B.

Could there be a better method still? For individual values of n we can indeed make some improvement. When $n = 15$ *Algorithm B* requires six multiplications, whereas five will suffice:

$$m \times m = m^2, \ m \times m^2 = m^3, \ m^2 \times m^3 = m^5,$$
$$m^5 \times m^5 = m^{10}, \ m^5 \times m^{10} = m^{15}.$$

However, it is quite easy to see that there can be no significant improvement on *Algorithm B* for all values of n. This is because a single multiplication cannot do more than double the largest exponent so far obtained, and so r multiplications cannot get us beyond m^{2^r}. Consequently, when $n = 2^r + 1$ the number of multiplications required to compute m^n is at least $r + 1$, which is approximately $\log_2 n$. It follows that the efficiency of any algorithm for this problem is at least $O(\log n)$.

Exercises 7.7

1 How many multiplications are required for the calculation of m^{55} by *Algorithms A* and *B*? Find a method which requires only eight multiplications.

2 In Fig. 7.6, let $P(k)$ denote the assertion: if the program reaches (*) with $y = k$ and any values of x and z, then it will eventually end with $A = zx^k$. Show that $P(n)$ is true for all $n \geq 0$, and deduce the validity of *Algorithm B*.

3 The rule for multiplying 2×2 matrices is

$$\begin{bmatrix} a & b \\ c & d \end{bmatrix} \begin{bmatrix} x & y \\ z & t \end{bmatrix} = \begin{bmatrix} ax + bz & ay + bt \\ cx + dz & cy + dt \end{bmatrix}.$$

This involves 8 multiplications and 4 additions. Show that the calculation can be done by using only 7 multiplications:

$$m_1 = (a + d) \times (x + t) \qquad m_5 = (a + b) \times t$$
$$m_2 = (c + d) \times x \qquad m_6 = (c - a) \times (x + y)$$
$$m_3 = \qquad a \times (y - t) \qquad m_7 = (b - d) \times (z + t).$$
$$m_4 = \qquad d \times (z - x)$$

(You must show that each entry of the product matrix can be expressed as a sum of terms $\pm m_i$ ($1 \leq i \leq 7$).) How many additions are involved in this method?

7.8 Introduction to sorting algorithms

Suppose we have a list of students in alphabetical order and we wish to arrange them in a new order according to the marks obtained in an examination. This is an example of a *sorting* problem. For convenience we shall discuss only the case where the marks are all distinct, since the modifications required when some marks are equal are quite straightforward.

We shall consider the problem in the following general form. A sequence of distinct integers $x_1, x_2, \ldots, x_n$ is given and it is desired to arrange them in increasing order. In other words we require the permutation π of the set $\{1, 2, \ldots, n\}$ for which

$$x_{\pi(1)} < x_{\pi(2)} < x_{\pi(3)} < \cdots < x_{\pi(n)}.$$

In practice, we do not determine π explicitly; we simply arrange the given integers in the required order. For example, the sequence 7, 2, 9, 3, 5, becomes 2, 3, 5, 7, 9.

One of the simplest sorting algorithms is known as *bubble sort*. The idea is to compare adjacent items in the list and switch them if they are in the wrong order. In the 'first pass' of the algorithm we compare the first and second items, then the second and third, and so on. Each comparison refers to the items in the order resulting from all previous comparisons. For example, the first pass operates on the initial order 4, 7, 3, 1, 5, 8, 2, 6 as shown in Table 7.8.1.

Table 7.8.1

Initial order:	4	7	3	1	5	8	2	6
After 1st comparison:	4	7						
After 2nd comparison:		3	7					
After 3rd comparison:			1	7				
After 4th comparison:				5	7			
After 5th comparison:					7	8		
After 6th comparison:						2	8	
After 7th comparison:							6	8
After first pass:	4	3	1	5	7	2	6	8

At the conclusion of the first pass, we can be sure that the largest integer is in the last place. In the 'second pass' the first $n-1$ integers are compared in the same way, and afterwards we can be sure that the two largest integers are in their correct places. We continue in this way for $n-1$ passes, after which the sorting is complete.

One of the main advantages of bubble sorting is that the programming is very simple. Given the sequence of distinct integers $x_1, x_2, \ldots, x_n$, the following program arranges them in increasing order $y_1, y_2, \ldots, y_n$, by the bubble sort method.

> **for** $j = 1$ **to** $n - 1$ **do**
>> **for** $i = 2$ **to** $n - j + 1$ **do**
>>> **if** $x_{i-1} > x_i$ **then** switch x_{i-1} and x_i
>>
>> $y_{n-j+1} \leftarrow x_{n-j+1}$
>
> $y_1 \leftarrow x_1$

The operations involved in bubble sort are comparisons and switches. At the jth pass $n - j$ comparisons are made, and so the total number of comparisons is

$$(n-1) + (n-2) + \cdots + 2 + 1 = \tfrac{1}{2}n(n-1),$$

which is $O(n^2)$. The number of switches is also $O(n^2)$, since in the worst-case (when the integers are given in reverse order) each comparison must be followed by a switch.

In Chapter 9 we shall show that the number of comparisons required by *any* algorithm for the sorting problem is at least $O(n \log n)$. If an algorithm with efficiency $O(n \log n)$ can be found then it will be better, in the sense of Section 7.7, than the bubble sort algorithm, whose efficiency is $O(n^2)$. In fact, several $O(n \log n)$ algorithms for the sorting problem are known. One of the simplest is a version of *insertion* sorting, and this is the method we shall now describe.

The basic idea of insertion sorting is to begin with the list $L = (x_1)$ and to insert x_i in its correct place in the list for $i = 2, 3, \ldots, n$. For example, if $x_1, x_2, \ldots, x_8$ are the integers 47, 73, 21, 45, 28, 69, 19, 23, the list is built up as shown in Table 7.8.2.

Table 7.8.2

47							
47	73						
21	47	73					
21	45	47	73				
21	28	45	47	73			
21	28	45	47	69	73		
19	21	28	45	47	69	73	
19	21	23	28	45	47	69	73

In insertion sorting the number of comparisons needed to find the correct place for x_i depends upon which of the several possible methods is adopted. If we use the 'sequential' method, in which we compare x_i with each member of the partial list in turn (starting at the left), then we could need as many as $i-1$ comparisons before we find the correct place. So the total number of comparisons could be $(n-1)+\cdots+1$, which is $O(n^2)$ again. But the insertion can be done more cunningly by the 'bisection' method: find which half of the list x_i should be in, then which half of that half, and so on. If there are between 2^{r-1} and 2^r items, then r comparisons will be required by this method (since the partial list is already in the correct order). In other words about $\log_2 i$ comparisons are needed for a list of length $i-1$. Hence the total number of comparisons is approximately

$$\log_2 2 + \log_2 3 + \cdots + \log_2 n,$$

which is an expression with $n-1$ terms, each term being not greater than $\log_2 n$. Consequently, the number of comparisons is $O(n \log n)$.

Although insertion sorting by the bisection method achieves the best possible order of magnitude for the number of comparisons, it is much harder to program than bubble sort. Also the number of data transfers is still $O(n^2)$, since when x_i is inserted all the $i-1$ members of the partial list may be shifted. However, there are algorithms which achieve $O(n \log n)$ comparisons and $O(n \log n)$ transfers and we shall describe one of them, *heapsort*, in Section 9.2.

Exercises 7.8

1 Describe the result of each 'pass' of the bubble sort algorithm when the given list is 4, 3, 9, 5, 1, 2, 7, 8, 6.

2 Use insertion sort (by hand!) to arrange the following list in increasing order: 516, 207, 321, 581, 762, 163, 921, 105, 721, 813, 316, 188, 733, 909, 281, 312, 871, 950, 135, 888, 417.

3 Write a program which, given a sequence $z_1, z_2, \ldots, z_n$ of distinct integers in increasing order, will insert a given integer q in its correct place and return a list $w_1, w_2, \ldots, w_{n+1}$. (You may assume that q is different from all the z_i, and you will find that the sequential method is much easier to program than the bisection method.)

4 Modify the bubble sort program to include a location c with the following properties.

(i) The value of c is the number of switches made during the current pass.

(ii) c is set equal to zero at the beginning of each pass.

(iii) If c is zero at the end of a pass the program will not execute any more passes.

5 Compute the number of comparisons and the number of switches required by the modified bubble sort program obtained in Ex. 4 when the initial list is in

(i) the correct order $x_1 < x_2 < \cdots < x_n$;

(ii) the reverse order $x_1 > x_2 > \cdots > x_n$.

6 Compute the number of comparisons and the number of data transfers required by the insertion sort algorithm (sequential method) in the two cases specified in the previous exercise.

7.9 Miscellaneous Exercises

1 Write out a formal proof of the correctness of the algorithm for addition in base 10, as given in Section 7.1 (*Example*). In other words, show that if

$$a = (a_{n-1}a_{n-2} \ldots a_1 a_0)_{10}, \qquad b = (b_{n-1}b_{n-2} \ldots b_1 b_0)_{10},$$

and the digits s_i $(0 \leqslant i \leqslant n)$ are defined by the given formulae, then

$$a + b = (s_n s_{n-1} \cdots s_1 s_0)_{10}.$$

2 Repeat the previous exercise for the algorithm described in Ex. 7.1.1.

3 Describe (in ordinary language) the usual method of 'long division', and explain why it is not, strictly speaking, an algorithm.

4 Formulate an algorithm for the subtraction of one n-digit integer from another (larger) one, in base 10 notation.

5 Suppose x and y are integers which have $2n$ digits in base 10 notation. Write

$$x = 10^n a + b, \qquad y = 10^n c + d,$$

where a, b, c, d are n-digit integers. Verify the formula

$$xy = (10^{2n} + 10^n)ac + 10^n(a - b)(d - c) + (10^n + 1)bd,$$

and explain how it could be used to improve the straightforward long multiplication algorithm for computing xy.

6 Construct a flowchart for the program which follows, and determine the final value of b.

$$x \leftarrow n; \quad z \leftarrow 1$$
$$\textbf{while} \quad x > 1 \quad \textbf{do}$$
$$z \leftarrow xz$$
$$x \leftarrow x - 1$$
$$b \leftarrow z$$

7 Write out a formal proof that your assertion about the final value of b in the previous exercise is correct.

8 Let n be a positive integer and p a prime. What is the significance of the value c calculated by the following program?

$$x \leftarrow n; \quad y \leftarrow p; \quad t \leftarrow p; \quad z \leftarrow 0$$
$$\textbf{while} \quad y \mathbin{!} x \quad \textbf{do}$$
$$\qquad y \leftarrow ty$$
$$\qquad z \leftarrow z + 1$$
$$c \leftarrow z$$

9 Write out the multiplications required in the calculation of m^n by Algorithm B (Section 7.7) when

$$\text{(i) } n = 59, \qquad \text{(ii) } n = 77.$$

Show that in both cases there is a method requiring only eight multiplications.

10 Write a program which, given two positive integers a and b in base 10 notation, will assign to q the values -1, 0, or 1 according as $a < b$, $a = b$, or $a > b$.

11 How many additions and multiplications are needed to compute the product of two $n \times n$ matrices, using the standard formulae?

12 Find a method of multiplying two 4×4 matrices which needs less than 64 multiplications. [Hint: $2 + 2 = 4$.]

13 Let the sequence (f_r) be defined by

$$f_1 = 1, \qquad f_2 = 2, \qquad f_r = f_{r-1} + f_{r-2} \quad (r \geqslant 2).$$

Show that if $m \geqslant n$, and the Euclidean algorithm requires k steps to compute $\gcd(m, n)$, then $n \geqslant f_k$. Deduce that the number of steps required is $O(\log n)$.

14 Prove that the following program assigns to d the gcd of the positive integers m and n.

$$x \leftarrow m; \qquad y \leftarrow n$$
$$\textbf{while} \quad x \neq y \quad \textbf{do}$$
$$\qquad \textbf{if} \quad x > y$$
$$\qquad\qquad \textbf{then} \quad x \leftarrow x - y$$
$$\qquad\qquad \textbf{else} \quad y \leftarrow y - x$$
$$d \leftarrow x$$

15 Use the principle of induction to prove that after k 'passes' of the bubble sort algorithm the last k integers are in their correct places.

16 Write a program for computing the gcd of three given positive integers r, s, t.

17   Let σ be a permutation of $\{1, 2, \ldots, n\}$. Write a program to calculate sgn σ, using the fact that sgn $\sigma = (-1)^{n-c}$, where c is the number of cycles in σ.

18   For any positive integer i let $\alpha(i)$ be the largest value of k such that 2^k divides i. Define a sequence of words $\mathbf{w}_1, \mathbf{w}_2, \ldots$, of length n in the alphabet $\{0, 1\}$, as follows. Take $\mathbf{w}_1$ to be the all-zero word, and for $i \geq 2$ take $\mathbf{w}_i$ to be the word obtained from $\mathbf{w}_{i-1}$ by altering the digit in position $\alpha(i) + 1$. Verify that when $n = 3$ we obtain the sequence

$$000, 100, 110, 010, 011, 111, 101, 001.$$

Prove that for any positive integer n this algorithm generates all 2^n words of length n. Explain how this result can be used as the basis of an algorithm for listing all the subsets of a given finite set.

8
Graphs

8.1 Graphs and their representation

The objects which we shall call *graphs* are very useful throughout Discrete Mathematics. Their name is derived from the fact that they can be regarded as a form of graphical (or pictorial) notation, and in this respect alone they resemble the familiar graphs of functions which are studied in elementary mathematics. But our *graphs* are quite different from graphs of functions, and correspond more closely with the objects which, in everyday language we call 'networks'.

Definition   A **graph** G consists of a finite set V, whose members are called **vertices**, and a set E of 2-subsets of V, whose members are called **edges**. We usually write $G = (V, E)$ and say that V is the **vertex-set** and E is the **edge-set**.

The restriction to finite sets is not essential, but it is convenient for us because we shall not consider infinite 'graphs' in this book.

A typical example of a graph $G = (V, E)$ is given by the sets

$$V = \{a, b, c, d, z\}, \qquad E = \{\{a, b\}, \{a, d\}, \{b, z\}, \{c, d\}, \{d, z\}\}.$$

This example and the definition itself are not very illuminating, and it is only when we turn to the *pictorial representation* of a graph that

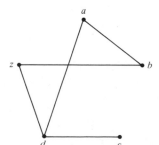

Fig. 8.1
A pictorial representation
of a graph.

the light dawns. We represent the vertices as points, and join two points by a line whenever the corresponding pair of vertices is an edge. Thus Fig. 8.1 is a pictorial representation of the graph given in the example above. This kind of representation is extremely convenient for working 'by hand' with relatively small graphs. Furthermore, its intuitive familiarity is of great assistance in formulating and understanding abstract arguments. We shall give a frivolous example.

Example   Professor McBrain and his wife April give a party at which there are four other married couples. Some pairs of people shake hands when they meet, but naturally no couple shake hands with each other. At the end of the party the Professor asks everyone else how many people they have shaken hands with, and he receives nine different answers. How many people shook hands with April?

Solution   We construct a graph whose vertices are the people at the party, and there is an edge $\{x, y\}$ whenever x and y shook hands. Since there are nine people apart from Professor McBrain, and the maximum number of handshakes in which any one person can be involved is eight, it follows that the nine different answers received by the Professor must be 0, 1, 2, 3, 4, 5, 6, 7, 8. We denote the vertices by these numbers and use M for McBrain himself. So we have a pictorial representation as in Fig. 8.2.

Now, vertex 8 is joined to all the other vertices except one, which must therefore represent the spouse of 8. This vertex must be 0, since it is certainly not joined to 8 (or any other vertex, for that matter). Thus 8 and 0 are a married couple, and 8 is joined to 1, 2, ..., 7 and M. In particular 1 is joined to 8 and this is the only edge from 1. Hence vertex 7 is not joined to 0 and 1 (only), and the spouse of 7 must be 1, since 0 is married to 8. Continuing in the

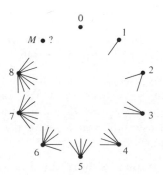

Fig. 8.2
April's party.

same way, we see that 6 and 2, and 5 and 3 are married couples. It follows that M and 4 are married, so vertex 4 represents April, who shook hands with four people. □

Although the pictorial representation of graphs is intuitively appealing to human beings, it is clearly useless when we wish to communicate with a computer. For that purpose we must represent a graph by some kind of list or table. Let us say that two vertices x and y of a graph are **adjacent** whenever $\{x, y\}$ is an edge. (We also say that x and y are **neighbours**.) Then we can represent a graph $G = (V, E)$ by its **adjacency list**, wherein each vertex v heads a list of those vertices adjacent to v. The graph in Fig. 8.1 has the adjacency list

a	b	c	d	z
b	a	d	a	b
d	z		c	d
			z	

Exercises 8.1

1   Three houses A, B, C each has to be connected to the gas, water, and electricity supplies: G, W, E. Write down the adjacency list for the graph which represents this problem, and construct a pictorial representation of it. Can you find a picture in which the lines representing the edges do not cross?

2   The pathways in a formal garden are to be laid out in the form of a **wheel graph** W_n, whose vertex-set is $V = \{0, 1, 2, \dots, n\}$ and whose edges are

$$\{0, 1\}, \quad \{0, 2\}, \dots, \{0, n\},$$
$$\{1, 2\}, \quad \{2, 3\}, \dots, \{n-1, n\}, \quad \{n, 1\}.$$

Describe a route around the pathways which starts and ends at vertex 0 and visits every vertex once only.

3   For each positive integer n we define the **complete graph** K_n to be the graph with n vertices in which each pair of vertices is adjacent. How many edges has K_n? For which values of n can you find a pictorial representation of K_n with the property that the lines representing the edges do not cross?

4   A *3-cycle* in a graph is a set of three mutually adjacent vertices.

Construct a graph with five vertices and six edges which contains no 3-cycles.

8.2 Isomorphism of graphs

At this point it must be emphasized that a graph is defined as an abstract mathematical entity. It is in this light that we shall discuss the important question of what we mean by saying that two graphs are 'the same'.

Clearly the important thing about a graph is not the names of the vertices, nor is it their representation pictorially or in any other concrete way. The characteristic property of a graph is the way in which the vertices are linked by its edges. This motivates the following definition.

Definition Two graphs G_1 and G_2 are said to be **isomorphic** when there is a bijection α from the vertex-set of G_1 to the vertex-set of G_2 such that $\{\alpha(x), \alpha(y)\}$ is an edge of G_2 if and only if $\{x, y\}$ is an edge of G_1. The bijection α is said to be an **isomorphism**.

For example, consider the two graphs depicted in Fig. 8.3. In this case there is a bijection from the vertex-set of G_1 to the vertex-set of G_2 which has the required property; it is given by

$$\alpha(a) = t, \qquad \alpha(b) = v, \qquad \alpha(c) = w, \qquad \alpha(d) = u.$$

We can check that each edge of G_1 corresponds uniquely to an edge of G_2, and conversely. For instance, the edge bc of G_1 corresponds to the edge vw of G_2, and so on. (We shall customarily use the abbreviation xy for an edge $\{x, y\}$, remembering that an edge is an unordered pair, so that xy and yx mean the same thing.)

When, as in Fig. 8.3, two graphs G_1 and G_2 are isomorphic we usually regard them as being 'the same' graph. In order to show that

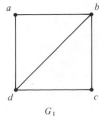
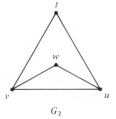

Fig. 8.3
G_1 and G_2 are isomorphic.

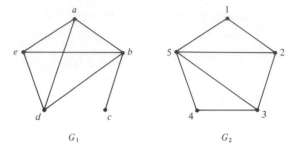

Fig. 8.4
G_1 and G_2 are not
isomorphic.

two graphs are *not* isomorphic, we must demonstrate that there can
be no bijection from the vertex-set of one to the vertex-set of the
other which takes edges to edges. If the two graphs have different
numbers of vertices, then no bijection is possible, and the graphs
cannot be isomorphic. If the graphs have the same number of
vertices, but different numbers of edges, then there are bijections
but none of them can be an isomorphism (Ex. 8.8.10). Even if the
graphs have the same numbers of vertices and edges, they need not
be isomorphic. For example, the two graphs in Fig. 8.4 both have
five vertices and seven edges, but they are not isomorphic. One way
to show this is to remark that the vertices a, b, d, e form a complete
subgraph of G_1 (each pair of them is linked by an edge). Any
isomorphism must take these vertices to four vertices of G_2 which
have the same property, and since there is no such set of vertices in
G_2 there can be no isomorphism.

**Exercises
8.2**

1 Prove that the graphs shown in Fig. 8.5 are not isomorphic.

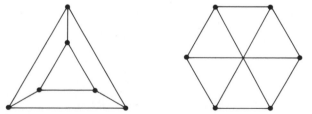

Fig. 8.5
Show that these graphs
are not isomorphic.

2 Find an isomorphism between the graphs defined by the following adjacency lists. (Both lists specify versions of a famous graph,
known as **Petersen's graph**. See also Ex. 8.8.3.)

a	b	c	d	e	f	g	h	i	j	0	1	2	3	4	5	6	7	8	9
b	a	b	c	d	a	b	c	d	e	1	2	3	4	5	0	1	0	2	6
e	c	d	e	a	h	i	j	f	g	5	0	1	2	3	4	4	3	5	7
f	g	h	i	j	i	j	f	g	h	7	6	8	7	6	8	9	9	9	8

3   Let $G = (V, E)$ be the graph defined as follows. The vertex-set V is the set of all words of length 3 in the alphabet $\{0, 1\}$, and the edge-set E contains those pairs of words which differ in exactly one position. Show that G is isomorphic to the graph formed by the corners and edges of an ordinary cube.

8.3 Valency

The **valency** of a vertex v in a graph $G = (V, E)$ is the number of edges of G which contain v. We shall use the notation $\delta(v)$ for the valency of v, so formally

$$\delta(v) - |D_v|, \quad \text{where} \quad D_v - \{e \in E \mid v \in e\}.$$

The graph depicted in Fig. 8.1 has $\delta(a) = 2$, $\delta(b) = 2$, $\delta(c) = 1$, $\delta(d) = 3$, $\delta(z) = 2$. The first theorem of graph theory tells us that the sum of these numbers is twice the number of edges; it is a simple application of the method for counting sets of pairs given in Section 3.2.

Theorem 8.3   The sum of the values of the valency $\delta(v)$, taken over all the vertices v of a graph $G = (V, E)$, is equal to twice the number of edges:

$$\sum_{v \in V} \delta(v) = 2|E|.$$

Proof   Let S denote the subset of $V \times E$ consisting of those pairs (v, e) for which v belongs to e. For each v in V the 'row total' $r_v(S)$ is the number of edges containing v, and so it is equal to $\delta(v)$. For each e in E the 'column total' $c_e(S)$ is the number of vertices in e, which is 2. Hence, by Theorem 3.2

$$\sum_{v \in V} \delta(v) = 2 + 2 + \cdots + 2 = 2|E|,$$

as required.   □

There is a useful corollary of this result. Let us say that a vertex of G is **odd** if its valency is odd, and **even** if its valency is even. Denote by V_o and V_e the sets of odd and even vertices respectively, so that $V = V_o \cup V_e$ is a partition of V. By Theorem 8.3, we have

$$\sum_{v \in V_o} \delta(v) + \sum_{v \in V_e} \delta(v) = 2\,|E|.$$

Now each term in the second sum is even, and so this sum is an even number. Since the right-hand side is an even number, the first sum must be even also. But a sum of odd numbers can only be even if there is an even number of them. In other words:

the number of odd vertices is even.

This result is sometimes known as the handshaking lemma, in view of its interpretation in terms of people and handshakes: given any set of people, the number of people who have shaken hands with an odd number of other members of the set is even.

A graph in which all vertices have the same valency r is said to be **regular** (with valency r), or **r-valent**. In this case, the result of Theorem 8.3 becomes

$$r\,|V| = 2\,|E|.$$

In fact, an r-valent graph is just another name for a 1-design with parameters $(|V|, 2, r)$, in the terminology of Section 4.6. But this viewpoint is not particularly useful in practice.

Many of the graphs which occur in applications are regular. We have already met the complete graphs K_n (Ex. 8.1.3); they are regular, with valency $n-1$. In elementary geometry we discuss the n-sided polygons, which correspond in graph theory to **cycle graphs** C_n. Formally, we can say that the vertex-set of C_n is $\mathbb{Z}_n$, the vertices i and j being joined by an edge whenever $j = i+1$ or $j = i-1$ in $\mathbb{Z}_n$. Clearly, C_n is a regular graph with valency 2, provided $n \geqslant 3$.

An important application of the notion of valency is to the problem of testing whether or not two graphs are isomorphic. If $\alpha: V_1 \to V_2$ is an isomorphism between G_1 and G_2, and $\alpha(v) = w$, then each edge containing v is transformed by α into an edge containing w. Consequently, $\delta(v) = \delta(w)$. On the other hand, if G_1 has a vertex x, with $\delta(x) = \delta_0$ say, and G_2 has no vertex with valency δ_0, then G_1 and G_2 cannot be isomorphic. This gives us another way of distinguishing between the two graphs in Fig. 8.4, since the first graph has a vertex of valency 1 and the second graph has no such vertex.

A further extension of this idea is given in Ex. 8.3.4.

1 Is it possible that the following lists are the valencies of all the vertices of a graph? If so, give a pictorial representation of such a graph. (Remember that there is at most one edge joining each pair of vertices.)

   (i) 2, 2, 2, 3.   (ii) 1, 2, 2, 3, 4.
   (iii) 2, 2, 4, 4, 4. (iv) 1, 2, 3, 4.

2 If $G = (V, E)$ is a graph, the **complement** $\bar{G}$ of G is the graph whose vertex-set is V and whose edges join those pairs of vertices which are not joined in G. If G has n vertices and their valencies are $d_1, d_2, \ldots, d_n$, what are the valencies of the vertices of $\bar{G}$?

3 Find as many different (non-isomorphic) regular 4-valent graphs with seven vertices as you can. [Hint: consider the complement of such a graph.]

4 Suppose G_1 and G_2 are isomorphic graphs. For each $k \geq 0$ let $n_i(k)$ be the number of vertices of G_i which have valency k $(i = 1, 2)$. Show that $n_1(k) = n_2(k)$.

5 Show that if G is a graph with at least two vertices then G has two vertices with the same valency.

8.4 Paths and cycles

Frequently we use graphs as models of practical situations involving routes: the vertices represent towns or junctions, and each edge represents a road or some other form of communication link. The definitions in this section are best conceived with that kind of picture in mind.

Definition A **walk** in a graph G is a sequence of vertices

$$v_1, v_2, \ldots, v_k,$$

such that v_i and v_{i+1} are adjacent $(1 \leq i \leq k - 1)$. If all its vertices are distinct, a walk is called a **path**.

 Thus a walk specifies a route in G which proceeds from a vertex to an adjacent one, and so on. A walk may visit any vertex several times, and in particular, it may reverse its direction by going from x to y and immediately back to x again. In a path, each vertex is visited at most once.

Let us write $x \sim y$ whenever vertices x and y of G can be joined by a path in G: strictly speaking, this means that there is a path $v_1, v_2, \ldots, v_k$ in G with $x = v_1$ and $y = v_k$. It is a simple matter to verify that $\sim$ is an equivalence relation on the vertex-set V of G, and so V is partitioned into disjoint equivalence classes. Two vertices are in the same class if they can be joined by a path, and in different classes if there is no such path.

Definition   Suppose $G = (V, E)$ is a graph and the partition of V corresponding to the equivalence relation $\sim$ is

$$V = V_1 \cup V_2 \cup \cdots \cup V_r.$$

Let E_i $(1 \leq i \leq r)$ denote the subset of E comprising those edges whose ends are both in V_i. Then the graphs $G_i = (V_i, E_i)$ are called the **components** of G. If G has just one component, it is said to be **connected**.

The terminology is almost self-explanatory. The graph shown in Fig. 8.6 has two components, and is therefore not connected. The decomposition of a graph into components is very useful, since many properties of graphs can be established by considering each component separately. For this reason, theorems about graphs are often proved only for the class of connected graphs.

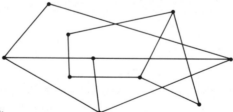

Fig. 8.6
A graph with two components.

When a fairly small graph is given by a pictorial representation it is quite easy to spot whether it is connected or not. However, when a graph is given by an adjacency list we shall need an efficient algorithm to decide if it is connected. This problem will be studied in the next chapter.

A walk $v_1, v_2, \ldots, v_{r+1}$ whose vertices are all distinct except that $v_1 = v_{r+1}$ is called a **cycle**. It has r distinct vertices and r edges, and we often speak of an **r-cycle**, or cycle of **length** r.

Example   Two senior members of the Mathematics Department at the University of Folornia plan to spend their vacation on the island of Wanda.

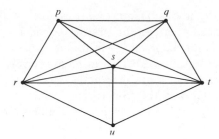

Fig. 8.7
The grand tour.

Figure 8.7 represents the interesting places on the island and the roads linking them. Dr Elsie Chunner is a tourist by nature, and wishes to visit each place once and return to her starting point. Dr Bob Dodder is an explorer, and wishes to traverse every road just once, in either direction; he is prepared to start and finish in different places. Can suitable routes for Drs Chunner and Dodder be found?

Solution   Dr C can use several routes: one possibility is the cycle p, q, t, s, u, r, p.

However, Dr D has a problem. Let us label his starting vertex x and his finishing vertex y, and suppose for the moment that $x \neq y$. Then he uses one edge at x when he starts and each time he returns to x he must arrive and depart by new edges. In this way, he uses an odd number of edges at x, and so x must be an odd vertex. Similarly, y must be an odd vertex also, since he uses two edges each time he passes through y, and one edge to finish at y. All the remaining vertices must be even, since every time he arrives at an intermediate vertex he also departs, and thereby uses two edges.

In summary, a route for Dr D starting and finishing at distinct vertices x and y is possible only if x and y are odd vertices, and the rest are even. But in the given graph the valencies are as follows:

$$v: \quad p \quad q \quad r \quad s \quad t \quad u$$
$$\delta(v): \quad 4 \quad 4 \quad 5 \quad 5 \quad 5 \quad 3 .$$

So there are too many odd vertices, and consequently no route for Dr D. If we allow the possibility that $x = y$, the situation is worse still, for then all the vertices would have to be even.   □

In general, Dr C's route is a cycle which contains all vertices of the given graph. Such cycles were studied by the Irish mathematician W. R. Hamilton (1805–65), and consequently a cycle with this property is known as a **Hamiltonian cycle**. In our example, it was very easy to find a Hamiltonian cycle, but this was a misleading special case. For a specific graph, it may be a difficult problem to decide whether or not a Hamiltonian cycle exists.

On the other hand, Dr D's problem is easily settled. A walk which uses each edge of a graph exactly once is called an **Eulerian walk**, because Euler was the first to study such walks. He found that if $x \neq y$ a necessary condition for an Eulerian walk starting at x and finishing at y is that x and y are odd vertices and the rest are even, while if $x = y$ the condition is that all vertices are even. Thus a necessary condition for the existence of an Eulerian walk in a graph G is that G has at most two odd vertices. Furthermore, it can be shown that this condition is also sufficient. Since it is easy to compute the valencies of vertices in any graph, it is correspondingly easy to decide if a given graph has an Eulerian walk.

Exercises 8.4

1 Find the number of components of the graph whose adjacency list is

a	b	c	d	e	f	g	h	i	j
f	c	b	h	c	a	b	d	a	a
i	g	e		g	i	c		f	f
j		g			j	e			

2 How many components are there in the graph of April's party (Section 8.1)?

3 Find a Hamiltonian cycle in the graph formed by the vertices and edges of an ordinary cube.

4 Next year Dr Chunner and Dr Dodder intend to visit the island of Meanda, where the interesting places and the roads joining them are represented by the graph whose adjacency list is

0	1	2	3	4	5	6	7	8
1	0	1	0	3	0	1	0	1
3	2	3	2	5	4	5	2	3
5	6	7	4		6	7	6	5
7	8		8		8		8	7

Is it possible to find routes for them which satisfy the requirements set out in the *Example*?

5 A mouse intends to eat a $3 \times 3 \times 3$ cube of cheese. Being tidy-minded, it begins at a corner and eats the whole of a $1 \times 1 \times 1$ cube before going on to an adjacent one. Can the mouse end in the centre?

8.5 Trees

Definition We say that a graph T is a **tree** if it has two properties:

(T1) T is connected;
(T2) there are no cycles in T.

Some typical trees are depicted in Fig. 8.8. Because of their special structure and properties, trees occur in many different applications of mathematics, especially in operations research and computer science. We begin our study of them by establishing some simple properties.

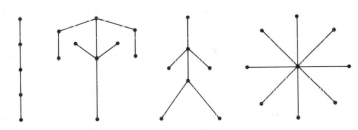

Fig. 8.8
Some trees.

Theorem 8.5 If $T = (V, E)$ is a tree with at least two vertices, then:

(T3) for each pair x, y of vertices there is a unique path in T from x to y;
(T4) the graph obtained from T by removing any edge has two components, each of which is a tree;
(T5) $|E| = |V| - 1$.

Proof (T3) Since T is connected, there is a path from x to y, say

$$x = v_0, v_1, \ldots, v_r = y.$$

If there is a different path, say

$$x = u_0, u_1, \ldots, u_s = y,$$

then let i be the smallest subscript for which $u_{i+1} \neq v_{i+1}$ (Fig. 8.9).

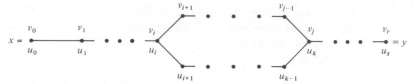

Fig. 8.9
Two distinct paths deter-
mine a cycle.

Since both paths finish at y they will meet again, and we can define j to be the smallest subscript such that

$$j > i \quad \text{and} \quad v_j = u_k \text{ for some } k.$$

Then $v_i, v_{i+1}, \ldots, v_j, u_{k-1}, u_{k-2}, \ldots, u_{i+1}, v_i$ is a cycle in T, contrary to hypothesis. Hence there is just one path in T from x to y.

(T4) Suppose uv is an edge of T, and let $S = (V, E')$ be the graph with the same vertex-set as T and edge-set $E' = E \backslash uv$. Let V_1 be the set of vertices x of T for which the unique path from x to v in T passes through u. Clearly the relevant path must end with the edge uv, otherwise T would have a cycle. Let V_2 be the complement of V_1 in V.

Each vertex in V_1 is joined by a path in S to u, and each vertex in V_2 is joined in S to v, but there is no path from u to v in S. It follows that V_1 and V_2 are the vertex-sets of the two components of S. Each component is (by definition) connected, and it contains no cycles, since there are no cycles in T. Hence the two components are trees.

(T5) The result is true when $|V| = 1$, since the only possible tree has no edges in that case.

Suppose it is true for all trees with k or fewer vertices. Let T be a tree with $|V| = k+1$, and let uv be any edge of T. If $T_1 = (V_1, E_1)$ and $T_2 = (V_2, E_2)$ are the trees obtained by removing uv from T, we have

$$|V_1| + |V_2| = |V|, \qquad |E_1| + |E_2| = |E| - 1.$$

Applying the induction hypothesis to T_1 and T_2 we have

$$|E| = |E_1| + |E_2| + 1 = |V_1| - 1 + |V_2| - 1 + 1 = |V| - 1,$$

as required. Hence the result is true for all positive integers k. $\square$

The properties (T1)–(T5) provide several alternative ways of defining a tree. For example, property (T3) alone can be taken as the defining property, instead of (T1) and (T2). We have already shown that (T3) is a consequence of (T1) and (T2), so it remains only to show that the converse holds (Ex. 8.5.3).

Exercises 8.5

1 There are six different (that is, mutually non-isomorphic) trees with six vertices: draw them.

2 Let $T = (V, E)$ be a tree with $|V| \geq 2$. Using property (T5) and Theorem 8.3 show that T has at least two vertices with valency 1.

3 Show that property (T3) implies (T1) and (T2).

4 A **forest** is a graph satisfying (T2) but not necessarily (T1). Prove that if $F = (V, E)$ is a forest with c components then

$$|E| = |V| - c.$$

8.6 Colouring the vertices of a graph

A problem which occurs frequently in modern life is that of time-tabling a set of events in such a way as to avoid clashes. We shall consider a very simple case as an example of how the theory of graphs can help us to study this problem.

Suppose we wish to schedule six one-hour lectures, v_1, v_2, v_3, v_4, v_5, v_6. Among the potential audience there are people who wish to hear both v_1 and v_2, v_1 and v_4, v_3 and v_5, v_2 and v_6, v_4 and v_5, v_5 and v_6, and v_1 and v_6. How many hours are necessary in order that the lectures can be given without clashes?

We can represent the situation by a graph (Fig. 8.10). The vertices correspond to the six lectures, and the edges signify the potential

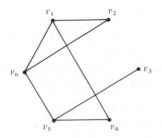

Fig. 8.10
The graph for a timetable problem.

clashes. A timetable which achieves the object of avoiding clashes is as follows:

Hour 1	Hour 2	Hour 3	Hour 4
v_1 and v_3	v_2 and v_4	v_5	v_6 .

In mathematical terms, we have a partition of the vertex-set into four parts, with the property that no part contains a pair of adjacent vertices of the graph. A more graphic description utilizes the function

$$c: \{v_1, v_2, v_3, v_4, v_5, v_6\} \rightarrow \{1, 2, 3, 4\}$$

which assigns to each vertex (lecture) the hour scheduled for it. Usually, we speak of colours assigned to the vertices, rather than hours, but clearly the exact nature of the objects 1, 2, 3, 4 is unimportant. We can use the names of actual colours, red, green, blue, yellow, or we can speak of colour 1, colour 2, and so on. The important point is that vertices which are adjacent in the graph must be given different colours.

Definition   A **vertex-colouring** of a graph $G = (V, E)$ is a function $c: V \rightarrow \mathbb{N}$ with the property that

$$c(x) \neq c(y) \quad \text{whenever} \quad \{x, y\} \in E.$$

The **chromatic number** of G, written $\chi(G)$, is defined to be the least integer k for which there is a vertex-colouring of G using k colours. In other words, $\chi(G) = k$ if and only if there is a vertex-colouring c which is a function from V to $\mathbb{N}_k$, and k is the least integer with this property.

Returning to the example depicted in Fig. 8.10, we see that our first attempt at a timetable is equivalent to a vertex-colouring using four colours. The smallest number of hours needed is the chromatic number of the graph, and we have now to ask if this number is less than 4. A quick trial with three colours will lead us to a solution:

Colour 1	Colour 2	Colour 3
v_1	v_2 and v_5	v_3, v_4, and v_6.

Furthermore, at least three colours are needed, since v_1, v_2, and v_6 are mutually adjacent and must have different colours. So we conclude that the chromatic number of this graph is 3.

In general, in order to show that the chromatic number of a given graph is k, we have to do two things:

(i) find a vertex-colouring using k colours;
(ii) show that no vertex-colouring uses fewer than k colours.

**Exercises
8.6**

1 Find the chromatic numbers of the following graphs:

    (i) a complete graph K_n;

    (ii) a cycle graph C_{2r} with an even number of vertices;

    (iii) a cycle graph C_{2r+1} with an odd number of vertices.

2 Determine the chromatic numbers of the graphs depicted in Fig. 8.11.

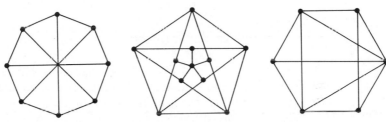

Fig. 8.11
Find the chromatic numbers.

3 Describe all graphs G for which $\chi(G) = 1$.

8.7 The greedy algorithm for vertex-colouring

The problem of finding the chromatic number of a given graph is a difficult one. Indeed, there is no known algorithm for the problem which works in 'polynomial-time', and most people believe that no such algorithm exists. However there is a simple method of constructing a vertex-colouring using a 'reasonable' number of colours.

The method is to assign colours to the vertices in order, in such a way that each vertex receives the first colour which has not already been assigned to one of its neighbours. In this algorithm we insist on making the best choice we can at each step, without looking ahead to see if that choice will create problems later on. An algorithm of this kind is usually referred to as a **greedy algorithm**.

The greedy algorithm for vertex-colouring is easy to program. Suppose we have arranged the vertices in some order $v_1, v_2, \ldots, v_n$. We assign colour 1 to v_1; for each v_i $(2 \le i \le n)$ we form the set S of colours assigned to vertices v_j $(1 \le j < i)$ which are adjacent to v_i; and we give v_i the first colour not in S. (In practice, more sophisticated methods of handling the data may be used.)

Greedy vertex-colouring algorithm

$c(v_1) \leftarrow 1$
for $i = 2$ **to** n **do**
    for $j = 1$ **to** $i - 1$ **do**
        $S \leftarrow \varnothing$
        if v_j is adjacent to v_i
            then $S \leftarrow S \cup \{c(v_j)\}$
    $k \leftarrow 1$
    while $k \in S$ **do**
        $k \leftarrow k + 1$
    $c(v_i) \leftarrow k$

Because the greedy strategy is a shortsighted one, the number of colours which it uses will normally be greater than the minimum possible. For example, the greedy algorithm applied to the graph in Fig. 8.10 gives precisely the vertex-colouring with four colours which we first proposed as our 'timetable', whereas we later found a vertex-colouring using only three colours. Of course, a great deal depends upon the order initially given to the vertices. It is quite easy to see that if we hit on the right order, then the greedy algorithm will give us a best possible colouring (Ex. 8.7.2). But there are $n!$ orders altogether, and if we have to check each one of them, the algorithm will require 'exponential-time'.

Despite its wastefulness, the greedy algorithm is useful both in theory and in practice. We shall prove two theorems by using the greedy strategy.

Theorem 8.7.1
If G is a graph with maximum valency k, then

    (i) $\chi(G) \leqslant k + 1$,
    (ii) if G is connected and not regular, $\chi(G) \leqslant k$.

Proof
(i) Let $v_1, v_2, \ldots, v_n$ be any ordering of the vertices of G. Each vertex v_i has at most k neighbours, and so the set S of colours assigned by the greedy algorithm to vertices v_j which are adjacent to v_i $(1 \leqslant j < i)$ has cardinality k at most. Hence at least one of the colours $1, 2, \ldots, k + 1$ is not in S, and the greedy algorithm will assign the first of these to v_i. In this way the greedy algorithm produces a vertex-colouring of G using at most $k + 1$ colours, and so $\chi(G) \leqslant k + 1$.

(ii) For this part, we arrange the vertices in a special order, starting with v_n and working backwards. Since G has maximum valency k and is not regular, there is at least one vertex of G whose valency is less than k: call it v_n. List the neighbours of v_n as $v_{n-1}, v_{n-2}, \ldots, v_{n-r}$; there are at most $k-1$ of them. Next, list the neighbours of v_{n-1} (except for v_n), remarking that since the valency of v_{n-1} is at most k, there are at most $k-1$ such vertices. Next list all the neighbours of v_{n-2} which have not already been listed, and so on. Since G is connected, the list will eventually contain all vertices of G. Furthermore, the method of construction ensures that every vertex is adjacent to at most $k-1$ of its predecessors in the order $v_1, v_2, \ldots, v_n$.

It follows from the same argument as used in part (i) that (for this ordering) the greedy algorithm will require at most k colours. Hence $\chi(G) \leqslant k$. □

Part (ii) of the theorem is false if we allow G to be regular. The reader who has correctly answered Ex. 8.6.1 will be able to supply two examples of this fact: the complete graphs, and the odd cycle graphs, both of which require $k+1$ colours. However, it can be shown that these are the only counter-examples.

Another useful consequence of the greedy algorithm concerns graphs G for which $\chi(G) = 2$. For such a graph, the sets V_1 and V_2 of vertices assigned the colours 1 and 2 respectively form a partition of V, with the property that every edge of G has one vertex in V_1 and the other in V_2. For this reason, when $\chi(G) = 2$ we say that G is **bipartite**. A vertex-colouring of the cube with two colours is illustrated in Fig. 8.12, together with an alternative picture which emphasizes the bipartite nature of the graph. We frequently use the latter kind of illustration in dealing with bipartite graphs.

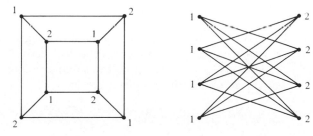

Fig. 8.12
The cube as a bipartite graph.

Theorem 8.7.2   A graph is bipartite if and only if it contains no cycles with odd length.

Proof If there is a cycle with an odd number of vertices then three colours are required for a vertex-colouring of that cycle alone, and the chromatic number of the graph is at least 3. Hence if the graph is bipartite it can have no odd cycles.

Conversely, suppose G is a graph with no odd cycles. We shall construct an ordering of G for which the greedy algorithm produces a vertex-colouring with two colours. Choose any vertex and call it v_1; we shall say that v_1 is at *level 0*. Next, list the neighbours of v_1, calling them $v_2, v_3, \ldots, v_r$; we shall say that these vertices are at *level 1*. Next, list the neighbours of the level 1 vertices (except v_1); we shall say that these vertices are at *level 2*. Continue in this way, listing at *level l* all those vertices adjacent to the *level l − 1* vertices, except for those previously listed at *level l − 2*. When no new vertices can be added in this way, we have a component G_0 of G (if G is connected, $G_0 = G$).

The crucial feature of this ordering is that a vertex at level l can be adjacent only to vertices at levels $l − 1$ and $l + 1$, not to vertices at the same level. For suppose x and y are vertices at the same level; then they are joined by paths of equal length m to some vertex z at a previous level, and the paths can be chosen so that z is the only common vertex (Fig. 8.13). If x and y were adjacent, there would be a cycle of odd length $2m + 1$ in G_0, contrary to hypothesis.

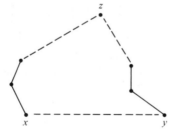

Fig. 8.13
Adjacent vertices at the same level yield an odd cycle.

It follows that the greedy algorithm assigns colour 1 to vertices at levels $0, 2, 4, \ldots$, and colour 2 to vertices at levels $1, 3, 5, \ldots$. Hence $\chi(G_0) = 2$. Repeating the same argument for each component of G we obtain the required result. □

Exercises 8.7 1 Find the orderings of the vertices of the cube graph (Fig. 8.12) for which the greedy algorithm requires 2, 3, and 4 colours respectively.

2 Show that for any graph G there is an ordering of the vertices for which the greedy algorithm requires $\chi(G)$ colours. [Hint: use a vertex-colouring with $\chi(G)$ colours to define the required ordering.]

3 Let $e_i(G)$ denote the number of vertices of a graph G whose valency is strictly greater than i. Use the greedy algorithm to show that if $e_i(G) \leqslant i+1$ for some i, then $\chi(G) \leqslant i+1$.

4 The graph M_r ($r \geqslant 2$) is obtained from the cycle graph C_{2r} by adding extra edges joining each pair of opposite vertices. Show that

    (i) M_r is bipartite when r is odd,
    (ii) $\chi(M_r) = 3$ when r is even and $r \neq 2$,
    (iii) $\chi(M_2) = 4$.

8.8 Miscellaneous Exercises

1 For which values of n is it true that the complete graph K_n has an Eulerian walk?

2 Use the principle of induction to show that if $G = (V, E)$ is a graph with $|V| = 2m$, and G has no 3-cycles, then $|E| \leqslant m^2$.

3 Let $X = \{1, 2, 3, 4, 5\}$ and let V denote the set of all 2-subsets of X. Let E denote the set of pairs of members of V which are disjoint (as subsets of X). Show that the graph $G = (V, E)$ is isomorphic with the graph depicted in Fig. 8.14. Show also that this is just another version of Petersen's graph, introduced in Ex. 8.2.2.

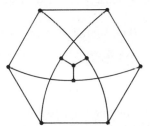

Fig. 8.14
Petersen's graph.

4 Let G be a bipartite graph with an odd number of vertices. Show that G cannot have a Hamiltonian cycle.

5 The **k-cube** Q_k is the graph whose vertices are the words of length k in the alphabet $\{0, 1\}$ and whose edges join words which differ in exactly one position. Show that

    (i) Q_k is a regular graph with valency k,
    (ii) Q_k is bipartite.

6 Prove that the graph Q_k defined in Ex. 8.8.5 has a Hamiltonian cycle.

7 Show that Petersen's graph does not have a Hamiltonian cycle.

8 In a game of dominos (Ex. 3.7.2) the rules require that the dominos be placed in a line so that adjacent dominos have matching numbers: $[x|y]$ is next to $[y|z]$, and so on. By regarding the dominos for which $x \neq y$ as the edges of the complete graph K_7, show that it is possible to have a game in which all the dominos are used.

9 Calculate the number of Eulerian walks in K_7 and the number of complete games of dominos.

10 Show that if $\alpha: V_1 \rightarrow V_2$ is an isomorphism of the graphs $G_1 = (V_1, E_1)$ and $G_2 = (V_2, E_2)$ then the function $\beta: E_1 \rightarrow E_2$ defined by

$$\beta\{x, y\} = \{\alpha(x), \alpha(y)\} \quad (\{x, y\} \in E_1)$$

is a bijection

11 If G is a regular k-valent graph with n vertices show that

$$\chi(G) \leq \frac{n}{n-k}.$$

12 Construct five mutually non-isomorphic connected regular graphs with valency 3 and eight vertices.

13 Show that the complete graph K_{2n+1} is the union of n Hamiltonian cycles, no two of which have a common edge.

14 Is it possible for a knight to visit all the squares of a chessboard exactly once and return to its starting square? Interpret your answer in terms of Hamiltonian cycles in a certain graph.

15 The **odd graph** O_k is defined as follows (when $k \geq 2$): the vertices are the $(k-1)$-subsets of a $(2k-1)$-set, and the edges join disjoint subsets. (Thus O_3 is Petersen's graph.) Show that $\chi(O_k) = 3$ for all $k \geq 2$.

16 Show that if G is a graph with n vertices, m edges, and c components then

$$n - c \leq m \leq \tfrac{1}{2}(n-c)(n-c+1)$$

Construct examples showing that both bounds can be attained for all values of n and c such that $n \geq c$.

17 A sequence $d_1, d_2, \ldots, d_n$ is **graphic** if there is a graph whose n vertices can be labelled $v_1, v_2, \ldots, v_n$ in such a way that $\delta(v_i) = d_i$ $(1 \leq i \leq n)$. Show that if the sequence $d_1, d_2, \ldots, d_n$ is graphic and $d_1 \geq d_2 \geq \cdots \geq d_n$ then

$$d_1 + d_2 + \cdots + d_k \leq k(k-1) + \sum_{i=k+1}^{n} \min(k, d_i)$$

for $1 \leq k \leq n$.

18   The **girth** of a graph G is the least value of g for which G contains a g-cycle. Show that a regular graph with valency k and girth $2m+1$ must have at least

$$1+k+k(k-1)+\cdots+k(k-1)^{m-1}$$

vertices, and that a regular graph with valency k and girth $2m$ must have at least

$$2[1+(k-1)+(k-1)^2+\cdots+(k-1)^{m-1}]$$

vertices.

19   Make a table of the lower bounds obtained in the previous exercise when $k=3$ and the girth is 3, 4, 5, 6, 7. Show that there is a graph which attains the lower bound in the first four cases, but not in the fifth case.

20   Let **B** be the set of blocks of 2-design on a set X, with parameters $(r^2+r+1, r+1, 1)$, and let G be the graph whose vertex-set is $X \cup \mathbf{B}$ and whose edges join the vertices x and B whenever the object x is in the block B. Show that G is a regular graph with girth 6 which attains the lower bound given in Ex. 18.

21   Let $G=(V, E)$ be a graph with at least three vertices such that

$$\delta(v) \geqslant \tfrac{1}{2}|V| \quad (v \in V).$$

Show that G has a Hamiltonian cycle.

22   Show that if $\bar{G}$ is the complement of the graph G (as defined in Ex. 8.3.2) then $\chi(G)\chi(\bar{G}) \leqslant n$, where n is the number of vertices of G.

9
Trees, sorting, and searching

9.1 Counting the leaves on a rooted tree

Recall that a *tree* is a connected graph which contains no cycles. Trees occur in many different contexts, and frequently one vertex of the tree is distinguished in some way. For example, in a 'family tree' which traces the descendants of King Henry VIII, we might emphasize the special position of the King by putting him at the 'top' of the tree. In general, we refer to the distinguished vertex as the **root**, and a tree with a specified root is said to be a **rooted tree**. (This terminology, although standard, has the unfortunate defect that, in a pictorial representation, the root will often be at the top of the tree, and the tree will appear to grow downwards.)

In order to study a rooted tree it is natural to arrange the vertices in levels, just as we did for bipartite graphs in Section 8.7. We say that the root vertex r is at *level 0*, and that the neighbours of r are at *level 1*. For each $k \geq 2$, *level k* contains those vertices adjacent to vertices at level $k-1$, except for those which have already been assigned to level $k-2$. The rooted tree depicted on the left in Fig. 9.1 can be redrawn as shown on the right in order to display the arrangement of the vertices in levels.

A vertex in a rooted tree is said to be a **leaf** if it is at level i ($i \geq 0$) and it is not adjacent to any vertices at level $i+1$. A vertex which is

Fig. 9.1
A rooted tree and its
levels.

not a leaf is an **internal** vertex. The **height** of a rooted tree is the maximum value of k for which level k is not empty. Thus the tree illustrated in Fig. 9.1 has six leaves and four internal vertices, and its height is 3.

Exercises
9.1

1 In the following table $n_5(h)$ is the number of non-isomorphic *rooted* trees which have five vertices and height h. (Two rooted trees are said to be isomorphic if there is an isomorphism between them (considered as unrooted trees) which also takes the root of the first tree to the root of the second.) Verify the table by sketching the required number of examples in each case.

$$h: \quad 1 \quad 2 \quad 3 \quad 4$$
$$n_5(h): \quad 1 \quad 4 \quad 3 \quad 1$$

2 If we consider ordinary (unrooted) trees, what is the number of non-isomorphic types with five vertices? Make a list of these types and hence check that the list obtained in the previous exercise is complete.

3 Construct two non-isomorphic rooted trees both having twelve vertices, six leaves, and height 4.

The two properties which we used in Section 8.5 to define a tree have obvious consequences when we arrange the vertices by levels. Since a tree is a connected graph (property T1), every vertex is contained in one of the levels. More significantly, since a tree has no cycles (property T2) each vertex v at level i $(i>0)$ is adjacent to exactly one vertex u at level $i-1$. We sometimes emphasize this point by referring to u as the 'father' of v, and v as a 'son' of u. Each vertex except the root has a unique father, but a vertex may have any number of sons (including zero). Clearly, a vertex is a leaf if and only if it has no sons.

In many applications it happens that every father (internal vertex) has the same number of sons. When each father has m sons we speak of an **m-ary** rooted tree; in particular when $m=2$ we use the word 'binary' and when $m=3$ we use the word 'ternary'.

Theorem
9.1

The height of an m-ary rooted tree with l leaves is at least $\log_m l$.

Proof   Since we have

$$h \geqslant \log_m l \quad \Leftrightarrow \quad m^h \geqslant l$$

it is sufficient to prove the equivalent assertion that an m-ary rooted tree with height h has at most m^h leaves. The proof is by induction on h.

Clearly the assertion is true when $h = 0$ since in this case the tree has just one vertex, the root, which is a leaf. Suppose the assertion is true whenever $0 \leqslant h \leqslant h_0$, and let T be an m-ary rooted tree with height $h_0 + 1$. If we delete the root and the edges containing it from T we obtain m trees $T_1, \ldots, T_m$, and we may specify as their roots the vertices at level 1 in T. Each T_i is now a rooted tree with height at most h_0, and so by the induction hypothesis it has at most m^{h_0} leaves. But the leaves of T are precisely the leaves of the trees $T_1, \ldots, T_m$, and so the number of leaves of T is at most $m \times m^{h_0} = m^{h_0 + 1}$.

By the strong form of the principle of induction, it follows that the result is true for all $h \geqslant 0$. $\square$

Since $\log_m l$ is generally not an integer, whereas the height h is, we can improve the statement of Theorem 9.1 slightly. For example if $m = 3$ and $l = 10$ then the inequality

$$h \geqslant \log_m l = 2.0959 \ldots$$

implies that $h \geqslant 3$. In general we can say that

$$h \geqslant \lceil \log_m l \rceil,$$

where $\lceil x \rceil$ stands for the least integer z such that $z \geqslant x$. (This should be compared with the notation $\lfloor x \rfloor$ introduced in Section 7.6).

A frequent application of Theorem 9.1 is in the study of *decision trees*. Each internal vertex of a decision tree represents a decision, and the possible results of that decision are represented by the edges leading to the vertices at the next level. The final outcomes of the procedure are represented by the leaves of the tree. If the result of each decision is simply that a statement is true or false, then we have a binary tree; this is a very common situation, which we shall discuss in the next section. Here we give an example involving a ternary tree.

Example   (The false coin problem). Suppose we have a genuine coin labelled 0, and r other coins, indistinguishable from 0 by appearance except for being labelled $1, 2, \ldots, r$. It is suspected that one coin may be false—either too light or too heavy. Show that at least $\lceil \log_3 (2r + 1) \rceil$

weighings on a balance are necessary to decide which coin (if any) is false, and whether it is light or heavy. Devise a procedure which uses exactly this number of weighings when $r = 4$.

Solution There are $2r + 1$ final outcomes, or leaves on the decision tree,

$$G, 1H, 1L, \ldots, rH, rL;$$

meaning that all coins are good, coin 1 is heavy, coin 1 is light, and so on. The decision tree is ternary, since there are three possible results of each decision (weighing one set of coins against another). They are

$$< \ : \ \text{left-hand set lighter}$$
$$= \ : \ \text{two sets equal}$$
$$> \ : \ \text{left-hand set heavier.}$$

Hence the height of the decision tree is at least $\lceil \log_3 (2r + 1) \rceil$.

When $r = 4$, $\lceil \log_3 (2r + 1) \rceil = 2$, and a solution with two weighings only is depicted in Fig. 9.2.

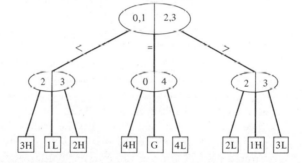

Fig. 9.2
Solution of the false-coin problem when $r = 4$.

**Exercises
9.1
(continued)**

4 Suppose that 20 teams have entered for the University of Folornia five-a-side soccer tournament. The tournament is organized on the 'knockout' principle, so that the winners of the matches in Round 1 proceed to Round 2, and so on. There are no drawn matches. Construct a scheme for the tournament based on a rooted tree, and show that at least five rounds are needed. Can it be done in such a way that all byes are in Round 1?

5 The FA Cup is a knockout football competition which can be represented as a rooted tree. How many rounds of the competition are needed if 4090 teams enter and none of them has a bye for more than one round? How many rounds are needed if 90 teams have byes until Round 6?

6   What is the lower bound for the number of weighings needed in the false coin problem (as stated in the *Example* above) when there are six coins given? Devise a scheme which achieves this number of weighings.

7   Consider the following variant of the false coin problem. There are eight coins, *exactly one of which is known to be light*. All the others are genuine, but *no genuine coin labelled 0 is given*. Find a theoretical lower bound for the number of weighings needed to find the light coin, and show that this number can be achieved.

9.2   Trees and sorting algorithms

In Section 7.8 we discussed the problem of how to arrange a given list $x_1, x_2, \ldots, x_n$ of distinct integers in increasing order. The algorithms we used for solving this problem involved the comparison of one integer with another and consequent transfers of the data, and this kind of of procedure can be represented by a decision tree, in the following way.

Each vertex of the decision tree represents a comparison of two integers, for instance those currently labelled x_i and x_j. So there are two possible results, $x_i < x_j$ or $x_i > x_j$, and the decision tree is a *binary* tree. In the bubble sort algorithm each comparison involves 'adjacent' integers x_i and x_{i+1}, and the rules which tell us which pair to compare at any stage do not depend on the results of previous comparisons. (Of course, the current values of x_i and x_{i+1} do depend on previous comparisons.) The decision tree for bubble sort when $n = 3$ is illustrated in Fig. 9.3. There are $3! = 6$ final outcomes, corresponding to the permutations of the initial order $\alpha\beta\gamma$; the leaves on the decision tree represent these outcomes, together with some 'impossible' outcomes.

For a general value of n the number of leaves on the decision tree is at least $n!$, and this is so whatever algorithm is used. Also, the height of the decision tree is equal to the number $s(n)$ of comparisons required. Thus it follows from Theorem 9.1 that

$$s(n) \geqslant \log_2{(n!)}.$$

Now $\log_2{(n!)}$ is $O(n \log n)$, since it is the sum of the n terms $\log_2 i$ $(1 \leqslant i \leqslant n)$, and each term is not greater than $\log_2 n$. Further-

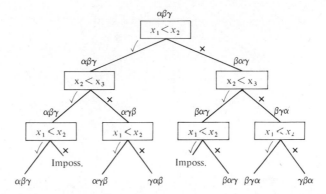

Fig. 9.3
The decision tree for bubble sort (three objects).

more, no better approximation is possible (Ex. 9.2.5). Hence the number of comparisons required by any algorithm for the sorting problem is at least $O(n \log n)$.

In Section 7.8 we showed that bubble sort is an $O(n^2)$ algorithm, while insertion sorting using the bisection method requires $O(n \log n)$ comparisons but $O(n^2)$ data transfers. Now we shall describe an algorithm for which the number of comparisons and the number of data transfers are both $O(n \log n)$. It is known as *heapsort*.

The heapsort algorithm uses rooted trees as an integral part of the method, and these trees should not be confused with the decision trees which can be used to analyse any sorting algorithm. We begin by assigning the members of the list $x_1, x_2, \ldots, x_n$ to the vertices of a rooted tree, in the following way (Fig. 9.4). We assign x_1 to the root (level 0), x_2 and x_3 to level 1, x_4, x_5, x_6, x_7 to level 2, and so on. The vertex labelled x_r is the father of x_{2r} and x_{2r+1}, provided $2r+1 \le n$. Thus the last level is usually incomplete, and when n is even the vertex $x_{\frac{1}{2}n}$ has just one son x_n. Apart from this, the tree is a binary rooted tree, as illustrated in Fig. 9.4 for the case $n = 12$.

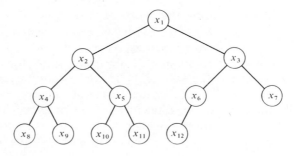

Fig. 9.4
A tree labelled for heapsort.

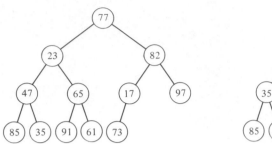 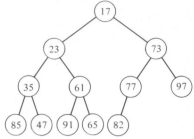

Fig. 9.5
Tree representations of an
unsorted list and the corresponding heap.

The heapsort algorithm is in two stages. First, the unsorted list is transformed into a special kind of list known as a **heap**, (Fig. 9.5), and secondly, the heap is transformed into the sorted list. The characteristic property of a heap is that each father is smaller than his sons; in other words

$$x_r < x_{2r} \quad \text{and} \quad x_r < x_{2r+1}.$$

Thus the unsorted list

$$77, 23, 82, 47, 65, 17, 97, 85, 35, 91, 61, 73,$$

will be transformed into the heap

$$17, 23, 73, 35, 61, 77, 97, 85, 47, 91, 65, 82,$$

(as in Fig. 9.5), and thence into the sorted list

$$17, 23, 35, 47, 61, 65, 73, 77, 82, 85, 91, 97.$$

The method for transforming an unsorted list into a heap pro-ceeds by dealing with the fathers (internal vertices) in reverse order. Let us suppose that when we come to deal with x_r the two subtrees rooted at x_{2r} and x_{2r+1} have already been made into heaps. If x_r is smaller than x_{2r} and x_{2r+1} we need do nothing, since the subtree rooted at x_r is a heap. If not, we store x_r temporarily and move the smaller of x_{2r} and x_{2r+1} to the vacant vertex. This creates a new vacancy; if x_r is smaller than the sons of the vacant vertex, or if there are no sons, then we fill the vacancy with x_r; if not, we fill the vacancy with the smaller of the sons, and continue. We shall find a place for x_r when we come to a leaf, if not before.

The rules stated above form the basis of a procedure **heap** (k, n) which, given a vertex x_k with the property that the subtrees rooted at x_{2k} and x_{2k+1} are heaps, makes the subtree rooted at x_k into a heap. Applying the procedure in turn for $k = \lfloor \frac{1}{2}n \rfloor, \lfloor \frac{1}{2}n \rfloor - 1, \ldots, 2, 1$ will transform the whole tree into a heap. The reader should verify that the heap shown in Fig. 9.5 is the result of applying this technique to the unsorted list given there.

The method for transforming a heap into a sorted list can also be expressed in terms of the **heap** procedure. In any heap, the root always has the smallest value, and so it goes into the first place y_1 in the sorted list. The vacancy at the root is then filled by the last value x_n, and the old x_n leaf is removed altogether. We now have a tree with $n - 1$ vertices, and the subtrees rooted at x_2 and x_3 are heaps, so the procedure **heap** $(1, n - 1)$ will restore the heap property of the whole tree. Now the root has the smallest remaining value, and this is assigned to y_2. The vacancy is filled by x_{n-1}, the old x_{n-1} leaf is removed, the heap property is restored by **heap** $(1, n - 2)$, and so on.

The entire heapsort algorithm can be expressed very concisely in terms of the **heap** procedure.

Heapsort algorithm

---

for $j = 0$ **to** $\lfloor \frac{1}{2}n \rfloor - 1$ **do**
    heap $(\lfloor \frac{1}{2}n \rfloor - j, n)$
for $i = 1$ **to** $n - 2$ **do**
    $y_i \leftarrow x_1$; $x_1 \leftarrow x_{n-i+1}$
    heap $(1, n - i)$
$y_{n-1} \leftarrow x_1$; $y_n \leftarrow x_2$

How many comparisons does the heapsort algorithm require? In the **heap** (k, n) procedure, we need to find the smaller of x_{2k} and x_{2k+1}, and compare it with x_k; this involves two comparisons. Then we repeat this operation at each level, possibly until a leaf is reached. Since x_k is at level $\lfloor \log_2 k \rfloor$, and a leaf is at level $\lfloor \log_2 n \rfloor$ or $\lfloor \log_2 n \rfloor - 1$, the number of comparisons required by **heap** (k, n) is approximately

$$2(\log_2 n - \log_2 k) = 2 \log_2 (n/k).$$

In the first phase of the algorithm the **heap** (k, n) procedure is executed for each value of k in the range $1 \leq k \leq \frac{1}{2}n$, approximately. In the second phase the **heap** $(1, j)$ procedure is executed for each value of j in the range $2 \leq j \leq n - 1$. Hence the total number of comparisons required is approximately

$$2 \sum_{k=1}^{n/2} \log_2 (n/k) + 2 \sum_{j=2}^{n-1} \log_2 (j).$$

Both sums have less than n terms, and each term is not greater than $\log_2 n$. It follows that the number of comparisons involved in the heapsort algorithm is $O(n \log n)$. Furthermore, the number of data transfers required in heapsort is proportional to the number of comparisons and so this number is also $O(n \log n)$.

Exercises 9.2

1   What is the smallest possible height of the decision tree of an algorithm for sorting four objects using binary comparisons?

2   Calculate the number of binary comparisons needed (in the worst case) when four objects are sorted

  (i) by bubble sort;
  (ii) by insertion (sequential method);
  (iii) by insertion (bisection method).

3   Use the heapsort method to form a heap from the following unsorted lists. In each case illustrate the resulting heap in tree form and write out the corresponding list.

  (i) 63, 55, 33, 16, 81, 76.
  (ii) 73, 21, 17, 28, 32, 56, 19, 84, 38, 49, 77, 51, 12.

4   The following is a program for the **heap** (k, n) procedure outlined in the text. Explain in ordinary language the purpose of each line of the program. (For example, line 2 checks that x_j is not a leaf.)

The **heap** (k, n) algorithm

```
1  j ← k
2  while  j ≤ ⌊½n⌋   do
3            if   2j < n and x₂ⱼ > x₂ⱼ₊₁   then   a ← 2j + 1
4                                          else   a ← 2j
5            if   xₐ < xⱼ   then   switch xₐ and xⱼ
6                                  j ← a
7                           else   j ← n
```

5 Prove that if i and n are integers such that $1 \leqslant i \leqslant n$ then

$$i(n-i+1) \geqslant n.$$

Deduce that, if n is even, $n! \geqslant n^{\frac{1}{2}n}$, and hence

$$\tfrac{1}{2}n \log_2 n \leqslant \log_2 (n!) \leqslant n \log_2 n.$$

9.3 Spanning trees and the MST problem

Suppose that $G = (V, E)$ is a connected graph, and T is a subset of E such that

(i) every vertex of G belongs to an edge in T;
(ii) the edges in T form a tree.

In this case we say that T is a **spanning tree** for G. For example, a spanning tree for the graph in Fig. 9.6 is indicated by the heavy lines.

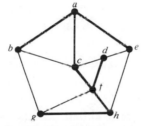

Fig. 9.6
A graph and one of its spanning trees.

It is easy to 'grow' a spanning tree in the following way. Take any vertex v as an initial 'partial tree', and add edges one by one so that each edge joins a new vertex to the partial tree. The spanning tree in Fig. 9.6 could be grown by starting at the vertex a, and linking up the other vertices in the order b, c, e, f, d, h, g, by means of the edges ab, ac, ae, cf, fd, fh, hg. In general, if there are n vertices, we shall continue for $n-1$ steps, after which we shall have $1 + (n-1) = n$ vertices and $n-1$ edges (which is the correct number according to Theorem 8.5.)

In order to show that the method always works, let S be the set of vertices in the partial tree at any intermediate stage, so that S is not empty or the whole of V. If there were no edges having one vertex in S and the other vertex in the complementary set $\bar{S}$, then there could be no path from any vertex in S to any vertex in $\bar{S}$, and G would be disconnected, contrary to hypothesis. Hence there is always an edge available at each stage of the construction.

1   Find spanning trees for the cube graph (Fig. 8.12) and Petersen's graph (Fig. 8.14).

2   Sketch all the spanning trees for the complete graph K_4 (there are 16 of them).

Spanning trees are useful in many contexts. For example, let us suppose that a number of towns must be linked in pairs by pipelines so as to form a connected network. Some pairs of towns may be impossible to link for geographical reasons; and for each possible link there will be an associated cost of construction. Formally, we have a graph $G = (V, E)$ whose vertices are the towns and whose edges are the possible links, and a function w from E to $\mathbb{N}$ so that $w(e)$ represents the cost of constructing the edge e. We say that G and w constitute a **weighted graph**, and w is a **weight function**.

In the pipeline problem, the aim in practice is to provide a connecting network at the least possible cost. Such a network corresponds to a spanning tree T for G whose total weight

$$w(T) = \sum_{e \in T} w(e)$$

is as small as possible. We shall refer to this as the **MST problem** (**minimum spanning tree problem**) for the weighted graph G.

Given that the values of w are positive integers, it is clear that there must be a solution to the MST problem, since there are only a finite number of spanning trees T for G and each one gives a positive integral value for $w(T)$. In other words, there is a minimum spanning tree T_0 such that

$$w(T_0) \leq w(T)$$

for all spanning trees T. However, we should notice that there may be several different trees with the same property.

A simple algorithm for the MST problem is based on applying the greedy principle to the tree-growing method given above. Specifically: at each stage we add the *cheapest* edge joining a new vertex to the partial tree. (If several edges with the same weight are available we can select any one of them.) For example, in Fig. 9.7 if we start at u, then we must add the edges in the order uv, ux, uy, yz. On the other hand, if we start at y, then we must add the edges in the order yz, yu, uv, ux.

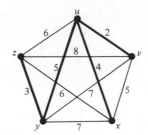

Fig. 9.7
A minimum spanning
tree.

At first sight it is rather surprising that the greedy algorithm for the MST problem really does work, especially when we recall that the greedy algorithm for the vertex-colouring problem does not always produce a colouring with the smallest possible number of colours. But in the case of the MST problem we are in luck.

Theorem 9.3

Let $G = (V, E)$ be a connected graph with weight function $w: E \to \mathbb{N}$, and suppose that T is a spanning tree for G constructed by the greedy algorithm. Then

$$w(T) \leq w(U)$$

for any spanning tree U of G.

Proof

Let the edges of T be denoted by $e_1, e_2, \ldots, e_{n-1}$, in the order of construction by the greedy algorithm. If $U = T$, the result is obviously true. If $U \neq T$ there are some edges of T which are not in U, and we shall suppose e_k is the first such edge in the given order. Denote by S the set of vertices in the partial tree immediately before the addition of e_k, and let $e_k = xy$ where x is in S and y is not in S.

Since U is a spanning tree there is a path in U from x to y, and as we travel along this path we shall encounter an edge e^* which has one vertex in S and the other not in S. Now, when e_k is selected for T by the greedy algorithm, e^* is also a candidate, but it is not selected. Hence we must have $w(e^*) \geq w(e_k)$. Also, if e^* is in T it must be selected after e_k, and so it comes after e_k in the given order.

The result of removing e^* from U and replacing it with e_k is a spanning tree U_1, for which

$$w(U_1) = w(U) - w(e^*) + w(e_k) \leq w(U).$$

Furthermore, the first edge of T which is not in U_1 occurs after e_k in the given order. We can thus repeat the procedure, obtaining a sequence of spanning trees $U_1, U_2, \ldots$, with the property that each one has a longer initial segment of the sequence $e_1, e_2, \ldots, e_{n-1}$ in common with T than its predecessor. The process ends when we obtain a spanning tree U_r identical with T, and we have

$$w(T) = w(U_r) \leqslant w(U_{r-1}) \leqslant \ldots \leqslant w(U_1) \leqslant w(U),$$

as required. $\qquad\square$

There is a rather neat way of keeping a record of the progress of this algorithm, by means of a table with three columns.

I	II	III
x	y	$w(xy)$
.	.	.
.	.	.

Column I is a list of all the vertices not in S, the set of vertices already linked by the partial tree. For each such x, the corresponding entry y in Column II is a vertex in S such that the edge xy is one of the cheapest edges joining x to a vertex in S. Column III contains the value of $w(xy)$.

At the ith step in the construction, $|S| = i$ and there are $n - i$ vertices in Column I. We have to select one of the smallest entries in column III, say $w(x_0 y_0)$, and this entails $n - i - 1$ comparisons. Then we must update the table as a result of adding x_0 to S by means of the edge $x_0 y_0$. This involves deleting the row x_0, and checking to see if x_0 (which is now in S) can replace any of the previous entries in Column II: that is, checking if $w(xx_0) < w(xy)$ for any of the remaining $n - i - 1$ vertices x, which entails another $n - i - 1$ comparisons. The total number of comparisons required is

$$\sum_{i=1}^{n-1} 2(n - i - 1) = (n - 1)(n - 2),$$

which is $O(n^2)$.

Exercises 9.3 (continued) 3 Use the greedy algorithm to find a minimum spanning tree for the weighted graph in Fig. 9.8. Is the minimum spanning tree unique in this case?

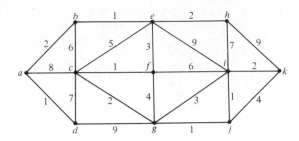

Fig. 9.8
Find the MST.

4 Let G be the weighted graph whose vertex-set is $\{x, a, b, c, d, e, f\}$ and whose edges and weights are given by the table:

xa	xb	xc	xd	xe	xf	ab	bc	cd	de	ef	fa
6	3	2	4	3	7	6	2	3	1	8	6.

Find all minimum spanning trees for G.

5 Suppose that T is a minimum spanning tree in a weighted graph K and e^* is an edge of K not in T. Let e by any edge of T belonging to the unique path in T which joins the vertices of e^*. Show that $w(e) \leqslant w(e^*)$.

6 Construct a flowchart for the greedy algorithm based on the tabular method outlined above.

9.4 Depth-first search

Suppose we wish to carry out a search of the vertices of a graph, beginning at a specified vertex. Roughly speaking, there are two kinds of strategy which we could employ. We could 'forge ahead', always moving on to a new vertex whenever one is available; or we could 'spread out', checking all the vertices at each 'level' before moving on to the next level. The first strategy, technically known as **depth-first search** (DFS), will be considered in this section.

The diagram in Fig. 9.9a represents a game of hide-and-seek. The seeker is initially at a, and wishes to check all the hiding places b, c, d, e, using the available routes. Clearly, we can use the graph shown in Fig. 9.9b to represent the essential features of the problem.

In the DFS strategy, the seeker starts out rather like an excited child. He goes from a to any adjacent vertex, say b, and thence immediately on to another new vertex c. From c he goes to d, but

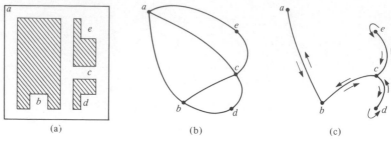

Fig. 9.9
Hide-and-seek.

there he is stuck, since the two adjacent vertices have already been visited. So he 'backtracks' from d to c, where he finds that a new vertex e is still available, and moves on to there.

Although the seeker cannot know it, there are no more hiding-places to be searched. In the DFS strategy, we may envisage that childlike excitement has been dissipated by this stage, and the rest of the procedure is a more careful double-check to see if any new vertices remain. Thus, since both neighbours, a and c, of e have been visited, the seeker backtracks to c (whence he first came to e). At c, again all the neighbouring vertices have been visited, so he backtracks to b (whence he first came to c). Similarly, from b he backtracks to a. At a, there are no new neighbours, and the seeker recognizes his starting point. The search is finished: if the procedure has been carried out correctly, all the vertices which can be reached from a will have been visited.

The DFS procedure can also be regarded as a special case of the general tree-growing method introduced at the beginning of the previous section. As each new vertex is visited, we add the edge leading to it to the partial tree, so that in the example the edges would be added in the order ab, bc, cd, ce (Fig. 9.9c). The special feature of the DFS method is that the new vertex is always chosen to be adjacent to the last possible one of the old vertices.

A flowchart for the DFS procedure is illustrated in Fig. 9.10. We start at any given vertex v and construct a partial tree W in the following way. Whenever the current vertex x has new neighbours, we choose y to be one of them, add xy to W, ADVANCE to y, and replace x by y as the new current vertex. Whenever there are no new vertices adjacent to x, we BACKTRACK to the vertex from which x was originally visited. Eventually we find ourselves at v again, with nowhere to go, so we put $T = W$ and stop. Each edge of T has been used twice in the procedure, once advancing and once back-tracking (Fig. 9.9c).

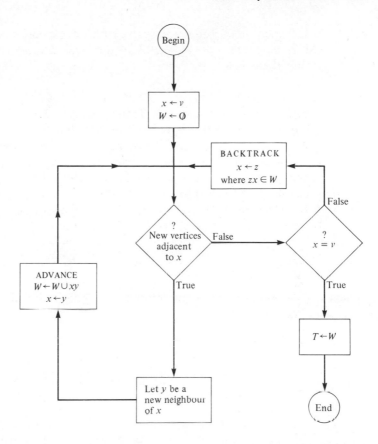

Fig. 9.10
Flowchart for DFS.

It is a simple matter to include in the procedure a list of the vertices in the order in which they are visited. However, there is one important difference between the tree-growing problems discussed in the previous section, and DFS regarded as a search procedure. In Section 9.3 we always assumed that G was known to be connected, so the list of vertices eventually contained all the vertices of the graph, and the tree was a spanning tree for G. (For this reason, we were able to stop after $n-1$ steps, without having to double-check that no more vertices could be reached.) But in a search problem we do not know that G is connected: indeed, the object of the search may be to decide that very point. What we do know is that DFS will find all the vertices which can be reached from the initial vertex v.

Theorem 9.4   Let v be a vertex of the graph G and let T be the subset of the edges of G constructed according to the DFS flowchart (Fig. 9.10). Then T is a spanning tree for the component of G which contains v.

Proof   It is a direct consequence of the rules of the construction that T is a tree. Suppose z is any vertex in the same component of G as v, so that there is a path

$$v = v_0, v_1, \ldots, v_k = z$$

in G. If z is not a vertex of T, then (since v is on T) there must be a pair v_i, v_{i+1} such that v_i is a vertex of T but v_{i+1} is not. Since v_i is on T, it will be the current vertex x at least once. Whenever we advance from v_i to a new vertex we must eventually return to v_i again, since we can only backtrack along edges of the tree. We cannot backtrack from v_i until all its neighbours have been visited. In particular, if v_{i+1} is not on T, then it cannot be a neighbour of v_i, contrary to hypothesis. Hence z must be on T.   □

Exercises 9.4

1   Let G be the graph defined by the adjacency list shown in Table 9.4.1.

Table 9.4.1

a	b	c	d	e	f	g	h
b	a	b	a	b	g	c	a
d	c	d	b			f	g
h	d	g	c			h	
	e						

Sketch the DFS tree in G, starting from g. (Whenever a choice is possible, pick the first vertex in alphabetical order.) Is G connected?

2   Using the DFS method in a systematic way, find the number of components of the graph whose adjacency list is shown in Table 9.4.2.

Table 9.4.2

0	1	2	3	4	5	6	7	8	9	10	11	12	13	14
2	3	0	1	11	0	11	0	1	10	9	4	1	2	4
5	8	5	12	14	2	14	5		12	12	6	3	5	6
7		13			7		13				14	9	7	11
					13							10		

3 The diagram (Fig. 9.11) represents a maze: the lines are walls and the spaces are passages.

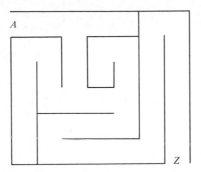

Fig. 9.11
Not a famous maze.

(i) Label the junctions and dead-ends. (The entrance A and exit Z are already labelled.)
(ii) Write down the adjacency list for the corresponding graph.
(iii) Use DFS to find a path from A to Z in the graph.
(iv) Sketch the corresponding route through the maze.

9.5 Breadth-first search

The main alternative to DFS is **breadth-first search** (BFS). Whereas in DFS we advance from the current vertex to a new one whenever possible, in BFS we check all the vertices adjacent to the current vertex before going on to the next one. Consequently there is no need for backtracking.

The flowchart for BFS (Fig. 9.12) is presented as a tree-growing method, although we can if we wish regard it purely as a search technique, and simply list the vertices in order of appearance. The following description of the flowchart should be compared and contrasted with that for DFS (Fig. 9.10).

We start at a given vertex v, and construct a partial tree W. Whenever the current vertex x has new neighbours, we choose y to be one of them, and ADJOIN xy to W. (We do *not* replace x by y.) Whenever there are no new vertices adjacent to x we move on to the NEXT vertex following x in the original order of appearance. Eventually we find ourselves at a vertex which has no new neighbours and for which there is no next vertex, so we put $T = W$ and stop. For example, the BFS procedure applied to the hide-and-seek graph

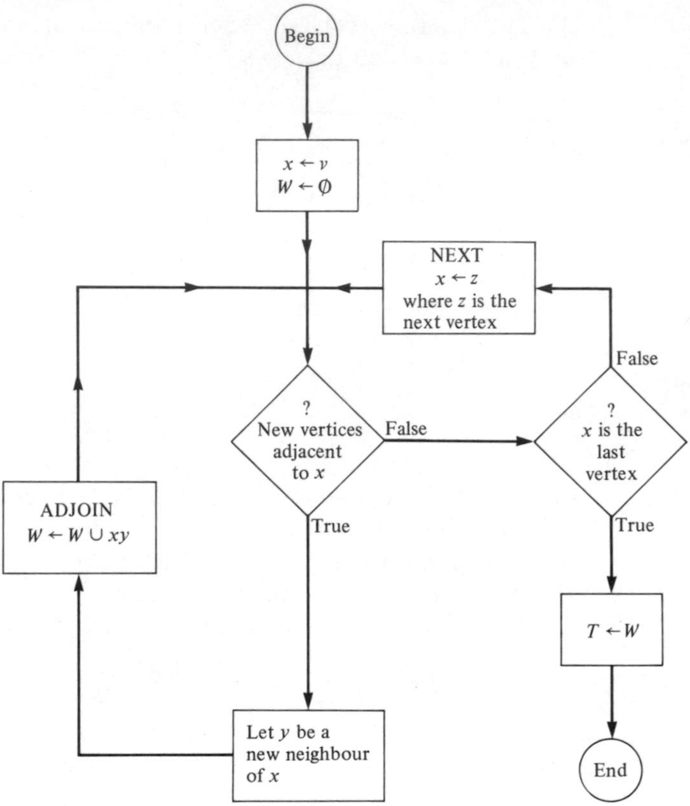

Fig. 9.12
Flowchart for BFS.

(Fig. 9.9) starting from a, would adjoin the edges in the order ab, ac, ae, bd.

The proof of the next theorem is very similar to that of Theorem 9.4, and is left as an exercise (Ex. 9.7.16).

Theorem 9.5
Let v be a vertex of the graph G, and let T be the subset of the edges of G constructed according to the BFS flowchart (Fig. 9.12). Then T is a spanning tree for the component of G which contains v.

☐

A good way of comparing the operation of BFS and DFS is to think of the vertices as forming a 'queue'. A vertex arrives in the queue when it first appears in the search, and departs (or is 'served') when it is no longer required. Thus, if *first* (Q) is the first member of a queue Q, the following procedure describes the formation of Q according to BFS, starting at the vertex v.

$$Q \leftarrow (v)$$
while $\quad Q \neq \varnothing \quad$ **do**
$\qquad x \leftarrow first\ (Q)$
$\qquad$ **if** $\quad x$ is adjacent to a new vertex y
$\qquad\qquad$ **then** $\quad Q \leftarrow (Q$ followed by $y)$
$\qquad\qquad$ **else** $\quad Q \leftarrow (Q$ with x deleted$)$

For example, the BFS method for the hide-and-seek graph proceeds as shown in Table 9.5.1.

Table 9.5.1

Q	Arrival	Departure
a	a	—
ab	b	—
abc	c	—
$abce$	e	—
bce	—	a
$bced$	d	—
ced	—	b
ed	—	c
d		e
$\varnothing$	—	d

Clearly, BFS deals with the queue of vertices in the way we expect an orderly queue to be served in everyday life: the vertex at the front of the queue is served first. This is sometimes known as the FIFO (first-in, first-out) mode of operation.

The DFS method differs only in the third line of the procedure. Instead of taking x to be the first member of Q, we must take

$$x \leftarrow last\ (Q);$$

where *last* (Q) is the last member of Q. This small change results in a totally different mode of operation; for example, applying it to the hide-and-seek graph we obtain the sequence of events shown in Table 9.5.2.

Table 9.5.2

Q	Arrival	Departure
a	a	—
ab	b	—
abc	c	—
$abce$	e	—
abc	—	e
$abcd$	d	—
abc	—	d
ab	—	c
a	—	b
$\varnothing$	—	a

In real life, this method of dealing with a queue would cause a riot. In this case it is more appropriate to use the term 'stack', suggested by the resemblance to the way in which plates or trays are stacked in a restaurant. The fact that a plate can only be removed from the top of the stack corresponds to the fact that, in DFS, a vertex can only depart when it is at the end of the list. This is known as the LIFO (last-in, first-out) mode of operation.

Exercises 9.5

1　Construct the BFS tree rooted at c for the graph G defined in Ex. 9.4.1.

2　Use BFS to test if the graph defined by Table 9.5.3 is connected.

Table 9.5.3

a	b	c	d	e	f	g	h	i
e	d	e	b	a	c	b	b	a
i	g	f	g	c	e	d	d	c
	h	i	h	f	i			f

3　Let v be a vertex of the complete graph K_n. Calculate the height of the DFS and BFS trees in K_n rooted at v. What are the respective heights when the graph is the cycle graph C_n?

4　Construct tables showing the formation of the queue and stack when BFS and DFS are applied to the graph in Ex. 9.4.1. (Start from the vertex a, and use alphabetical order if a choice is possible.)

5   Let G be a connected graph which is regular with valency k, and let d be the height of the BFS tree rooted at a vertex of G. Prove that the number of vertices of G is at most

$$1 + k + k(k-1) + k(k-1)^2 + \cdots + k(k-1)^{d-1}.$$

9.6 The shortest path problem

Many of the algorithms used in computer science and operations research are based upon BFS or DFS. In any given problem it may be difficult to decide which technique is appropriate, and the decision may depend on theoretical analysis, or practical experience, or a combination of both. Roughly speaking, DFS is usually preferred when the problem requires just one of many possible solutions: for example, when we are asked to find any path joining a given pair of vertices in a graph. On the other hand, BFS is better suited to problems where some kind of optimization is required, such as finding a path with the smallest possible number of edges. In this section we shall study a generalization of the latter problem.

Suppose we have a weighted graph, for example the one illustrated in Fig. 9.13, and we think of the weights as representing the 'lengths' of the edges in some practical interpretation. We ask for the shortest path from a given vertex v to another given vertex w, where the length of a path is measured by the sum of the lengths of its edges.

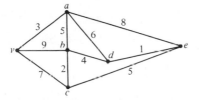

Fig. 9.13
Find the shortest path
from v to e.

Suppose we have established that the shortest path from v to a vertex p has length $l(p)$. Suppose also that y is a neighbour of p for which we have only an estimate $l(y)$ of the length of the shortest path from v to y (Fig. 9.14). The shortest route from v to y via p has length $l(p) + w(py)$ and, if this is less than $l(y)$, we can improve our estimate accordingly. That is, we assign a new value of $l(y)$, equal to

$$\min \{l(y), l(p) + w(py)\}.$$

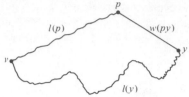

Fig. 9.14
$l(y) \leftarrow \min\{l(y), l(p) + w(py)\}$.

The idea behind the algorithm for the shortest path problem is to use a version of the BFS procedure to construct a tree, rooted at v, whose edges define the shortest path from v to any vertex. At each stage, every vertex x has a label $l(x)$, which may be either temporary or permanent. The label $l(x)$ becomes permanent, and we link x to the tree, when we are sure that $l(x)$ is the length of the shortest path from v to x.

We begin with $l(v) = 0$, which is already permanent, and for all other vertices x we set $l(x)$ equal to some large number L. Let p be the latest vertex to receive a permanent label (initially $p = v$). For each vertex y which is adjacent to p and has a temporary label, calculate the new value of $l(y)$ as above. Then find the vertex q which has the smallest temporary label: make its label permanent, and link it to the tree by means of the edge zq, where z is the vertex used to obtain the value $l(q)$. Now replace p by q, and repeat until all vertices have been permanently labelled.

Table 9.6.1 shows the labelling process and the formation of the tree for the graph in Fig. 9.13.

Table 9.6.1

p	v	a	b	c	d	e	q	zq
	0	L	L	L	L	L		
v		3	9	7	L	L	a	va
a			8	7	9	11	c	vc
c			8		9	11	b	$-ab$
b					9	11	d	ad
d						10	e	de

A suitable value for L in this case would be 50, which is the sum of all the weights; clearly, no path can be longer than this. The shortest path from v to any vertex is the unique path in the tree from v to that vertex; for example, the shortest path from v to e is v, a, d, e. If we require only the shortest path to a particular vertex, we can stop the procedure as soon as that vertex receives a permanent label.

1 Use the tabular method described above to find the shortest path from v to w in the weighted graph depicted in Fig. 9.15.

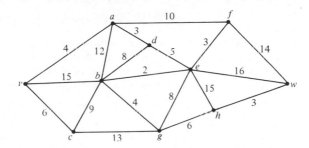

Fig. 9.15
Find the shortest path from v to w.

2 Find the shortest route from A to F in the weighted graph specified in Table 9.6.2.

Table 9.6.2

	A	B	C	D	E	F
A	—	5	8	3	4	9
B		—	6	1	5	4
C			—	3	9	2
D				—	4	6
E					—	3

3 Suppose we are given a weighted graph with n vertices. Define *Stage i* of the algorithm to be the phase which is executed when i of the vertices have been assigned permanent labels ($1 \leq i \leq n-1$). What is the maximum number of comparisons required in order to calculate the new labels at Stage i? What is the maximum number of comparisons required to decide which vertex is to be the next to receive a permanent label? Deduce that the total number of comparisons required by the algorithm is $O(n^2)$.

9.7 Miscellaneous Exercises

1 Let T be an m-ary rooted tree with n vertices, l leaves, and i internal vertices. Show that

$$n = mi + 1$$

and hence find an equation for l in terms of m and n.

2 For each one of the six different (unrooted) trees with six vertices (Ex. 8.5.1), find the number of essentially different ways of choosing one vertex as a root. Hence calculate the number of different rooted trees with six vertices.

3 Use the heapsort method to construct a heap from the list

$$29, 38, 71, 15, 32, 61, 83, 35, 47, 67, 78, 63, 91.$$

4 Suppose the vertices of an 'almost-binary' tree are labelled $x_1, x_2, \ldots, x_n$ in the manner of Fig. 9.4, and let r_i denote the number of vertices of the subtree rooted at x_i. Show that the number of ways of assigning n distinct integers to the vertices so as to form a heap is

$$\frac{n!}{r_1 r_2 \cdots r_n}.$$

5 Show that the number of different spanning trees in the complete graph K_5 is 125. (Do not attempt to list all the trees individually.)

6 Find all minimum spanning trees for the weighted graph depicted in Fig. 9.16.

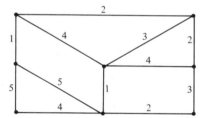

Fig. 9.16
Find all MSTs.

8 In the weighted graph shown in Fig. 9.8, find the shortest path from a to k.

9 Kruskal's version of the greedy algorithm is: choose the edges in order of increasing weight, rejecting those whose inclusion would complete a cycle. Show by an example that in Kruskal's method the set of edges constructed at an intermediate stage need not form a connected graph.

10 Prove the version of Theorem 9.3 relevant to Kruskal's method. (In other words, prove that Kruskal's method works.)

11 Devise a maze with a single 'gate' and a single 'centre', using a representation similar to that in Fig. 9.11. Describe how you would use DFS to find a method of escaping from the centre to the gate.

12 In 1895, G. Tarry gave the following rule for traversing a maze. *Do not return along the passage which has led to a junction for the first time unless you cannot do otherwise.* Explain the connection between Tarry's rule and the depth-first search algorithm.

13 Suppose T is a spanning tree of a graph G and r is a vertex of G.

Regarding T as a directed graph (by directing each edge away from r) we may say that vertices x and y are *T-related* if there is a directed path in T from x to y or from y to x.

Show that if T is a DFS tree rooted at r, and xy is an edge of G not in T, then x and y are T-related.

14 Give an example to show that the conclusion of the previous exercise may be false when T is a BFS tree rooted at r.

15 A vertex is a **cut-vertex** of a connected graph if its removal (together with the edges which contain it) results in a disconnected graph. Use the idea contained in Ex. 13 to formulate an algorithm, based on DFS, for finding the cut-vertices in a given connected graph.

16 Write out the proof of Theorem 9.5.

17 Let the vertices of the complete graph K_n be denoted by $1, 2, \ldots, n$, and for each spanning tree T of K_n define the **Prüfer symbol** $(p_1, p_2, \ldots, p_{n-2})$ as follows. The Prüfer symbol of a tree with two vertices is null. If $n > 2$, the Prüfer symbol of a tree T with n vertices is $(j, q_1, \ldots, q_{n-3})$ where

   (i) j is the unique vertex of T adjacent to the vertex i of valency one which comes first in numerical order,

   (ii) $(q_1, \ldots, q_{n-3})$ is the Prüfer symbol of the tree obtained by deleting the edge ij from T.

Show that the Prüfer symbol construction defines a bijection from the set of spanning trees of K_n to the set of ordered $(n-2)$-tuples from the set $\{1, 2, \ldots, n\}$ and deduce that K_n has n^{n-2} spanning trees.

18 Suppose we have four coins, at most one of which is false (light or heavy). Show that the determination of the false coin, if there is one, requires at least two weighings theoretically, but that this number cannot be achieved. (In this problem, no true coin is given.)

19 Suppose the vertices of a weighted graph are denoted by $1, 2, \ldots, n$ and let w_{ij} denote the weight of the edge ij (if there is such an edge) or an arbitrary large number (if there is no such edge). Determine the final values of d_{ij} obtained by the following program.

$$d_{11} \leftarrow w_{11}; \quad d_{12} \leftarrow w_{12}; \quad \ldots; \quad d_{nn} \leftarrow w_{nn}$$

for $i = 1$ **to** n **do**

    for $j = 1$ **to** n **do**

        for $k = 1$ **to** n **do**

            if $d_{ij} > d_{ik} + d_{kj}$

                then $d_{ij} \leftarrow d_{ik} + d_{kj}$

10
Bipartite graphs and matching problems

10.1 Relations and bipartite graphs

In Section 3.2 we introduced the wide class of problems which can be expressed in terms of counting a given subset of a product set $X \times Y$. One way of describing such a subset is to say that a member x of X and a member y of Y are 'related' whenever the pair (x, y) belongs to the given subset. For example, if X is a set of students and Y is a set of courses, we could say that x and y are related whenever x is a student who is taking course y.

These remarks lead to the conclusion that a **relation** R between two sets X and Y is simply a subset of $X \times Y$, and so the statements

x and y are related (by R),

the pair (x, y) is in R,

mean precisely the same thing. It is possible that $X = Y$; for example, this is the case when we consider equivalence relations defined on X. However, in this chapter we shall study relations which are defined when X and Y are *disjoint* sets. We shall base our discussion on a representation of such a relation by a bipartite graph, which we shall now describe.

When R is a relation between disjoint sets X and Y (that is, when R is a subset of $X \times Y$), we define a bipartite graph G representing R as follows. The vertex-set of G is the union of X and Y, and the edge-set E contains those edges xy for which (x, y) is in R. Since every edge has one vertex in X and one vertex in Y it is clear that G is bipartite, and it is convenient to think of G in pictorial terms, as in Fig. 10.1 for example. In order to emphasize that G is bipartite,

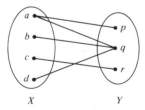

Fig. 10.1
The bipartite graph representing the relation $R = \{(a, p), (a, q), (b, q), (c, r), (d, q)\}$.

we shall use the notation $G = (X \cup Y, E)$ for a graph of this kind.

The basic theorem on counting sets of pairs has a simple interpretation in graphical terms.

Theorem 10.1

Let $G = (X \cup Y, E)$ be a bipartite graph and let $\delta(v)$ denote the valency of a vertex v in G. Then

$$\sum_{x \in X} \delta(x) = \sum_{y \in Y} \delta(y) = |E|.$$

Proof

Since each edge has exactly one vertex in X, the total number of edges is the sum of the valencies of the vertices in X. Similarly, the total number of edges is equal to the sum of the valencies of the vertices in Y. Hence we have the required result.   □

Throughout the rest of this chapter we shall frame our results in terms of bipartite graphs, rather than relations. But all the results can be interpreted as results about relations, if we so wish.

Example

Suppose we are given a set of people and a set of jobs, such that each person is qualified to do exactly k of the jobs and there are exactly k people qualified to do each job. Show that

(i) the number of people is equal to the number of jobs;
(ii) given any n-subset A of the people there are at least n jobs for which some member of A is qualified.

Solution

Let X denote the set of people and Y the set of jobs. Define x and y to be related whenever person x is qualified to do job y. The given condition says that the bipartite graph $G = (X \cup Y, E)$ representing this relation is a regular graph with valency k. Hence, by Theorem 10.1, we have

$$|X|\, k = |Y|\, k = |E|,$$

so that $|X| = |Y|$, as claimed in part (i).

For part (ii), let A be any n-subset of X, and define $J(A)$ to be the set of jobs for which at least one member of A is qualified; that is

$$J(A) = \{y \in Y \mid xy \in E \text{ for some } x \in A\}.$$

Since each vertex belongs to exactly k edges of G, the set E_A of edges which have one vertex in A has cardinality $k\,|A| = kn$. By definition of $J(A)$, each of these edges has one vertex in $J(A)$, and the given condition says that the total number of edges with one vertex in $J(A)$ is $k |J(A)|$. Hence

$$|E_A| = kn \leqslant k |J(A)|,$$

and it follows that $|J(A)| \geqslant n$, as claimed in part (ii). $\qquad \square$

**Exercises
10.1** 1 Let $X = \{2, 3, 5, 7, 11\}$, $Y = \{99, 100, 101, 102, 103\}$, and define x and y to be related whenever x is a divisor of y. Draw the bipartite graph representing this relation, and verify that Theorem 10.1 is satisfied.

2 The **complete bipartite graph** $K_{r,s}$ is the bipartite graph $(X \cup Y, E)$ in which $|X| = r$, $|Y| = s$, and every pair xy with $x \in X$ and $y \in Y$ is an edge.

(i) What is the valency of each vertex in X?
(ii) What is the valency of each vertex in Y?
(iii) How many edges are there in $K_{r,s}$?
(iv) Describe in ordinary language the relation which $K_{r,s}$ represents.
(v) Show that for any $s \geqslant 1$ the graph $K_{1,s}$ is a tree.
(vi) Show that $K_{r,s}$ is not a tree whenever $r \geqslant s \geqslant 2$.

3 Find the number of 4-cycles in $K_{r,s}$ when $r \geqslant s \geqslant 2$. (Two 4-cycles are equal if they have the same set of vertices.)

10.2 Edge-colourings of graphs

There are many problems which can be interpreted in terms of a partition of the edge-set E of a graph, that is, a decomposition of the form

$$E = E_1 \cup E_2 \cup \cdots \cup E_r$$

where $E_1, E_2, \ldots, E_r$ are disjoint, non-empty sets. Intuitively, it is helpful to describe such a partition by means of a 'colouring' of the edges: the edges in E_1 are given a certain colour, those in E_2 a different colour, and so on. We shall use small Greek letters $\alpha, \beta, \gamma, \ldots$, for the names of the colours, and usually we shall require that the colouring satisfies a condition analogous to the one we imposed for a vertex-colouring.

Definition Let G be a graph with edge-set E. A colouring of E is said to be an **edge-colouring** of G if any two edges containing the same vertex have different colours.

The diagram (Fig. 10.2) illustrates two colourings of the edges of the same graph. One is an edge-colouring but the other is not.

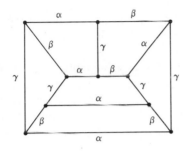 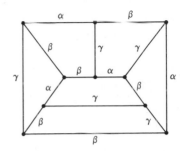

Fig. 10.2
Which one is an edge-colouring?

Exercises
10.2

1 What is the least number of colours required for an edge-colouring of:

(i) the complete graph K_4;
(ii) the complete graph K_5;
(iii) the cube graph (Fig. 8.12).

2 Suppose there is an edge-colouring of Petersen's graph (Fig. 8.14) using only three colours. Show that each colour must be used twice on the edges of the 'outer hexagon' in the Figure, and that there are only two essentially different ways of doing this. Deduce that no such edge-colouring of the whole graph is possible.

3 Prove that for any positive integer n the complete bipartite graph $K_{n,n}$ has an edge-colouring with n colours.

If v is a vertex with valency k, then there are k edges containing v. In order that these edges shall receive different colours, it is clear that at least k colours must be available. Hence, if $\hat{k}$ is the maximum valency of G, at least $\hat{k}$ colours are needed for an edge-colouring of G. In general, $\hat{k}$ colours will not be enough (as in

Ex. 10.2.2); however, we can prove that when G is *bipartite* $\hat{k}$ colours are always sufficient.

Theorem 10.2 If $G = (X \cup Y, E)$ is a bipartite graph, then the minimum number of colours needed for an edge-colouring of G is equal to the maximum valency of G.

Proof We shall proceed by induction on m, the number of edges. If $m = 1$, then G has maximum valency 1 and clearly one colour is sufficient to colour the one edge.

Suppose the result is true for any bipartite graph with m edges, and suppose G has $m + 1$ edges and maximum valency $\hat{k}$. Remove any edge xy from G to get a bipartite graph G' with m edges. Since the maximum valency of G' is $\hat{k}$ or $\hat{k} - 1$, it follows from the induction hypothesis that there is an edge-colouring of G' using at most $\hat{k}$ colours $\alpha, \beta, \ldots$.

The valency of x in G' is at most $\hat{k} - 1$ (since xy has been removed) and so there must be a colour, say α, not used on the edges at x. Similarly there must be colour β not used at y. If we can choose α and β to be the same then we can give this colour to xy and thereby obtain an edge-colouring of G as required. We shall call this the *easy case*. Thus we need only consider the situation where $\alpha \neq \beta$, and we shall show how, in this situation, we can modify the given colouring of G' so that the easy case applies.

Suppose then that $\alpha \neq \beta$. Define a path $x, y_1, x_1, y_2, x_2, \ldots$ as follows:

(1) xy_1 is the edge at x coloured β;
(2) if there is an edge at y_1 coloured α, call it y_1x_1, otherwise stop;
(3) if there is an edge at x_1 coloured β, call it x_1y_2, otherwise stop;
(4) continue with edges coloured α and β alternately until forced to stop.

The path is illustrated in Fig. 10.3. It must stop eventually, since the graph is finite. Also the path does not contain y, since it arrives at each vertex in Y by an edge coloured β, and β is defined to be a colour not used at y.

Now we alter the edge-colouring of G' by interchanging the colours α and β on the path, leaving the colours on other edges unchanged (Fig. 10.3). The result is an edge-colouring of G' in the easy case: no edge at x is coloured β. From this colouring we obtain the required edge-colouring of G by giving xy the colour β.

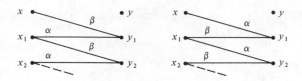

Fig. 10.3
An alternating path and
its recolouring.

By the principle of induction the result holds for all bipartite graphs. □

The crucial part of the proof is the construction of the 'alternating path' $x, y_1, x_1, y_2, \ldots$, which depends critically on the bipartite character of G. We shall make more use of this construction in later sections.

**Exercises
10.2
(continued)**

4 Show that the graph illustrated in Fig. 10.4 is bipartite and construct an edge-colouring of it using only three colours.

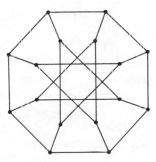

Fig. 10.4
A bipartite graph.

5 According to Ex. 8.2.3, the graph Q_3 formed by the corners and edges of a cube can be represented in the following way: the vertices are the words of length 3 in the alphabet $\{0, 1\}$, and edges join words which differ in just one letter. Use this representation to show that Q_3 is bipartite and to construct an edge-colouring of Q_3 using only three colours.

6 Generalize the results of Ex. 5 to the graph Q_k obtained by replacing 3 by k throughout.

10.3   Application of edge-colouring to latin squares

There is a simple relationship between latin squares and edge-colourings of bipartite graphs. We may describe an $n \times n$ latin square in terms of the rows $r_1, r_2, \ldots, r_n$, the columns $c_1, c_2, \ldots, c_n$, and the symbols $s_1, s_2, \ldots, s_n$, arranged so that each symbol occurs just once in each row and column. In practice, of course, we often use the same labels for the names of the rows and columns, and for the symbols, but the amplified notation will make the ensuing discussion clearer.

If we are given a latin square, for example the one in Fig. 10.5a, then we can use it to assign colours to the edges of a complete bipartite graph, as in Fig. 10.5b. The vertices of the graph are regarded as the rows and columns of the latin square and the edge $r_i c_j$ is assigned the 'colour' s_k, where s_k is the symbol in row r_i and column c_j of the square. The defining property of the latin square ensures that this assignment of colours really is an edge-colouring. so any latin square of order n defines an edge-colouring of $K_{n,n}$.

In mathematics, it sometimes pays to take a twisted view, and that is true here. Instead of constructing an edge-colouring of $K_{n,n}$ by the obvious method described above, we shall use the latin square in a different way. We take the vertices to be the symbols $s_1, s_2, \ldots, s_n$ and the columns $c_1, c_2, \ldots, c_n$, and assign the edge $s_i c_j$ the 'colour' r_k, where r_k is the row such that s_i appears in row r_k and column c_j of the square (Fig. 10.5c). Here again, the defining property of the latin square ensures that we have an edge-colouring of $K_{n,n}$.

We shall adopt the twisted view to investigate problems about the construction of latin squares. Suppose we set out to construct an $n \times n$ latin square by filling in one row at a time. Naturally, we make sure that each row contains every symbol once only, and that no symbol occurs more than once in a column. The result of filling in m

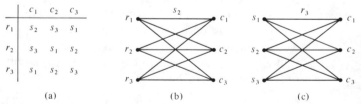

	c_1	c_2	c_3
r_1	s_2	s_3	s_1
r_2	s_3	s_1	s_2
r_3	s_1	s_2	s_3

(a)   (b)   (c)

Fig. 10.5
A 3×3 latin square and
two ways of edge-
colouring $K_{3,3}$.

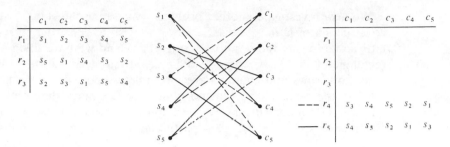

Fig. 10.6
A latin rectangle and how to complete it.

rows in this way is an $m \times n$ **latin rectangle** $(1 \leq m < n)$. An example of a 3×5 latin rectangle is given in Fig. 10.6.

Given an $m \times n$ latin rectangle, can we fill in the remaining $n - m$ rows to get an $n \times n$ latin square? Rather surprisingly, the answer is yes, without any extra conditions on the latin rectangle. In other words, if we construct a latin square one row at a time then, provided we observe the obvious constraints at each stage, we shall never get stuck.

Theorem 10.3.1

Any $m \times n$ latin rectangle with $1 \leq m < n$ can be completed to form an $n \times n$ latin square.

Proof

The m rows of the latin rectangle enable us to colour some of the edges of $K_{n,n}$, using the twisted rule given above. Let E denote the set of edges which remain as yet uncoloured; that is

$$E = \{s_i c_j \mid \text{symbol } s_i \text{ does } not \text{ appear in column } c_j\}.$$

The graph formed by the uncoloured edges is $G = (S \cup C, E)$, where S and C denote the sets of symbols and columns respectively. The graph G is regular, with valency $n - m$, and so (by Theorem 10.2) it has an edge-colouring using $n - m$ colours. Call these colours $r_{m+1}, r_{m+2}, \ldots, r_n$.

Now we can fill in the remaining rows of the square, by putting s_i in row r_k and column c_j whenever the edge $s_i c_j$ is coloured r_k $(m + 1 \leq k \leq n)$. $\qquad \square$

The procedure for a 3×5 latin rectangle is illustrated in Fig. 10.6. The graph formed by 'uncoloured' edges has valency $5 - 3 = 2$, and can be edge-coloured with two colours as shown. The last two rows of the square can then be filled in.

We now investigate another problem concerning the step-by-step construction of latin squares. Suppose we wish to construct an $n \times n$ latin square using the symbols $s_1, s_2, \ldots, s_n$, and we have filled in a rectangle of size $k \times l$, where k and l are strictly less than n. In this case, both rows and columns are incomplete, and even if we have filled in the rectangle so that no symbol occurs more than once in any row or column, it may yet be impossible to complete the latin square. For example, the 3×4 partial latin rectangle

$$A \quad C \quad D \quad E$$
$$C \quad E \quad A \quad B$$
$$E \quad A \quad C \quad D$$

cannot be completed to give a 5×5 latin square. To see this we need only remark that B occurs just once in the rectangle, and it can occur only three times more in the square (once in the last column and twice in the last two rows); hence there are only four possible occurrences of B, whereas five are needed. The following theorem shows that the condition that each symbol occurs often enough in the rectangle is both necessary and sufficient for a completion.

Theorem 10.3.2

Let R be a partial $m \times p$ latin rectangle in which the symbols $\{s_1, s_2, \ldots, s_n\}$ are used, and let $n_R(s_i)$ denote the number of times s_i occurs in R $(1 \le i \le n)$. Then R can be completed to an $n \times n$ latin square if and only if

$$n_R(s_i) \ge m + p - n \qquad (1 \le i \le n).$$

Proof

Suppose that R can be completed. Since there are $n - m$ rows and $n - p$ columns to be filled in, there are at most $(n - m) + (n - p)$ further occurrences of any symbol. Hence

$$n_R(s_i) + (n - m) + (n - p) \ge n,$$

which leads to the stated condition.

Conversely, suppose the condition holds. Construct the bipartite graph whose vertices are the rows $\{r_1, \ldots, r_m\}$ and the symbols $\{s_1, \ldots, s_n\}$, and whose edge-set is

$$E = \{r_i s_j \mid s_j \text{ does } not \text{ occur in row } r_i\}.$$

Since each row of R contains p different symbols, the valency of each r_i vertex is

$$\delta(r_i) = n - p \qquad (1 \le i \le n).$$

Since each symbol s_j occurs at least $m + p - n$ times in R, and each occurrence is in a different row, s_j does not occur in at most $m-(m+p-n)$ rows. That is,

$$\delta(s_j) \leqslant n-p.$$

Since the maximum valency of the bipartite graph is $n-p$, it follows from Theorem 10.2 that it has an edge-colouring using $n-p$ colours, which we shall call $c_{p+1}, c_{p+2}, \ldots, c_n$.

Using this colouring we can complete the m rows of R, by putting s_j in row r_i and column c_k when the edge $r_i s_j$ is coloured c_k. We now have an $m \times n$ latin rectangle with complete rows and, by Theorem 10.3.1, this can be completed to form an $n \times n$ latin square. $\square$

Exercises 10.3

1 Use the edge-colouring method to extend the following latin rectangle to a 5×5 latin square.

$$A \quad B \quad C \quad D \quad E$$
$$C \quad D \quad B \quad E \quad A$$
$$B \quad C \quad E \quad A \quad D$$

2 Find all values of Q for which the rectangle R_1 can be extended to a 6×6 latin square. Show that R_2 cannot be so extended, whatever value Q has.

$$
\begin{array}{cccc@{\qquad}cccc}
R_1: & A & B & C & D & R_2: & A & B & C & D \\
& F & E & A & B & & F & E & A & B \\
& C & D & F & A & & B & D & F & A \\
& D & A & B & Q & & D & A & B & Q \\
\end{array}
$$

3 Show that any $n \times n$ latin square can be used as the 'top left quarter' of a $2n \times 2n$ latin square.

10.4 Matchings

In Section 10.1 (*Example*) we discussed a special case of the situation where we have a set X of people and a set Y of jobs, and each person is qualified to do some of the jobs. A question with obvious practical implications is the following. How shall we assign people to jobs, so that the maximum number of people get jobs for which they are qualified?

We shall translate the question into the language of bipartite graphs. The relation of 'being qualified' enables us to set up a bipartite graph $G = (X \cup Y, E)$ in the usual way: xy is an edge if and only if x is qualified to do job y. An assignment of people to jobs for which they are qualified corresponds to a 'matching' in G, in the technical sense defined below.

Definition   A **matching** in a bipartite graph $G = (X \cup Y, E)$ is a subset M of E with the property that no two edges in M have a common vertex.

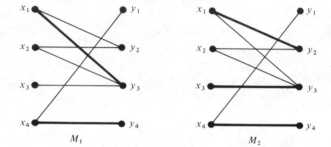

Fig. 10.7
A matching M_1 and a
maximum matching M_2.

In Fig. 10.7 two matchings M_1 and M_2 in the same graph are illustrated; the edges which belong to the matchings are indicated by heavy lines. Using the people-and-jobs terminology, M_1 yields an assignment of jobs for the two people x_1 and x_4 and M_2 for the three people x_1, x_3, x_4. In fact, M_2 cannot be bettered, since it is impossible for all four people to get jobs for which they are qualified. To see this, we simply remark that the three people $\{x_1, x_2, x_3\}$ are collectively qualified only for the two jobs y_2 and y_3, and so one of these people must be disappointed however the jobs are filled.

We shall say that a matching M is a **maximum matching** for $G = (X \cup Y, E)$ if no other matching has a greater cardinality. If $|M| = |X|$ (all the people get jobs), then we say that M is a **complete matching**. In the example, M_2 is a maximum matching, but not a complete matching.

The first step in the study of matchings is to decide when a complete matching is possible. A necessary condition has already appeared in our discussion of Fig. 10.7, where we remarked that three people are collectively qualified for only two jobs. More generally, if $G = (X \cup Y, E)$ and A is a subset of X, let

$$J(A) = \{y \in Y \mid xy \in E \text{ for some } x \in A\},$$

so that $J(A)$ is the set of jobs for which the people in A are collectively qualified. Our remark about Fig. 10.7 amounts to saying that, if $|J(A)| < |A|$, then someone in A is bound to be disappointed. So, if there is a complete matching, it must be true that $|J(A)| \geq |A|$ for all $A \subseteq X$. This is known as **Hall's condition**, after the mathematician Philip Hall, who studied a similar problem in 1935. (See Section 10.6.)

The fundamental theorem about complete matchings says that Hall's condition is both necessary and sufficient.

Theorem 10.4

The bipartite graph $G = (X \cup Y, E)$ has a complete matching if and only if Hall's condition is satisfied, that is

$$|J(A)| \geq |A| \quad \text{for all } A \subseteq X.$$

Proof

Suppose there is a complete matching. For any $A \subseteq X$ the vertices in Y matched with those in A form a subset of $J(A)$ with size $|A|$. Hence $|J(A)| \geq |A|$.

Conversely, suppose Hall's condition holds. Given any matching M with $|M| < |X|$ we shall show how to construct a matching M' with $|M'| = |M| + 1$.

Let x_0 be any vertex not matched by M. Since $|J\{x_0\}| \geq |\{x_0\}| = 1$, there is at least one edge $x_0 y_1$. If y_1 is unmatched then we can add $x_0 y_1$ to M and obtain the required M'.

If y_1 is matched, with x_1 say, then

$$|J\{x_0, x_1\}| \geq |\{x_0, x_1\}| = 2,$$

and so there is another vertex y_2 apart from y_1 adjacent to x_0 or x_1. If y_2 is unmatched, stop. If y_2 is matched, to x_2 say, repeat the argument and select a new vertex y_3 adjacent to at least one of x_0, x_1, x_2. Continuing in this way, we must eventually stop at an unmatched vertex y_r, since G is finite.

Each vertex y_i $(1 \leq i \leq r)$ is adjacent to at least one of $x_0, x_1, \ldots, x_{i-1}$. So, retracing our steps, we have a path

$$y_r, x_s, y_s, x_t, y_t, \ldots, y_w, x_0$$

in which the edges $x_i y_i$ are in M, and the alternate edges are not in M. We construct the new matching M' such that the edges $x_i y_i$ of the path are not in M', but the alternate edges are in M'. Since the terminal edges $y_r x_s$ and $y_w x_0$ are both in M', we have $|M'| = |M| + 1$, as required. $\qquad\square$

The key idea in the proof is the construction of a path whose edges are alternately in M and not in M. In general, suppose $G = (X \cup Y, E)$ is a bipartite graph and M is a matching in G. We say that the path

$$x_0, y_1, x_1, y_2, x_2, \ldots, x_{k-1}, y_k$$

is an **alternating path** (for M) if the edges $y_i x_i$ are in M, the edges $x_{i-1} y_i$ are not in M ($1 \leqslant i \leqslant k$), and x_0 and y_k do not belong to any edge of M. Notice that the first and last edges are not in M, so that the path has one edge fewer in M than it has not in M. The proof of the theorem shows that if Hall's condition is satisfied, and M is an incomplete matching, then an alternating path for M exists. Switching the status of the edges on this path yields a new matching M' with one more edge.

The idea is not only the basis of the proof, it is also a practical device for the construction of matchings. In Fig. 10.8 the path $ABCD$ is an alternating path, and switching the status of the edges on this path yields the complete matching shown.

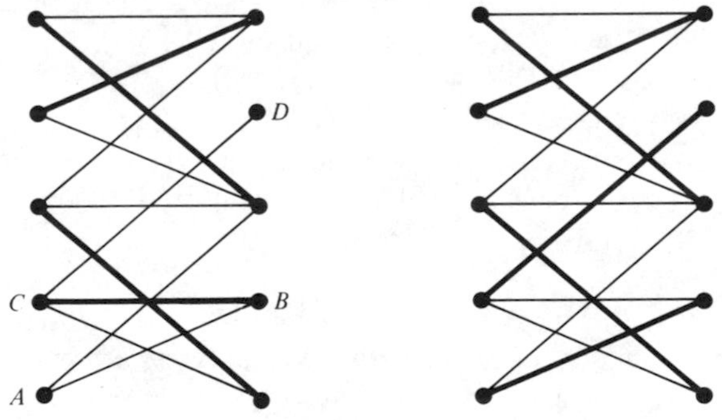

Fig. 10.8
An alternating path
$ABCD$ and the result of
switching.

In Section 10.5 we shall explain how this approach leads to an algorithm for finding a maximum matching in any bipartite graph.

**Exercises
10.4**

1   Use Hall's condition to show that the graph in Fig. 10.9 has no complete matching.

2   Let M be the matching denoted by heavy lines in Fig. 10.9.

(i) Find an alternating path for M beginning at x_2.
(ii) Use it to construct a matching M' with $|M'| = 4$.
(iii) Check that there is no alternating path for M'.
(iv) Is M' a maximum matching?

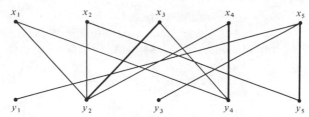

Fig. 10.9
Illustration for Ex. 10.4.2.

3 Suppose that each member of a set of people has a list of k books which he or she wishes to borrow from a library. Suppose also that each book appears on exactly k lists. Show that it is possible for everyone to borrow one of the books on his or her list at the same time. [Hint: use the result established in the *Example*, Section 10.1.]

10.5 Maximum matchings

In general, a bipartite graph will not have a complete matching. This remark leads us back to our original question of finding the maximum size of a matching. In the people-and-jobs terminology, we are trying to find the assignment which results in the largest possible number of people getting suitable jobs. The solution to this problem is, in fact, a fairly straightforward deduction from the result on complete matchings, Theorem 10.4.

The crucial point is the remark that, if a set of people A are collectively qualified for the set of jobs $J(A)$, and $|A| > |J(A)|$, then some of the people will be disappointed. Indeed, at least $|A| - |J(A)|$ people will be disappointed.

Definition The **deficiency** d of a bipartite graph $G = (X \cup Y, E)$ is defined to be

$$d = \max_{A \subseteq X} \{|A| - |J(A)|\}.$$

We remark that the empty set $\varnothing$ is a subset of X and $|\varnothing| = |J(\varnothing)| = 0$, so that $d \geq 0$ in all cases. Theorem 10.4 asserts that G has a complete matching if and only if $d = 0$, and the next theorem deals with the size of a maximum matching in the general case.

Theorem 10.5.1 The size of a maximum matching M in a bipartite graph $G = (X \cup Y, E)$ is

$$|M| = |X| - d$$

where d is the deficiency of G.

Proof   By the definition of d, there is a set $A_0 \subseteq X$ for which $|A_0| - |J(A_0)| = d$. In any matching, at least d members of A_0 remain unmatched, and so $|M| \leq |X| - d$. We have to show that there is a matching with size $|X| - d$.

Let D be a new set of size d, and let G^* be the graph $(X^* \cup Y^*, E^*)$ given by

$$X^* = X, \qquad Y^* = Y \cup D, \qquad E^* = E \cup K,$$

where K is the set of all possible edges linking X and D. The set of vertices $J^*(A)$ linked to a subset A of X in G^* is $D \cup J(A)$, and so

$$|J^*(A)| - |A| = d + |J(A)| - |A| \geq 0,$$

by the definition of d. Hence G^* satisfies Hall's condition, and it has a complete matching M^*. Removing from M^* the d edges which have a vertex in D, we obtain the required matching in G. $\square$

Theorem 10.5.1 does not tell us how to find a maximum matching. Indeed, it is not even the basis of a good practical method for finding the *size* of a maximum matching, since in order to calculate d we must examine all the $2^{|X|}$ subsets of X.

A more practical approach is based on the fact that if we have an alternating path for a matching M then we can construct a better matching M'. In order to make this idea work, we need the following result.

Theorem 10.5.2   If the matching M in a bipartite graph G is not a maximum matching, then G contains an alternating path for M.

Proof   Let M^* be a maximum matching, and let F denote the set of edges which are in one of M or M^*, but not both. (F is the 'symmetric difference' of M and M^*.) The edges in F and the vertices they contain form a graph in which every vertex has valency 1 or 2, so the components of this graph are paths and cycles. In each path or cycle the edges in M alternate with edges not in M, and so in any cycle the number of edges in M is equal to the number not in M. Since $|M^*| > |M|$, there must be at least one component which is a path, and this is an alternating path for M. $\square$

On the basis of Theorem 10.5.2 we can outline a strategy for finding a maximum matching.

(1) Begin with any matching M (one edge alone will do).
(2) Search for an alternating path for M.

(3) If an alternating path is found, construct a better matching M' in the usual way, and return to (2) with M' replacing M.

(4) If no alternating path can be found, stop: M is a maximum matching.

The search for an alternating path can be carried out by a modified BFS procedure. We choose an unmatched vertex x_0 and construct a tree of 'partial' alternating paths starting from x_0 as follows.

(1) At level 1 insert all the vertices $y_1, y_2, \ldots, y_k$ adjacent to x_0. If any one of these vertices y_i is unmatched, stop: $x_0 y_i$ is an alternating path.

(2) If all level 1 vertices are matched, insert the vertices $x_1, x_2, \ldots, x_k$ with which they are matched at level 2.

(3) At level 3, insert all new vertices adjacent to the level 2 vertices. If any one of them is unmatched, stop: the path leading from this vertex to x_0 is an alternating path.

(4) If all level 3 vertices are matched, insert the vertices with which they are matched at level 4.

... and so on.

It remains only to remark that the construction may be halted because there are no new vertices to insert at an odd-numbered level. When this happens there is no alternating path beginning at the chosen vertex x_0. We must however repeat the procedure for every unmatched vertex in X before we can be sure that no alternating path whatsoever can be found in G.

Example   Let $G = (X \cup Y, E)$ be the bipartite graph with $X = \{x_1, x_2, x_3, x_4, x_5\}$, $Y = \{y_1, y_2, y_3, y_4, y_5\}$, and E specified by Table 10.5.1. Let M denote the matching $\{x_1 y_3, x_2 y_1, x_3 y_5, x_4 y_4\}$. Construct the tree of partial alternating paths rooted at x_5 and use it to find a complete matching in G.

Table 10.5.1

x_1	x_2	x_3	x_4	x_5
y_1	y_1	y_1	y_2	y_3
y_3	y_3	y_3	y_4	y_4
		y_5	y_5	

Solution   The tree is illustrated in Fig. 10.10. At level 3 we see that y_2 is

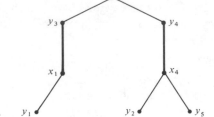

Fig. 10.10
The tree of partial alternating paths rooted at x_5.

unmatched, so that $x_5 y_4 x_4 y_2$ is an alternating path. Switching the
status of the edges on this path gives the complete matching
$\{x_1 y_3, x_2 y_1, x_3 y_5, x_4 y_2, x_5 y_4\}$. □

**Exercises
10.5**

1   Let $G = (X \cup Y, E)$ be the bipartite graph with $X = \{a, b, c, d, e\}$,
$Y = \{v, w, x, y, z\}$, and $E = \{av, ax, bv, bz, cw, cy, cz, dy, dz, ez\}$. Use
the algorithmic method to find a complete matching in G, starting
from the matching $M = \{av, bz, cy\}$.

2   Let $G = (X \cup Y, E)$ be the graph depicted in Fig. 10.7. For
which 3-subsets of X is it possible to find a maximum matching in G
such that the three given vertices are matched?

3   Suppose $G = (X \cup Y, E)$ is a bipartite graph with $|X| = |Y| = n$.
Show that if δ is the minimum valency of G then

$$|A| - |J(A)| \leq n - \delta \quad \text{for all } A \subseteq X.$$

Deduce that if $|E| > (m-1)n$ then G has a matching with at least m
edges.

10.6   Transversals for families of finite sets

At the University of Folornia, the Mathematics Department is run
by committees. There are only six members of the department
(Professor McBrain, Dr Angst, Dr Blott, Dr Chunner, Dr Dodder,
and Dr Elder), and they have organized themselves into four
committees:

> Teaching: {McBrain, Angst},
> Administration: {McBrain, Blott},
> Research: {McBrain, Angst, Blott},
> Car Parking: {Chunner, Dodder, Elder}.

It has been decided that each committee must select a representative to serve on the department's new Committee for Committees. No one is allowed to represent more than one committee. Can this be done?

Given the stated membership of the committees, there are several ways to select distinct representatives: one way is to select Angst, Blott, McBrain, and Chunner to represent the respective committees in the order listed above. However, if the Car Parking committee contained only Angst and Blott, then the selection would be impossible. (Why?)

The general form of this problem is best expressed by using the notion of a family of sets, as introduced in Section 5.1. We are given a family

$$\mathcal{S} = \{S_i \mid i \in I\}$$

of sets, not necessarily distinct, and we wish to choose representatives s_i $(i \in I)$ such that

$$s_i \in S_i \quad \text{and} \quad i \neq j \Rightarrow s_i \neq s_j.$$

Such a set of distinct representatives is usually called a **transversal** for $\mathcal{S}$. The basic problem is to find conditions which will ensure that a given family $\mathcal{S}$ has a transversal.

In fact, this problem is merely a disguised form of the problem of finding a sufficient condition for the existence of a complete matching in a bipartite graph. To see this, we construct the bipartite graph whose two parts correspond to the names of the sets and the members of the sets respectively, and whose edges signify which sets contain which members. (The situation at the University of Folornia is illustrated in Fig. 10.11.)

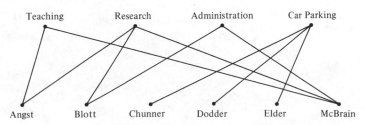

Fig. 10.11
Democracy at the University of Folornia.

Generally, we define $G = (X \cup Y, E)$ as follows:

$$X = I \qquad \text{(the names of the sets),}$$

$$Y = \bigcup_{i \in I} S_i \qquad \text{(the union of the sets),}$$

and put the edge iy in E whenever y is a member of the set S_i. Then a transversal for $\mathcal{S}$ is simply a complete matching for G; if y_i represents S_i then the edge joining i to y_i is in the matching. Now, Hall's condition is easily expressed in the transversal terminology. a subset H of I is simply a subfamily of sets in $\mathcal{S}$, and $J(H)$ is the total membership of those sets, that is

$$\bigcup_{i \in H} S_i.$$

Translating Hall's condition into this language leads to the following version of Theorem 10.4.

Theorem 10.6

The finite family of finite sets

$$\mathcal{S} = \{S_i \mid i \in I\}$$

has a transversal if and only if

$$\left| \bigcup_{i \in H} S_i \right| \geq |H| \quad \text{for all } H \subseteq I. \qquad \square$$

A useful way of expressing the condition is to say that any k of the sets must have at least k members collectively ($k \geq 1$).

Exercises 10.6

1 Let $\mathcal{S}$ be the family of sets $\{a, b, l, e\}$, $\{l, e, s, t\}$, $\{s, t, a, b\}$, $\{s, a, l, e\}$, $\{t, a, l, e\}$, $\{s, a, l, t\}$. Find a transversal for $\mathcal{S}$.

2 Show that there is no transversal for the family $\mathcal{S}$ given in Ex. 1 in which the first three sets are represented by e, l, s, respectively.

3 Prove that the family of sets $\{a, m\}$, $\{a, r, e\}$, $\{m, a, r, e\}$, $\{m, a, s, t, e, r\}$, $\{m, e\}$, $\{r, a, m\}$ has no transversal, by showing explicitly that Hall's condition does not hold.

4 Let $\{X_1, X_2, \ldots, X_n\}$ be a family of sets and let X denote the union of the sets. Show that if the family has a transversal then, given any x in X, there is a transversal containing x.

10.7 Miscellaneous Exercises

1 Show that if a regular graph with valency 3 has a Hamiltonian cycle, then it has an edge-colouring with three colours.

2 There are n married couples at a party. Conversations only take place between two people of opposite sex who are not married. Represent this situation by means of a bipartite graph and show explicitly that your graph has an edge-colouring using $n-1$ colours.

3 Let $S = \{a, d, i, m, o, r, s, t\}$ and let $\mathcal{S}$ be the family of subsets $\{r, o, a, d\}$, $\{r, i, o, t\}$, $\{r, i, d, s\}$, $\{s, t, a, r\}$, $\{m, o, a, t\}$, $\{d, a, m, s\}$, $\{m, i, s, t\}$. Show that any 7-subset of S is a transversal for $\mathcal{S}$.

4 Let X denote the union of the family of sets $\{X_1, X_2, \ldots, X_n\}$, and suppose x and y are members of X. Show by an example that the family may have a transversal, but not a transversal which contains both x and y.

5 Let the elements of $\mathbb{Z}_{14}$ represent the vertices of the cycle graph C_{14}, and let G be the graph obtained from C_{14} by adding the edges $\{i, i+5\}$ ($i = 0, 2, 4, 6, 8, 10, 12$). ($G$ is known as **Heawood's graph**.) Show that G is bipartite and construct an edge-colouring of G which uses the smallest possible number of colours.

6 Suppose there are five committees: $C_1 = \{a, c, e\}$, $C_2 = \{b, c\}$, $C_3 = \{a, b, d\}$, $C_4 = \{d, e, f\}$, $C_5 = \{e, f\}$. Each committee must send a different representative to the Annual Congress of Committees, and C_1 wishes to nominate e, C_2 wishes to nominate b, C_3 wishes to nominate a, and C_4 wishes to nominate f.

(i) Show that it is not possible to respect the wishes of C_1, C_2, C_3, and C_4.
(ii) Use the alternating path method and the associated graph to find a complete system of distinct representatives.
(iii) Is it possible to find a complete system of distinct representatives if committee C_1 refuses to change its nomination?

7 A bipartite graph $G = (V \cup W, E)$ may be represented by an $m \times n$ matrix $B = (b_{ij})$ where $m = |V|$, $n = |W|$ and

$$b_{ij} = \begin{cases} 1 & \text{if } \{v_i, w_j\} \in E; \\ 0 & \text{if not.} \end{cases}$$

Describe the alternating path algorithm for finding a maximum matching in G, in terms of operations on B.

8 Let us say that a *step* of the alternating path algorithm for finding a maximum matching comprises the operations required to augment the partial matching by one edge. Show that if $G = (V \cup W, E)$ satisfies $\max(|V|, |W|) = n$, then the number of operations required for each step is $O(n^2)$. Deduce that the efficiency of the algorithm is $O(n^3)$.

9 Prove that the complete graph K_{2m} has an edge-colouring with $2m-1$ colours.

10   In 'projective geometry' two triangles $A_1B_1C_1$ and $A_2B_2C_2$ are said to be **in perspective** if A_1A_2, B_1B_2, C_1C_2 have a common point X. Desargues's theorem asserts that if the triangles are in perspective then the points P (intersection of A_1B_1 and A_2B_2), Q (intersection of B_1C_1 and B_2C_2), and R (intersection of A_1C_1 and A_2C_2) are collinear.

Let G be the bipartite graph whose vertices represent the ten points and ten lines occurring in the theorem, two vertices being adjacent if and only if they represent a point and a line containing that point. Show that G is Hamiltonian and construct an edge-colouring of G using only three colours.

11   Let B denote the graph used in the proof of Theorem 10.3.1 to show that any $r \times n$ latin rectangle can be completed. Show that if $r = n - 2$ then the rectangle can be completed in an essentially unique way if and only if B is connected.

12   Let p, q and n be integers such that $1 \le p \le n$ and $1 \le q \le n$, and let $s_1, s_2, \ldots, s_n$ be n symbols. Let M be an $n \times n$ array of cells such that q cells in each of the first p rows are occupied by a symbol, and the remaining $n^2 - pq$ cells are empty. Suppose also that no symbol occurs more than once in any row or column. Show that each of the remaining cells can be assigned a symbol s_i ($1 \le i \le n$) in such a way as to complete a latin square only if the number $N(i)$ of occupied cells in column i satisfies

$$N(i) \ge p + q - n \qquad (1 \le i \le n).$$

13   Show that the following infinite family of sets satisfies Hall's condition but does not have a transversal.

$$X_0 = \{1, 2, 3, \ldots\}, \qquad X_1 = \{1\}, \qquad X_2 = \{1, 2\}, \ldots$$
$$\ldots \quad X_i = \{1, 2, \ldots, i\}, \ldots \quad .$$

14   Let X be the union of the family of sets $X_1, X_2, \ldots, X_n$, and suppose the family has a transversal. Show that there is a unique transversal if and only if $|X| = n$.

15   Suppose that we are given two partitions of a set X:

$$X = A_1 \cup A_2 \cup \ldots \cup A_n = B_1 \cup B_2 \cup \ldots \cup B_n.$$

A **simultaneous transversal** is a set $\{x_1, x_2, \ldots, x_n\}$ of distinct elements of X such that each part of either partition contains one x_i. Prove that there is a simultaneous transversal if and only if no k of the parts A_i are contained in fewer than k of the parts B_j ($1 \le k \le n - 1$).

16   Let m and n be integers such that $m \ge n$. Construct an explicit edge-colouring of $K_{m,n}$ which uses m colours.

17   Show that if a regular graph with valency k has an odd number of vertices, then it cannot have an edge-colouring with k colours.

18   Suppose that G is a graph with n vertices, m edges, and maximum valency $\hat{k}$. Show that if $m > \hat{k} \lfloor n/2 \rfloor$ then G does not have an edge-colouring with $\hat{k}$ colours.

19   A **vertex-cover** in a graph G is a set of vertices C such that each edge of G contains at least one vertex in C. Show that if G is bipartite then the size of a maximum matching is equal to the size of a minimum vertex-cover.

20   Show that there are at least $(n-r)!$ ways of adding one more row to an $r \times n$ latin rectangle, in such a way that the latin property is preserved $(1 \leqslant r \leqslant n-1)$.

11
Digraphs, networks, and flows

11.1   Digraphs

A **digraph** (or **directed graph**) consists of a finite set V, whose members are called **vertices**, and a subset A of $V \times V$, whose members are called **arcs**. We shall use the notation $D = (V, A)$ for the digraph D defined in this way. Digraphs can be represented by pictures in a manner similar to that used for graphs: the only difference is that an arc is an ordered pair (v, w), whereas an edge of a graph is an unordered pair $\{v, w\}$. In the picture of a digraph, we indicate the order of the vertices v and w by putting an arrow pointing from v to w on the line representing the arc (v, w) (see Fig. 11.1). If (v, v) is an arc, we indicate this by drawing a loop at v (the direction of the arrow is immaterial). If (v, w) and (w, v) are both arcs, we draw two lines joining v and w, with arrows in the appropriate directions.

Fig. 11.1
Two digraphs.

Formally, a digraph is just another way of describing a relation R between members of the *same* set. Instead of saying that v is related to w by the relation R, we can say that (v, w) is an arc of the digraph whose set of arcs is R. (This should be distinguished from the bipartite graph representation introduced in Section 10.1, which is relevant when R is a relation between two *disjoint* sets.) The properties of relations can be easily translated into properties of digraphs. For instance, if a relation is symmetric, the corresponding digraph has the property that its arcs (except for loops) occur in pairs; either both (v, w) and (w, v) are arcs, or neither of them is an arc.

The definitions of walk, path, and cycle carry over from graphs to digraphs in a fairly obvious way. Thus, a **directed walk** in a digraph $D = (V, A)$ is a sequence of vertices $v_1, v_2, \ldots, v_k$, with the property that (v_i, v_{i+1}) is in A for $1 \leq i \leq k - 1$. A directed walk is a **directed path** if all its vertices are distinct, and it is a **directed cycle** if all its vertices are distinct except that $v_1 = v_k$.

A simple illustration of some of these ideas occurs in the analysis of a 'round-robin tournament'. This is a competition in which each competitor plays against all the others, and each match has a definite result: either x beats y or y beats x. In the corresponding digraph, the vertices represent the competitors and, for each pair of distinct vertices x and y, there is precisely one of the arcs (x, y), (y, x), according as x beats y or y beats x. Such a digraph may be thought of as a complete graph, in which each edge has been converted into an arc by the insertion of an arrow. Because of the interpretation in terms of a round-robin tournament, any digraph of this kind is known as a **tournament**.

A round-robin tournament is great fun, but it is not particularly useful for deciding the relative merits of the players. It frequently happens that although x beats y and y beats z, we find that z has beaten x. So we have a directed cycle (x, y, z) of three players, each of whom beats, and is beaten by, one of the others.

In view of this, the following result may appear rather surprising at first sight.

Theorem 11.1　In any tournament there is a directed path containing all the vertices.

Proof　We shall show that any directed path $y_1, y_2, \ldots, y_l$ which does not contain all the vertices can be extended by the addition of an extra vertex. Thus we can begin with any arc (y_1, y_2) and successively extend until we have a directed path containing all the vertices.

Let z be any vertex not on the directed path $y_1, y_2, \ldots, y_l$. If (z, y_1) is an arc, insert z at the beginning. If (z, y_1) is not an arc then, since we have a tournament, (y_1, z) is an arc. Let r be the greatest integer for which $(y_1, z), (y_2, z), \ldots, (y_r, z)$ are arcs. If $r < l$, we have arcs (y_r, z) and (z, y_{r+1}), and so we can insert z between y_r and y_{r+1}. If $r = l$, we can insert z at the end.　□

The directed path guaranteed by Theorem 11.1 has the desirable property that it arranges the 'competitors' in a sequence so that y_1

beats y_2, y_2 beats y_3, and so on. Unfortunately, this does not mean that each competitor beats all those that follow him in the sequence, since (for example) y_1 may lose to y_3.

Exercises 11.1

1 In the adjacency list for a *digraph* we put y in column x whenever (x, y) is an arc. Sketch the digraph whose adjacency list is

a	*b*	*c*	*d*	*e*	*f*
d	*a*	*b*	*b*	*f*	*a*
e			*c*		
			e		

.

Find a directed path from c to f, and a directed cycle starting and ending at d.

2 In Table 11.1.1, there is a + in row i and column j if i beats j, and a − if j beats i. Find a directed path containing all the vertices of the tournament.

Table 11.1.1

	j								
i	1	2	3	4	5	6	7	8	9
1		+	+	−	−	+	−	+	−
2			−	+	+	−	+	−	−
3				+	−	−	+	−	+
4					+	−	−	+	−
5						−	+	−	−
6							+	−	−
7								+	+
8									−

3 Write a program for finding a directed path through all the vertices in a tournament, using the method in the proof of Theorem 11.1.

4 The **out-degree** $\vec{\delta}(v)$ of a vertex in a digraph is the number of arcs of the form (v, w), and the **in-degree** $\overleftarrow{\delta}(v)$ is the number of arcs

of the form (x, v). Show that in general

$$\sum_{v \in V} \bar{\delta}(v) = \sum_{v \in V} \vec{\delta}(v),$$

and, if the digraph is a tournament, that

$$\sum_{v \in V} (\bar{\delta}(v))^2 = \sum_{v \in V} (\vec{\delta}(v))^2.$$

11.2 Networks and critical paths

In many practical situations it is appropriate to use a digraph, rather than a graph, as a model. This will be true, for example, when each arc represents a one-way street, or some other kind of link for which movement in only one direction is possible. Often the model will be such that the arcs of the digraph carry numbers representing costs or distances. With this kind of picture in mind we shall use the term **network** for a digraph $D = (V, A)$ together with a weight function $w : A \rightarrow \mathbb{N}$. The reason for restricting the values of w to be positive integers is to avoid complications about the existence of optimal solutions. In practice it is easy to circumvent any difficulties caused by this restriction.

A typical example of a network arising from a practical problem is the so-called *activity network*. A large construction project is usually broken down into a number of smaller activities, and these activities are related, in that some of them cannot start until others have been completed. For instance, building a house involves laying foundations, bricklaying, carpentry, roofing, electrical work and so forth, and an activity like electrical work cannot begin until some of the other activities have been completed. In planning such a project it is customary to use a network in which the arcs represent activities and the vertices represent 'events', each event being the completion of some activities. The weight of an arc represents the time required for that activity, and the problem is to schedule the activities so that the total time required for the whole project is as small as possible.

Example The table lists the activities $\alpha_1, \alpha_2, \alpha_3, \alpha_4, \alpha_5, \alpha_6, \alpha_7, \alpha_8$ involved in a project, with the time (in days) needed for each one, and the activities which must be completed before it can be started. What is the least number of days needed for the whole project?

Activity	α_1	α_2	α_3	α_4	α_5	α_6	α_7	α_8
Time needed	4	3	7	4	6	5	2	5
Prerequisites	–	–	α_1	α_1	α_2	α_4	α_3	α_4
						α_5	α_6	α_5

Solution   The first step is to construct the activity network, as in Fig. 11.2. The vertex s represents the start of the whole project and t represents its completion, while q represents the completion of activities α_4 and α_5, and so on. The full list of activities and their representative arcs is as follows:

$$\text{Activity: } \alpha_1 \quad \alpha_2 \quad \alpha_3 \quad \alpha_4 \quad \alpha_5 \quad \alpha_6 \quad \alpha_7 \quad \alpha_8$$
$$\text{Arc: } (s, r) \ (s, p) \ (r, z) \ (r, q) \ (p, q) \ (q, z) \ (z, t) \ (q, t)$$

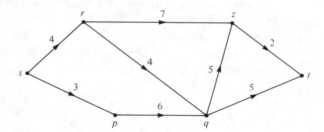

Fig. 11.2
An activity network.

For each vertex v we shall calculate $E(v)$, the earliest time for the corresponding event: that is, $E(v)$ is the earliest completion time for all the activities represented by arcs directed towards v. Initially, we set $E(s) = 0$. Clearly, $E(p) = 3$ since the only activity involved is $\alpha_2 = (s, p)$, which takes 3 days. Similarly $E(r) = 4$. At q, the activities (p, q) and (r, q) must both be completed. Now $E(p) = 3$ and (p, q) takes 6 days, while $E(r) = 4$ and (r, q) takes 4 days, so that

$$E(q) = \max{(3+6, 4+4)} = 9.$$

(Recall that *both* activities must be completed, so we must allow the *larger* number of days.) Continuing in this way, we obtain the following values of E:

$$v: s \quad p \quad q \quad r \quad z \quad t$$
$$E(v): 0 \quad 3 \quad 9 \quad 4 \quad 14 \quad 16.$$

Hence the total time needed for the whole project is at least 16 days. $\qquad\square$

The example deals with a special case of the problem of finding the *longest* path in a network. In general, we would use a version of the BFS strategy (appropriately modified for dealing with digraphs), together with the recursive calculation of the 'earliest time' function according to the rules

$$E(s) = 0, \qquad E(v) = \max_{u} \{E(u) + w(u, v)\},$$

where the maximum is taken over all vertices u for which (u, v) is an arc.

This method of studying an activity network is part of a technique called **critical path analysis**. The rest of the technique proceeds as follows. For each vertex v we calculate $L(v)$, the latest time by which all activities (v, x) must be started if the whole project is to be completed on time. This is done by means of the 'backwards' recursion

$$L(t) = E(t), \qquad L(v) = \min_{x} (L(x) - w(v, x)),$$

where the minimum is taken over all the vertices x for which (v, x) is an arc. Now consider what we know about a typical activity (y, z):

(i) it cannot start before time $E(y)$, at the earliest;
(ii) it must finish by time $L(z)$, at the latest;
(iii) it takes time $w(y, z)$.

It follows that if we define the **float time** $F(y, z)$ to be

$$F(y, z) = L(z) - E(y) - w(y, z),$$

then (y, z) can start at any time after $E(y)$ and before $E(y) + F(y, z)$ without delaying the project. An activity (y, z) for which the float time is zero is said to be **critical**: it must be started at the earliest possible time $E(y)$ if the project is to finish on time. In any activity network there will be at least one directed path from s to t consisting entirely of critical activities, and this is called a **critical path**.

Exercises 11.2

1   Calculate the latest times $L(v)$ and the float times $F(u, v)$ for the project described in the *Example* above. Find a critical path, and draw up a schedule for the project showing the alternative start times for the activities which are not critical.

2   Carry out the complete critical path analysis for the project described below.

Activity	α_1	α_2	α_3	α_4	α_5	α_6	α_7	α_8	α_9	α_{10}	α_{11}
Time needed	6	2	10	1	4	2	7	7	9	2	4
Prerequisites	–	–	α_1	α_1	α_1	α_5	α_2	α_3	α_2	α_7	α_8
							α_4	α_6	α_4		α_{10}

11.3   Flows and cuts

In the next three sections we shall think of the arcs of a network as 'pipelines', along which some commodity can flow. The numerical weights assigned to the arcs will be regarded as capacities, giving practical limits to the amounts which can flow along the arcs. In addition, there will always be one vertex s with the property that all arcs containing s are directed away from s, and one vertex t with the property that all arcs containing t are directed towards t. The vertices s and t are known as the **source** and **sink** respectively. In summary then, we shall be dealing with networks comprising

(i) a digraph $D = (V, A)$;
(ii) a capacity function $c: A \to \mathbb{N}$;
(iii) a source s and sink t.

An example is illustrated in Fig. 11.3.

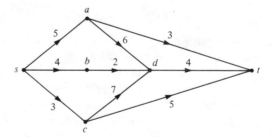

Fig. 11.3
A network, showing the capacity of each arc.

Suppose a commodity is flowing along the arcs of a network, and let $f(x, y)$ denote the amount which flows along the arc (x, y). We shall insist that, except at s and t, the amount of flow which arrives at a vertex v must equal the amount which leaves v. Thus, we define the **inflow** and **outflow** at v to be

$$\text{inflow}\,(v) = \sum_{(x,v)\in A} f(x, v), \qquad \text{outflow}\,(v) = \sum_{(v,y)\in A} f(v, y),$$

and require that the two quantities are equal except when $v = s$ or $v = t$. This is the 'conservation rule' for network flows. We shall also require the 'feasibility rule' that no arc carries flow exceeding its capacity. For convenience, we shall incorporate these two rules into our definition of a flow.

Definition   A **flow** from the source s to the sink t in a network is a function which assigns a non-negative number $f(x, y)$ to each arc (x, y), subject to the rules

$$conservation: \quad \text{inflow } (v) = \text{outflow } (v) \qquad (v \neq s, t);$$
$$feasibility: \qquad f(x, y) \leqslant c(x, y) \qquad ((x, y) \in A).$$

The reader should check that the function defined by the following table is a flow in the network of Fig. 11.3.

(x, y):	(s, a)	(s, b)	(s, c)	(a, d)	(b, d)	(c, d)	(a, t)	(c, t)	(d, t)
$f(x, y)$:	3	2	3	1	2	1	2	2	4

Since nothing is allowed to accumulate at intermediate vertices, it is clear that the total amount flowing out of s must equal the total amount flowing into t. (A formal proof is outlined in Ex. 11.3.3.) In other words,

$$\text{outflow } (s) = \text{inflow } (t)$$

for any flow from s to t. The common value of these two quantities measures the total amount flowing across the network, and is called the **value** of the flow, written val (f). In the example given above

$$\text{outflow } (s) = f(s, a) + f(s, b) + f(s, c) = 3 + 2 + 3 = 8$$
$$\text{inflow } (t) = f(a, t) + f(d, t) + f(c, t) = 2 + 4 + 2 = 8,$$

and so val $(f) = 8$.

Our problem is to calculate the maximum value of a flow for a given network. We begin by finding an upper bound for this value in terms of the capacities. In Fig. 11.3, we notice that the total capacity of the arcs directed away from s is $5 + 4 + 3 = 12$, so clearly no flow can have value exceeding 12. More generally, suppose we partition the vertex-set in two parts, one part S containing s, and the other part T containing t. Then the net flow from S to T is, by the conservation rule, the same as the flow from s to t, which is the value of f. That is

$$\text{val } (f) = \sum_{\substack{x \in S \\ y \in T}} f(x, y) - \sum_{\substack{u \in T \\ v \in S}} f(u, v).$$

The first sum measures the total flow from S to T, and the second sum measures the total flow in the reverse direction. Since each term in the second sum is non-negative, we certainly have

$$\text{val}\,(f) \leqslant \sum_{\substack{x \in S \\ y \in T}} f(x, y).$$

Furthermore, $f(x, y) \leqslant c(x, y)$ for all arcs, and so we may conclude that the sum

$$\sum_{\substack{x \in S \\ y \in T}} c(x, y)$$

is an upper bound for the value of any flow. For example, in Fig. 11.3, if we take $S = \{s, b\}$, $T = \{a, c, d, t\}$ then the arcs directed from S to T are (s, a), (s, c) and (b, d), with a total capacity of 10, and we deduce that val $(f) \leqslant 10$ for any flow in that network.

Formally, we define (S, T) to be a **cut** (separating s and t) if $S \cup T$ is a partition of the vertex-set such that s is in S and t is in T. The **capacity** of the cut is

$$\text{cap}\,(S, T) = \sum_{\substack{s \in S \\ y \in T}} c(x, y),$$

and, with this terminology, we have established the following result.

Theorem 11.3 Let s and t be the source and sink of a network. If f is any flow from s to t and (S, T) is any cut, then

$$\text{val}\,(f) \leqslant \text{cap}\,(S, T). \qquad \square$$

Suppose f_0 is a flow with the maximum possible value, and (S_0, T_0) is a cut with the minimum possible capacity. Theorem 11.3 tells us that val $(f_0) \leqslant \text{cap}\,(S_0, T_0)$ or, more expressively,

$$\text{max-flow} \leqslant \text{min-cut}.$$

In the next section we shall prove the fundamental 'max-flow min-cut theorem', which says that the two expressions are in fact equal.

Exercises 11.3 1 Find a flow f^* in Fig. 11.3 for which val $(f^*) = 10$. Why is this the maximum possible value?

2 Sketch the network whose vertices are s, a, b, c, d, t, and whose

arcs and capacities are

$$(x, y): \quad (s, a) \quad (s, b) \quad (a, b) \quad (a, c) \quad (b, d) \quad (d, c) \quad (c, t) \quad (d, t)$$
$$c(x, y): \quad 5 \quad\quad 3 \quad\quad 3 \quad\quad 3 \quad\quad 5 \quad\quad 2. \quad\quad 6 \quad\quad 2$$

Find a flow with value 7 and a cut with capacity 7. What is the value of a maximum flow, and why?

3   Let $D = (V, A)$ be a digraph, and suppose $\phi : A \to \mathbb{N}$ is a function, not necessarily a flow. If the outflow and inflow for ϕ are defined as above, show that

$$\sum_{v \in V} \text{outflow}(v) = \sum_{v \in V} \text{inflow}(v).$$

Deduce that if s and t are the source and sink of a network, and ϕ is a flow, then

$$\text{outflow}(s) = \text{inflow}(t).$$

11.4   The max-flow min-cut theorem

In this section we shall describe a method for increasing the value of a given flow, provided that the flow does not have the maximum possible value. This method is not only the basis for a practical algorithm (Section 11.5), but also leads to a proof of the fundamental max-flow min-cut theorem.

Let us look again at the network of Fig. 11.3 and the flow f with value 8 specified in the previous section. In Fig. 11.4 the values of f are encircled to distinguish them from the capacities. Consider the directed path s, a, t. Neither the arc (s, a) nor (a, t) is carrying flow to its full capacity, and so we can increase the flow on both arcs, until the capacity of one of them is reached. If we define $f_1(s, a) = 4$ and $f_1(a, t) = 3$ then (a, t) is now saturated; furthermore, since the flow on both arcs has been increased by the same amount, the

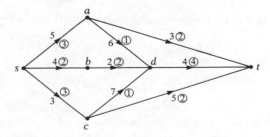

Fig. 11.4
A flow with value 8.

conservation rule still holds at the vertex a. Defining $f_1(x, y) = f(x, y)$ on all the remaining arcs, we have a new flow,

(x, y):	(s, a)	(s, b)	(s, c)	(a, d)	(b, d)	(c, d)	(a, t)	(c, t)	(d, t)
$f_1(x, y)$:	4	2	3	1	2	1	3	2	4

with val $(f_1) = $ val $(f) + 1 = 9$.

In order to make a further improvement, we have to use a more cunning approach, involving a 'path' like s, a, d, c, t. Unfortunately, we have not defined the meaning of a 'path' in a network, only a 'directed path', and this is certainly not a directed path, since the arc (c, d) is in the wrong direction. Strictly speaking therefore, we should refer to s, a, d, c, t as a path in the *graph* obtained by ignoring the directions on the arcs of the digraph and treating them as edges. However, we shall simply use the word 'path', without the cumbersome qualifications. The capacities and f_1-values on the path are

$$s \xrightarrow[\textcircled{4}]{5} a \xrightarrow[\textcircled{1}]{6} d \xleftarrow[\textcircled{1}]{7} c \xrightarrow[\textcircled{2}]{5} t.$$

Because the flow on (c, d) is contrary to the direction of the path we can reduce the flow on (c, d) by 1, and increase the flow on the other arcs by the same amount, without violating the conservation rule. In this way, we obtain a new flow f_2, whose values on the path are

$$s \xrightarrow[\textcircled{5}]{} a \xrightarrow[\textcircled{2}]{} d \xleftarrow[\textcircled{0}]{} c \xrightarrow[\textcircled{3}]{} t,$$

and whose values on the other arcs are the same as for f_1. Notice that we cannot make any greater change on this path, since (s, a) is now saturated, and the flow on (c, d) cannot be less than zero.

Now val $(f_2) = $ val $(f_1) + 1 = 10$. In the previous section we found a cut with capacity 10, and by Theorem 11.3 we know that no flow can have a value greater than this. Hence f_2 is a maximum flow.

The paths s, a, t and s, a, d, c, t, which we used to augment the flows f and f_1 respectively, are examples of what is called a 'flow-augmenting path'. Generally, if a flow f is given, a path

$$s = x_1, x_2, \ldots, x_{k-1}, x_k = t$$

is called an **f-augmenting path** provided that

$$f(x_i, x_{i+1}) < c(x_i, x_{i+1}) \quad \text{and} \quad (x_i, x_{i+1}) \in A,$$

or

$$f(x_{i+1}, x_i) > 0 \quad \text{and} \quad (x_{i+1}, x_i) \in A,$$

for $1 \le i \le k - 1$. In other words, the 'forward' arcs are not used to their full capacity, whereas the 'backward' arcs are carrying some 'contra-flow'. Given such a path, we can increase the flow on the forward arcs and decrease the flow on the backward arcs by the same amount, without violating the conservation rule. The greatest change which can be obtained in this way, without overloading the forward arcs or making the flow on the backward arcs negative, is the minimum in the range $1 \leq i \leq k-1$ of the quantities

$$c(x_i, x_{i+1}) - f(x_i, x_{i+1}) \quad \text{if} \quad (x_i, x_{i+1}) \in A,$$
$$f(x_{i+1}, x_i) \quad\quad\quad\quad \text{if} \quad (x_{i+1}, x_i) \in A.$$

If this quantity is denoted by α, then we add α to the flow on forward arcs and subtract it from the backward arcs, obtaining a new flow f^* with val $(f^*) =$ val $(f) + \alpha$.

The existence of an f-augmenting path from s to t enables us to find a new flow f^* with val $(f^*) >$ val (f). On the other hand, we know from Theorem 11.3 that the value of any flow cannot exceed the capacity of any cut. In the next theorem these two ideas are combined. As part of the proof we shall need to consider *incomplete* f-augmenting paths, which satisfy the conditions for an f-augmenting path except that the final vertex is not t.

Theorem 11.4 The maximum value of a flow from s to t in a network is equal to the minimum capacity of a cut separating s and t.

Proof Let f be a maximum flow. Define S to be the set of vertices x for which there is an incomplete f-augmenting path from s to x, and let T be the complementary set of vertices. The sink t must be in T, otherwise we should have an f-augmenting path from s to t, and f could be augmented, contrary to the hypothesis that it is a maximum flow. Hence (S, T) is a cut separating s and t.

We shall prove that cap $(S, T) =$ val (f). Let (x, y) be any arc with $x \in S$ and $y \in T$. By definition of S there is an incomplete f-augmenting path from s to x, and if $f(x, y) < c(x, y)$ we could extent it to y, contradicting the fact that y is in T. Hence $f(x, y) = c(x, y)$. Similarly, given an arc (u, v) with $u \in T$ and $v \in S$, there is an incomplete f-augmenting path from s to v, and if $f(u, v) > 0$ we could extend it to u, contradicting the fact that u is in T. Hence $f(u, v) = 0$. Thus

$$\text{val } (f) = \sum_{\substack{x \in S \\ y \in T}} f(x, y) - \sum_{\substack{u \in T \\ v \in S}} f(u, v) = \sum_{\substack{x \in S \\ y \in T}} c(x, y)$$
$$= \text{cap } (S, T).$$

Suppose (S', T') is any other cut. By Theorem 11.3 and the result just proved we have

$$\text{cap}\,(S', T') \geqslant \text{val}\,(f) = \text{cap}\,(S, T).$$

It follows that (S, T) is a minimum cut, as required. □

Exercises 11.4

1 The diagram (Fig. 11.5) represents a network, and the numbers on the arcs are their capacities. A flow f is defined as follows.

(x, y):	(s, a)	(s, b)	(s, c)	(a, b)	(a, d)	(b, c)	(b, d)	(b, e)	(c, e)	(d, t)	(e, t)
$f(x, y)$:	5	6	0	0	5	1	2	3	1	7	4

(i) What is the value of f?
(ii) Find an f-augmenting path and compute the value of the augmented flow.
(iii) Find a cut with capacity 12.
(iv) What can you deduce?

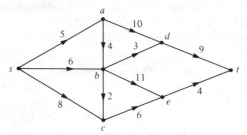

Fig. 11.5
A network, showing the capacity of each arc.

2 The vertex-set of a network is $\{s, a, b, c, d, e, t\}$, and the capacity of an arc (x, y), if there is such an arc, is given in row x and column y of Table 11.4.1. Find a maximum flow from s to t in the network and explain why it is a maximum.

Table 11.4.1

	s	a	b	c	d	e	t
s	—	14	—	14	—	12	—
a	—	—	8	14	—	3	—
b	—	—	—	—	—	2	15
c	—	—	—	—	19	—	—
d	—	—	6	—	—	1	10
e	—	—	—	—	—	—	14

11.5 The labelling algorithm for network flows

The foregoing theory points the way to a practical algorithm for the maximum flow problem. Roughly speaking, the strategy is as follows.

(A) Start with any flow (assigning zero to each arc will do).

(B) Use BFS to construct a tree of incomplete f-augmenting paths rooted at s.

(C1) If the tree reaches t, augment f accordingly and return to (B) with the new flow.

(C2) If the tree does not reach t, let S be the set of vertices it does reach, and let T be the complement of S. The flow f is a maximum flow and (S, T) is a minimum cut.

Example Find a maximum flow in the network illustrated in Fig. 11.6, starting with the flow f_1 which is equal to 4 on (s, b) and (b, t), and zero on the other arcs.

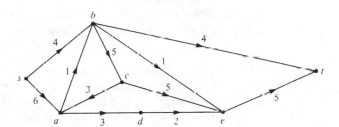

Fig. 11.6
The network discussed in the *Example*.

Solution We regard the network as a graph, by ignoring the direction of the arcs, and use BFS to construct a tree rooted at s. The edges of the tree must form (incomplete) f_1-augmenting paths, and when an edge xy is added to the tree, the new vertex y receives a label $l(y)$ equal to the greatest possible augmentation on the path from s to y.

For the purposes of exposition we shall record (see Table 11.5.1) the process in terms of the queue Q of vertices, formed according to BFS rules (Section 9.5). At each stage we 'scan' at x, the first vertex of Q, to see if there are any edges xy which extend the f_1-augmenting path from s to x; if there is a choice, we shall use alphabetical order to decide which edge is added first.

Table 11.5.1

Step	Queue Q	Scan at x	Arrival y	Label $l(y)$	Departure
1	s	s	a	6	—
2	sa	s	—	—	s
3	a	a	b	1	—
4	ab	a	d	3	—
5	abd	a	—	—	a
6	bd	b	c	1	—
7	bdc	b	e	1	—
8	$bdce$	b	—	—	b
9	dce	d	—	—	d
10	ce	c	—	—	c
11	e	e	t	1	—

Comments

Step 2: sb cannot be added since the arc (s, b) is saturated.

Step 3: ab can be added, $l(b) = c(a, b) - f_1(a, b) = 1$.

Step 4: ac cannot be added as a backward arc (c, a),
since $f_1(c, a) = 0$.

Step 9: de cannot be added, as e is already on the tree.

Step 11: t has been reached, so we stop.

Since $f(t) = 1$ we can augment f_1 by 1 on the arcs (s, a), (a, b),
(b, e), (e, t). For examples like this one, it is easy to mark the tree
edges and the labels on a diagram of the network. This is done in
Fig. 11.7, and the augmented flow f_2 is indicated below.

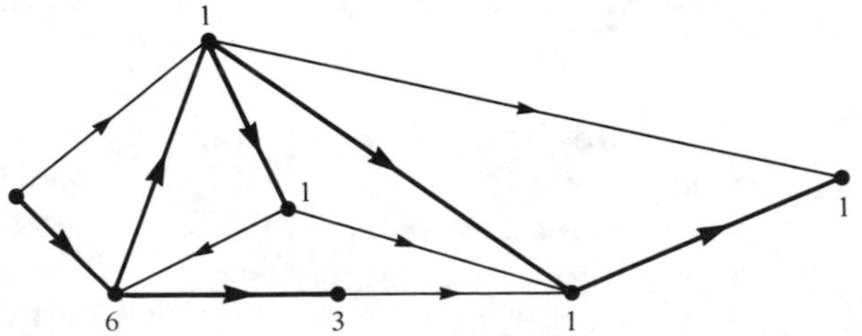

Fig. 11.7
First iteration of the algorithm.

(x, y):	(s, a)	(s, b)	(a, b)	(a, d)	(b, c)	(b, e)	(b, t)	(c, a)	(c, e)	(d, e)	(e, t)
$f_2(x, y)$:	1	4	1	0	0	1	4	0	0	0	1

Starting from f_2 and repeating the procedure we obtain the tree
and labels as shown in Fig. 11.8, and the augmented flow f_3.

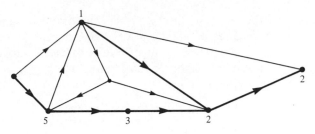

(x, y):	(s, a)	(s, b)	(a, b)	(a, d)	(b, c)	(b, e)	(b, t)	(c, a)	(c, e)	(d, e)	(e, t)
$f_2(x, y)$:	3	4	1	2	0	1	4	0	0	2	3

Finally, starting from f_3 and repeating the procedure yields a tree
which does not reach t (Fig. 11.9).

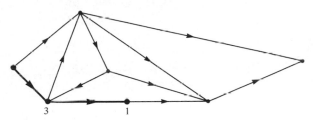

We conclude that f_3 is a maximum flow. Its value is 7 and the fact
that it is maximal can be verified by checking the cut (S, T), where S
is the set of vertices reached by the final tree. In this case $S =$
$\{s, a, d\}$, and

$$\text{cap } (S, T) = c(s, b) + c(a, b) + c(d, e) = 4 + 1 + 2 = 7,$$

so (S, T) is a minimum cut and f_3 is a maximum flow. □

The formal description of the labelling algorithm follows the lines
of the procedure outlined above, but there are some adjustments to
meet the requirements of dumb computers.

Let $F(x, y)$ stand for the statement '(x, y) is a forward arc and
$f(x, y) < c(x, y)$', and let $B(y, x)$ stand for the statement '(y, x) is a
backward arc and $f(y, x) > 0$'. When the edge xy is added to the
tree, the vertex y receives a double label $(l_1(y), l_2(y))$. The first label
records how y is reached, and the second is the $l(y)$ used in the
example above. Explicitly, if $F(x, y)$ holds we define

$$l_1(y) = x^+, \qquad l_2(y) = \min \{l_2(x), \ c(x, y) - f(x, y)\},$$

while if $B(y, x)$ holds we define

$$l_1(y) = x^-, \qquad l_2(y) = \min \{l_2(x), \ f(y, x)\}.$$

Initially, we take f to be the flow which is zero on all arcs, and the label $l_2(s)$ to be l_{big} (which, in practice, could be the sum of the capacities of all the arcs). The program set out hereafter indicates how the labelling algorithm is grafted onto a BFS procedure (enclosed in the box); it will stop when f_{max} is equal to the value of a maximum flow.

Labelling algorithm for maximum flow

$f_{max} \leftarrow l_{big}$

$f(v, w) \leftarrow 0$ for all $(v, w) \in A;$ val $(f) \leftarrow 0$

while   val $(f) \neq f_{max}$ **do**

     $Q \leftarrow (s);$ | $l_2(s) \leftarrow l_{big}$

     while $Q \neq \varnothing$ **do**

         $x \leftarrow first\,(Q)$

         if x is adjacent to a new vertex y

             then | **if** $F(x, y)$ or $B(x, y)$

                     then   label y

                 $Q \leftarrow (Q$ followed by $y)$

                 if $y = t$

                     then   augment f

                         val $(f) \leftarrow$ val $(f) + l_2(t)$

                         $Q \leftarrow \varnothing$

             else $Q \leftarrow (Q$ with x deleted$)$

         if $Q = \varnothing$

             then $f_{max} \leftarrow$ val (f)

Exercises 11.5

1   Starting from the zero flow, use the labelling algorithm (by hand!) to find the maximum flow in the network illustrated in Fig. 11.10.

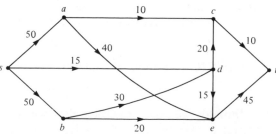

Fig. 11.10
Find the maximum flow.

2   A commodity produced at s_1 and s_2 is transported through the network illustrated in Fig. 11.11 to markets at t_1, t_2, t_3.

(i) Add a new 'supersource' and 'supersink' to make a network of the standard kind.
(ii) Find a 'good' initial flow by inspection.
(iii) Use the labelling algorithm to find a maximum flow.

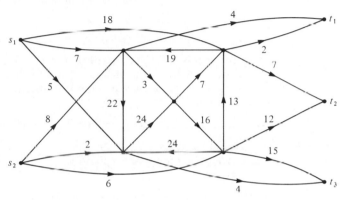

Fig. 11.11
Illustration for Ex. 11.5.2.

11.6   Miscellaneous Exercises

1   Find a directed path containing all the vertices of the tournament described by Table 11.6.1. (The notation is as in Ex. 11.1.2.)

Table 11.6.1

i	1	2	3	4	5	6	7	8
1		+	−	+	−	−	−	+
2			+	−	−	+	−	+
3				−	−	+	+	+
4					+	+	+	+
5						−	−	−
6							+	−
7								+

2 Define the **score** of a vertex in a tournament to be its out-degree, and the **score-sequence** of the tournament to be the sequence of scores arranged in non-decreasing order. Verify that the score sequence of the tournament described in Ex. 11.1.2. is $(6, 5, 4, 4, 4, 3, 3, 3)$, and show that in general if the score-sequence is $(s_1, s_2, \ldots, s_n)$ then

$$\sum_{i=1}^{n} s_i = \tfrac{1}{2}n(n-1).$$

3 Show that the score sequence of a tournament with n vertices satisfies

 (i) $s_1 + s_2 + \cdots + s_k \geqslant \tfrac{1}{2}k(k-1)$ $(1 \leqslant k \leqslant n-1)$;

 (ii) $\tfrac{1}{2}(k-1) \leqslant s_k \leqslant \tfrac{1}{2}(n+k-2)$ $(1 \leqslant k \leqslant n)$.

4 A tournament is said to be **transitive** if it represents a transitive relation; that is, if the existence of the arcs (x, y) and (y, z) implies the existence of the arc (x, z). Show that a tournament is transitive if and only if the kth term of its score sequence is $k-1$.

5 What is the number of tournaments with n vertices? How many of them are transitive?

6 Show that in any tournament there is a vertex s such that any vertex x can be reached from s by a directed path of length 0, 1, or 2.

7 Carry out the complete critical path analysis for the project described below.

Activity	α_1	α_2	α_3	α_4	α_5	α_6	α_7	α_8	α_9
Time needed	3	5	4	4	4	4	2	4	7
Prerequisites	—	—	—	α_1	α_3	α_1	α_2	α_6	α_3
							α_4	α_7	

8 Modify the values of the times needed by the activities in Ex. 7 so that there is more than one critical path. Show that, in general, any activity whose float time is zero is on some critical path.

9 Consider the network with vertex-set $\{s, a, b, c, t\}$ and arcs and capacities given by the following table.

(s, a)	(s, b)	(a, b)	(a, c)	(a, t)	(b, c)	(c, t)
5	2	3	1	3	3	4

Calculate the capacities of all cuts separating s and t, and find a maximum flow from s to t.

10 Use the labelling algorithm to verify that if the capacities of the arcs of a network are integers, then there is a maximum flow such that the flow on each arc is an integer.

11 Show that the maximum flow in Fig. 11.11 is not uniquely determined, and hence give an example of a maximum flow for which the flow on each arc is not an integer.

12   Let $(S_1, \bar{S}_1)$ and $(S_2, \bar{S}_2)$ be cuts separating s and t in a network, both of which are minimum cuts. Show that $(S_1 \cap S_2, \bar{S}_1 \cap \bar{S}_2)$ is also a minimum cut.

13   Write down the general form of the procedure for introducing a 'supersource' and a 'supersink' into a network with several sources and sinks, as exemplified in Ex. 11.5.2. State and prove the corresponding version of the max-flow min-cut theorem.

14   Consider the network with vertex-set $\{s, a, b, c, d, e, f, t\}$ and arcs and capacities given by the following table.

Arc:	(s, a)	(s, d)	(a, b)	(b, c)	(b, e)	(c, t)	(d, e)	(e, f)	(f, t)
Capacity:	n	n	n	n	1	n	n	n	n

Show that if we start with the zero flow and make appropriate choices in the use of the labelling algorithm, then $O(n)$ iterations of the algorithm may be required to construct the maximum flow. What does this tell us about the labelling algorithm?

15   Let $G = (V \cup W, E)$ be a bipartite graph. Construct a digraph associated with G by adding two new vertices s and t, together with arcs (s, v) for all v in V and (w, t) for all w in W, and regarding each edge vw of G as an arc (v, w). Assign capacities to the arcs so that the arcs (s, v) and (w, t) have large capacities and the arcs (v, w) have unit capacity.

Let $(S, \bar{S})$ be a minimum cut in this network, and define $A = S \cap V$, $B = S \cap W$. Show that the capacity of $(S, \bar{S})$ is $|V| - |A| + |B|$ and deduce that the size of a maximum matching in G is $|V| - d$, where d is the deficiency.

16   Prove the following generalization of Theorem 11.1. Let D be a digraph and let χ be the chromatic number of the graph obtained from D by ignoring the directions of the arcs. Show that D contains a directed path of length $\chi - 1$.

17   A 'bad triple' in a tournament is a set of three vertices $\{a, b, c\}$ such that a beats b, b beats c, and c beats a. Show that the number of bad triples is equal to

$$\binom{n}{3} - \sum_{i=1}^{n} \binom{s_i}{2},$$

where (s_i) is the score sequence. Deduce that the maximum number of bad triples is $\frac{1}{4}\binom{n+1}{3}$.

12
Recursive techniques

12.1 Generalities about recursion

In previous chapters we have encountered many problems whose solution can be expressed as a function u of a positive integer n. The values of u may be written as function-values $u(n)$, or as the terms of a sequence (u_n). Some typical examples are

$\phi(n)$: the number of integers k such that $1 \leqslant k \leqslant n$ and $\gcd(k, n) = 1$.

d_n: the number of derangements of $\{1, 2, \ldots, n\}$,

q_n: the number of partitions of an n-set.

In many cases we can set up an equation giving the value of u_n in terms of some of the values of u_r with $r < n$: this is the essence of the recursive method. Thus in Ex. 4.3.4 we showed that

$$d_n = (n-1)(d_{n-1} + d_{n-2}) \quad (n \geqslant 3).$$

This equation, together with the knowledge that $d_1 = 0$ and $d_2 = 1$, means that we can compute $d_3, d_4, d_5, \ldots$, as follows:

$$d_3 = 2(1+0) = 2, \qquad d_4 = 3(2+1) = 9, \qquad d_5 = 4(9+2) = 44,$$

and so on. In this way the existence of a suitable recursion equation enables us to calculate the value of d_n for any positive integer n.

For some purposes, it is useful to have an explicit formula for u_n. Indeed, when we have such a formula, it is conventional to say that we have 'solved' the recursion. In these terms the explicit formula

$$d_n = n!\left(1 - \frac{1}{1!} + \frac{1}{2!} - \cdots + (-1)^n \frac{1}{n!}\right)$$

obtained in Section 4.4 can be regarded as the solution of the recursion $d_n = (n-1)(d_{n-1} + d_{n-2})$ $(n \geqslant 3)$, given the initial values $d_1 = 0$ and $d_2 = 1$. But it must be stressed that the formula is frequently less useful than the equation itself. If we wish to calculate d_{10} (say) then it is usually more efficient to use the equation than to substitute values in the formula (Ex. 12.1.1).

One purpose for which a formula can be useful is to describe the behaviour of a sequence for large values of n. The reader who knows that

$$e^{-1} = 1 - \frac{1}{1!} + \frac{1}{2!} - \cdots + (-1)^r \frac{1}{r!} + \cdots$$

will deduce from the formula that d_n is approximately equal to $n!/e$. In other words, the proportion of the $n!$ permutations which are derangements is about $e^{-1} = 0.367 \ldots$.

Exercises 12.1

1 Show that the formula for d_n can be written as follows:

$$d_n = 3 \times 4 \times \cdots \times (n-1) \times n \; - \; 4 \times \cdots \times (n-1) \times n \; + \; \cdots$$
$$+ (-1)^{n-1} n + (-1)^n.$$

Show that the number of multiplications required to compute d_n by this formula is $O(n^2)$. What is the number of multiplications required if the recursion is used?

2 (i) Write a program to calculate values of d_n by means of the recursion.

(ii) Modify your program so that it finds the smallest value of n for which $d_n > 10^{10}$.

3 Show that the derangement numbers d_n also satisfy the recursion

$$d_1 = 0, \qquad d_n = nd_{n-1} + (-1)^n \quad (n \geqslant 2).$$

Is there any advantage in using this recursion (rather than the usual one) for the calculation of d_n?

4 Suppose that constants $c_0, c_1, \ldots, c_{k-1}$ and functions $f_k, f_{k+1}, \ldots$, are given. Show that there is a unique function u specified by the recursion

$$u(0) = c_0, \qquad u(1) = c_1, \quad \ldots \quad , \qquad u(k-1) = c_{k-1},$$
$$u(n+k) = f_{n+k}[u(0), u(1), \ldots, u(n-k+1)] \quad (n \geqslant 0).$$

[Hint: suppose there are two functions u and u', and let X be the set of n for which $u(n) \neq u'(n)$; now use the well-ordering axiom in the same way as it was used in Section 1.3.]

12.2 Linear recursion

A particular form of the general recursion given in Ex. 12.1.4 arises when $u(n+k)$ is specified by a constant linear combination of the values $u(n), u(n+1), \ldots, u(n+k-1)$. We shall use the sequence notation rather than the functional notation, and we shall rearrange the equations so that the typical case takes the form

$$u_0 = c_0, \qquad u_1 = c_1, \qquad \ldots \quad, \qquad u_{k-1} = c_{k-1},$$
$$u_{n+k} + a_1 u_{n+k-1} + a_2 u_{n+k-2} + \cdots + a_k u_n = 0 \quad (n \geqslant 0),$$

where $c_0, c_1, \ldots, c_{k-1}$ and $a_1, a_2, \ldots, a_k$ are constants. This is called a **linear recursion** of degree k. We shall find that there is always an explicit formula for the terms of a sequence given by a linear recursion, although it may not always be practicable (or sensible) to use it. Here we shall restrict our attention to the case $k = 2$; the general case will be discussed in Chapter 18.

Theorem 12.2

Let (u_n) be a sequence satisfying the linear recursion

$$u_0 = c_0, \qquad u_1 = c_1,$$
$$u_{n+2} + a_1 u_{n+1} + a_2 u_n = 0 \quad (n \geqslant 0),$$

and let α and β be the roots of the **auxiliary equation**

$$t^2 + a_1 t + a_2 = 0.$$

If $\alpha \neq \beta$ then there are constants A and B such that

$$u_n = A\alpha^n + B\beta^n \quad (n \geqslant 0),$$

while if $\alpha = \beta$ there are constants C and D such that

$$u_n = (Cn + D)\alpha^n \quad (n \geqslant 0).$$

The constants A and B (or C and D) are determined by the values of c_0 and c_1.

Proof

If $\alpha \neq \beta$ the equations

$$A + B = c_0, \qquad A\alpha + B\beta = c_1,$$

detemine A and B as follows:

$$A = \frac{c_1 - c_0 \beta}{\beta - \alpha}, \qquad B = \frac{c_1 - c_0 \alpha}{\alpha - \beta}.$$

Consequently, when A and B are assigned these values the result holds for u_0 and u_1, and we have the basis for a proof by means of the strong principle of induction.

For the induction hypothesis, suppose that the result holds for all u_r with $0 \leqslant r \leqslant n+1$ (where $n \geqslant 0$). Then using the recursion equation for u_{n+2} and the induction hypothesis we have

$$u_{n+2} = -(a_1 u_{n+1} + a_2 u_n)$$
$$= -[a_1(A\alpha^{n+1} + B\beta^{n+1}) + a_2(A\alpha^n + B\beta^n)]$$
$$= -A\alpha^n(a_1\alpha + a_2) - B\beta^n(a_1\beta + a_2)$$
$$= A\alpha^{n+2} + B\beta^{n+2}.$$

At the last step we use the fact that α and β are roots of the quadratic equation. Hence the result holds for u_{n+2}, and by the strong principle of induction it holds for all u_n with $n \geqslant 0$.

When $\alpha = \beta$ we apply the same method using the alternative formula. $\square$

The theorem provides a simple method of finding a formula for the terms of a sequence (u_n) which satisfies a linear recursion with $k = 2$. We have only to solve a quadratic equation and determine the constants A and B (or C and D) so that we get the specified values of u_0 and u_1.

ample Find an explicit formula for u_n when

$$u_0 = 0, \qquad u_1 = 1, \qquad u_{n+2} - 5u_{n+1} + 6u_n = 0 \quad (n \geqslant 0).$$

Solution The auxiliary equation is

$$t^2 - 5t + 6 = 0.$$

Its roots are $\alpha = 2$ and $\beta = 3$, so the formula for u_n takes the form $A(2^n) + B(3^n)$. The specified values of u_0 and u_1 yield the equations

$$A + B = 0, \qquad 2A + 3B = 1,$$

whence it follows that $A = -1$ and $B = 1$. So the formula for u_n is

$$u_n = 3^n - 2^n. \qquad \square$$

It should be noted that the roots α and β may be real or complex numbers rather than integers. When the initial values c_0 and c_1, and the coefficients a_1 and a_2, are integers, then it is clear from the form of the recursion itself that every u_n must be an integer. However, the *formula* for u_n may involve powers of non-integral numbers. This is the case in Ex. 12.2.2, for example, where the roots α and β are $\frac{1}{2}(1+\sqrt{5})$ and $\frac{1}{2}(1-\sqrt{5})$. When α and β belong to the set of complex numbers $\mathbb{C}$, we can often use De Moivre's theorem to manipulate the solution into a recognizable form. For example, suppose we are given that

$$u_0 = 2, \qquad u_1 = 0, \qquad u_{n+2} + u_n = 0 \quad (n \geqslant 0).$$

The auxiliary equation is $t^2 + 1 = 0$, and its roots are i and $-$i. From the given values of u_0 and u_1 we obtain the formula

$$u_n = i^n + (-i)^n,$$

which can be expressed in the form

$$u_n = 2 \cos \tfrac{1}{2} n \pi.$$

However, this is just a rather complicated way of saying that the terms of the sequence are $2, 0, -2, 0, 2, 0, -2, 0, \ldots$, which is an obvious direct consequence of the recursion equation. So here again is an illustration of the maxim that in many circumstances the equation itself is more useful than a formula.

Exercises 12.2

1   Find an explicit formula for u_n when

(i) $u_0 = 1$, $\qquad u_1 = 1$, $\qquad u_{n+2} - 3u_{n+1} - 4u_n = 0 \quad (n \geqslant 0)$;
(ii) $u_0 = -2$, $\qquad u_1 = 1$, $\qquad u_{n+2} - 2u_{n+1} + u_n = 0 \quad (n \geqslant 0)$.

2   The **Fibonacci numbers** F_n are defined by the recursion

$$F_0 = 1, \qquad F_1 = 1, \qquad F_{n+2} = F_{n+1} + F_n \quad (n \geqslant 0).$$

Show that

$$F_n = \frac{1}{\sqrt{5}} \left[\left(\frac{1 + \sqrt{5}}{2} \right)^{n+1} - \left(\frac{1 - \sqrt{5}}{2} \right)^{n+1} \right].$$

3   Let q_n denote the number of words of length n in the alphabet $\{0, 1\}$ which have the property that no two consecutive terms are 0. Show that

$$q_1 = 2, \qquad q_2 = 3, \qquad q_{n+2} = q_{n+1} + q_n \quad (n \geqslant 0).$$

4   What is the relationship between the Fibonacci numbers and the numbers q_n defined in the previous exercise? Use this relationship to give an explicit formula for q_n.

5   Without using the formula for the Fibonacci numbers F_n, show that

(i) $F_{n+2} = F_n + F_{n-1} + \cdots + F_1 + 2$, $\qquad$ (ii) $F_n F_{n+2} = F_{n+1}^2 + (-1)^n$.

12.3 Recursive bisection

In the discussion of insertion sorting (Section 7.8) we mentioned the method of repeated bisection. A similar method is used in a number of other algorithms, and it leads us to study recursions of the general form

$$u_{2n} = Pu_n + Q(n).$$

where P is a constant and Q is a function of n. Of course, such a recursion does not determine u_n for all values of n, but if we are given u_2 (for example) we can calculate u_4, u_8, u_{16} and so on. The behaviour of these terms may be sufficient to indicate the behaviour of the sequence (u_n). The method is sometimes known as 'divide and conquer'.

Example Let S be a set of $n = 2^k$ distinct integers $(k \geqslant 1)$. Show the maximum and minimum of S can be found by using

$$f_n = \tfrac{3}{2}n - 2$$

binary comparisons. What can be said if n is not a power of 2?

Solution By way of motivation, let us remark that in order to find the maximum of a set of n integers we need to make $n - 1$ comparisons (since the maximum may be the last integer to be compared). Similarly, we need $n - 1$ comparisons to find the minimum. So if we carry out the two searches independently, $2n - 2$ comparisons are needed.

In order to obtain the improvement claimed, we must proceed as follows. Partition S into two subsets S_1 and S_2 of size $n/2 = 2^{k-1}$, and suppose the maximum and minimum of S_i are σ_i and τ_i $(i = 1, 2)$. The maximum of S is the larger of σ_1 and σ_2, and the minimum of S is the smaller of τ_1 and τ_2. So, if $f_{n/2}$ is the number of comparisons needed to find the maximum and minimum of a set of size $n/2$ (such as S_1 or S_2), we have shown that

$$f_n = 2f_{n/2} + 2.$$

Now we shall use this recursion to prove that

$$f_{2^k} = 3 \cdot 2^{k-1} - 2 \quad (k \geqslant 1).$$

Clearly, only one comparison is needed when $|S| = 2$, so $f_2 = 1$ and the formula holds when $k = 1$. Suppose the formula holds for $f_{2^{k-1}}$. Then, using the recursion,

$$f_{2^k} = 2(3 \cdot 2^{k-2} - 2) + 2 = 3 \cdot 2^{k-1} - 2,$$

and so the formula holds for f_{2^k}. Hence the formula holds for f_n whenever n is a power of 2, and we obtain the original form by substituting $n = 2^k$.

The case when n is not a power of 2 can be dealt with as follows. First we partition S into a set S_1 of size 2^m and a set S_2 of size $n - 2^m$, where 2^m is the greatest power of 2 less than n. Then we repeat the process for S_2, and so on. For example, if $n = 26$ we have $26 = 16 + 10$ and $10 = 8 + 2$, and by the basic argument above

$$f_{26} = f_{16} + f_{10} + 2, \qquad f_{10} = f_8 + f_2 + 2.$$

Inserting the known formulae for f_2, f_8, f_{16} we find

$$f_{10} = (\tfrac{3}{2} \cdot 8 - 2) + (\tfrac{3}{2} \cdot 2 - 2) + 2 = \tfrac{3}{2} \cdot 10 - 2,$$
$$f_{26} = (\tfrac{3}{2} \cdot 16 - 2) + (\tfrac{3}{2} \cdot 10 - 2) + 2 = \tfrac{3}{2} \cdot 26 - 2.$$

In this way we can show that $f_n = \tfrac{3}{2}n - 2$, whenever n is even. (When n is odd we need a trivial modification to deal with the 'spare' one.) $\square$

Exercises 12.3

1 Prove that if

$$u_2 = 5, \qquad u_{2r} = 2u_r + 3 \quad (r = 2^k, \ k \geqslant 1)$$

then $u_n = 4n - 3$ whenever n is a power of 2.

2 Prove that if

$$u_2 = 5, \qquad u_{r+s} = u_r + u_s + 3 \quad (r \geqslant s \geqslant 2)$$

then $u_n = 4n - 3$ whenever n is even. [Hint: write n as a sum of powers of 2 and use Ex. 1.]

3 Find an explicit formula for a_n when n is a power of 4 and

$$a_1 = 2, \qquad a_{4n} = 2a_n + 4n \quad (n = 4^k, \ k \geqslant 0).$$

4 A version of the *merge sort* algorithm proceeds as follows. Suppose the given list is $x_1, x_2, \ldots, x_n$, where n is a power of 2. Compare x_1 and x_2 and put them in order, repeat with x_3 and x_4, x_5 and x_6, and so on. Then merge the ordered pairs (x_1, x_2) and (x_3, x_4) into a single ordered list, and so on. Show that the number of binary comparisons required satisfies

$$u_2 = 1, \qquad u_{2n} = 2u_n + 2n - 1.$$

Deduce that u_n is $O(n \log n)$.

12.4 Recursive optimization

In this section we shall show how a recursive method can be used to find the maximum value of a function of several variables. Roughly speaking, the idea is to solve the problem in stages, each stage involving a relatively simple optimization problem.

Example
A speculator has £5 m and is considering investing capital in three companies C_1, C_2, C_3 (in units of £1 m). He predicts that his return from investing x units ($0 \leqslant x \leqslant 5$) in the respective companies will be as follows:

	$x = 5$	$x = 4$	$x = 3$	$x = 2$	$x = 1$	$x = 0$
C_1	8	7	6	4	1	0
C_2	11	10	8	5	3	0
C_3	11	11	10	2	0	0 .

How should he invest his money in order to maximize his total return?

Solution
We shall attack the problem in stages, *Stage i* being the investment in C_i ($i = 1, 2, 3$).

Stage 1 If x_1 units are invested in C_1, the return $r_1(x_1)$ is given directly by the table. For uniformity, we introduce here the notation

$$y_1 = \text{total invested so far } (= x_1),$$
$$f_1(y_1) = \text{best return from this total } (= r_1(x_1)).$$

With this notation, we have the values

y_1	5	4	3	2	1	0
$f_1(y_1)$	8	7	6	4	1	0 .

Stage 2 Suppose a further x_2 units are invested in C_2, giving a return $r_2(x_2)$ as in the original table. Let

$$y_2 = \text{total invested so far } (= y_1 + x_2),$$
$$f_2(y_2) = \text{best return from this total.}$$

The calculation of $f_2(y_2)$ can be displayed in tabular form (Table 12.3.1), as explained below.

Since the total amount available is 5 units we need only calculate the best returns for $0 \leqslant y_2 \leqslant 5$. Each such value of y_2 arises from several pairs of values of x_2 and y_1, and in the table these values lie on a sloping line, as indicated. The value of $f_2(y_2)$ is the maximum

Table 12.3.1

x_2	$r_2(x_2)$	5 / 8	4 / 7	3 / 6	2 / 4	1 / 1	0 / 0	y_1 / $f_1(y_1)$
5	11						11	
4	10					11	10	
3	8				12	9	8	
2	5			11	9	6	5	
1	3		10	9	7	4	3	
0	0	8	7	6	4	1	0	
		5	4	3	2	1	0	y_2
		12	10	8	5	3	0	$f_2(y_2)$

on the relevant line; that is

$$f_2(y_2) = \max \{f_1(y_1) + r_2(x_2)\}$$

where the maximum is taken over those values of x_2 and y_1 for which $y_1 + x_2 = y_2$.

  Stage 3 Suppose an additional x_3 units are invested in C_3, giving a return $r_3(x_3)$ as in the original table. Let

$$y_3 = \text{total invested so far } (= y_2 + x_3),$$

$$f_3(y_3) = \text{best return from this total.}$$

Since this is the final stage, and we are given that 5 units are available, we need consider only the case $y_3 = 5$. For the sake of uniformity, we set out the calculation as in *Stage 2*; see Table 12.3.2.

Table 12.3.2

x_3	$r_3(x_3)$	5 / 12	4 / 10	3 / 8	2 / 5	1 / 3	0 / 0	y_2 / $f_2(y_2)$
5	11						11	
4	11					14		
3	10				15			
2	2			10				
1	0		10					
0	0	12						
		5						y_3
		15						$f_3(y_3)$

So we see that $f_3(5) = 15$, where

$$f_3(y_3) = \max \{f_2(y_2) + r_3(x_3)\},$$

the maximum being taken over those values of x_3 and y_2 for which $y_2 + x_3 = y_3$.

We conclude that the best return is 15 units. Checking back through the tables we see that $f_3(5) = 15$ is obtained when $x_3 = 3$ and $y_2 = 2$, and that the value $f_2(2) = 5$ is obtained when $x_2 = 2$ and $y_1 = x_1 = 0$. So the manager should put nothing into C_1, 2 units into C_2 and 3 units into C_3. □

In the solution of the example we have implicitly invoked a very useful principle. At each stage we calculated only the best return $f_i(y_i)$ for each feasible value of y_i; and the reason why this works is as follows. *If the best policy overall involves a particular value of y_i, then the policy as far as Stage i must surely be the best one for that value of y_i.*

This 'principle of optimality' is fundamental to the technique of recursive optimization, as outlined in the example. A practical illustration may be helpful. If the best route from London to Paris is via Dover, then it must surely use the best route from London to Dover; on the other hand, if the best route is via Newhaven, then it must surely use the best route from London to Newhaven.

Exercises 12.4

1 Suppose a manager has six units to invest in four companies C_1, C_2, C_3, C_4, but companies C_1 and C_2 allow investment only in multiples of two units. If the estimated returns are as given in Table 12.4.1, find the best total return. Is the best policy unique?

Table 12.4.1

	Company			
Investment	C_1	C_2	C_3	C_4
0	0	0	0	0
1	—	—	0	2
2	0	3	1	3
3	—	—	3	4
4	4	6	5	4
5	—	—	7	4
6	8	9	9	4

2   The owner of three stores has purchased five crates of fresh raspberries. The stores are in different areas, and the pattern of sales varies; for example, at Store B the customers will pay high prices but turnover is small and the raspberries may go bad before they are sold. For such reasons, the owner estimates the profit from allocating x crates to the stores shown in Table 12.4.2. Find the allocation which yields the largest profit. (Any store may get no raspberries.)

Table 12.4.2

		Store	
x	A	B	C
1	3	5	4
2	7	10	6
3	9	11	11
4	12	11	12
5	13	11	12

12.5   The framework of dynamic programming

The method which we used to solve the speculator's problem is typical of a general technique known as **dynamic programming**, or simply **DP**. In fact, the name is rather misleading, and it is advisable to remember that it is just a form of recursive optimization.

In general, the dynamic programming method depends on decomposing a problem into a finite number N of stages. The typical *Stage i* can be represented diagrammatically as in Fig. 12.1.

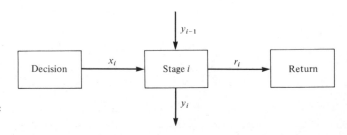

Fig. 12.1
A typical stage in dynamic programming.

The symbols y_{i-1} and y_i represent what are called *state variables*. The value of y_{i-1} depends upon the initial value of y_0 and all the decisions that have been made in Stages $1, 2, \ldots, i-1$. At Stage i, a new decision is made, represented by the value assigned to a *decision variable* x_i, and this together with y_{i-1} determines the value of y_i. We express the fact that y_i depends on x_i and y_{i-1} by using the functional notation and writing

$$y_i = y_i(x_i, y_{i-1}).$$

The rule which enables us to determine y_i, given x_i and y_{i-1}, is called the *stage transformation*. Finally, each stage produces a *stage return* r_i, which also depends on the decision x_i and the input y_{i-1}, so we write

$$r_i = r_i(x_i, y_{i-1}).$$

Referring back to the *Example* in Section 12.4, we can give concrete examples of this barrage of terminology. We have $N = 3$ stages, corresponding to investment in the three companies. The state variable y_i represents the total investment in the companies C_1 to C_i, while the decision variable x_i represents the amount invested in C_i. Clearly, the stage transformation is given by the equation

$$y_i = x_i + y_{i-1}.$$

Finally, the stage return r_i depends (in this case) only on x_i and is given explicitly by the table at the beginning of the *Example*.

The speculator's objective is to maximize the sum of the stage returns, given that the initial investment y_0 is zero and that y_N, the total number of units purchased, is 5. In general, a DP problem may involve maximization or minimization of a more complicated function of the stage returns, but for our purposes it will be sufficient to consider the following standard form.

Given the initial value $y_0 = y_0^*$,
maximize (or minimize) the total return $r_1 + r_2 + \cdots + r_N$,
over all values of $x_1, x_2, \ldots, x_N$ which yield a
given final value $y_N = y_N^*$.

The essence of the DP method is that it enables us to optimize in stages, instead of over all the variables at once. This will usually involve a dramatic improvement in efficiency. See Ex. 12.7.16.

The recursive method of dealing with the standard form of DP problem can be justified as follows. The problem is to find

$$\max \{r_1(x_1, y_0^*) + r_2(x_2, y_1) + \cdots + r_N(x_N, y_{N-1})\}$$

over all those values of $(x_1, x_2, \ldots, x_N)$ which yield $y_N = y_N^*$. This is equivalent to the problem of finding

$$\max_{\underline{N}} \{\ldots \max_{2} \{\max_{1} \{r_1(x_1, y_0^*)\} + r_2(x_2, y_1)\} + \cdots + r_N(x_N, y_{N-1})\},$$

where the successive maximizations are taking over appropriate values of the variables. The first maximization is taken over the values of x_1 which yield a specified value of y_1, and the result is a function of y_1:

$$f_1(y_1) = \max_{1} \{r_1(x_1, y_0^*)\}.$$

(Recall that y_1 is a known function of x_1 and y_0, and y_0 must take the specified initial value y_0^*.) The second maximization is taken over those values of x_2 and y_1 which yield a specified value of y_2, and the result is a function of y_2:

$$f_2(y_2) = \max_{2} \{f_1(y_1) + r_2(x_2, y_1)\}.$$

Continuing in this way, we see that the calculation at Stage i is given by

$$f_i(y_i) = \max_{i} \{f_{i-1}(y_{i-1}) + r_i(x_i, y_{i-1})\},$$

where the maximum is taken over those values of x_i and y_{i-1} which yield the stated value of y_i, according to the stage transformation. At each intermediate stage we must calculate $f_i(y_i)$ for all possible values of y_i, but at the final stage we need only find one value $f_N(y_N^*)$.

To summarize, the solution of the N-stage DP problem, in the standard form given above, may be found by a recursion involving N optimization problems of the following form:

$$f_1(y_1) = \max \{r_1(x_1, y_0^*)\},$$
$$f_i(y_i) = \max \{f_{i-1}(y_{i-1}) + r_i(x_i, y_{i-1})\} \quad (2 \leq i \leq N).$$

The required solution is $f_N(y_N^*)$.

Exercises 12.5

1 An *investment problem* is defined to be a DP problem of the standard form which can be decomposed into stages in the following way: the stage return r_i depends only on x_i, where x_i is the decision variable and represents an amount 'invested' at Stage i; and the state variable y_i represents the total amount invested up to and including Stage i.

Write down the stage transformations for an investment problem and show that if $y_0 = 0$ the recursion can be written in the form

$$f_1(y_1) = r_1(y_1), \qquad f_i(y_i) = \max_{y_{i-1}} \{f_{i-1}(y_{i-1}) + r_i(y_i - y_{i-1})\}.$$

2 The Eclectic Electric Company has six salesmen who sell its products in Berkshire, Hampshire, and Surrey. The managing director estimates that the monthly profit (in thousands of pounds) obtained by allocating a given number of salesmen to a particular county is as stated in Table 12.5.1.

Table 12.5.1

County	Number of salesmen						
	0	1	2	3	4	5	6
Berkshire	38	42	48	58	66	72	83
Hampshire	40	42	50	60	66	75	82
Surrey	60	64	68	78	87	102	104

The problem is to find the allocation of the six salesmen which yields the maximum profit.

Formulate this as an investment problem, identifying the state variables, decision variables, and stage returns. Solve the problem.

Suppose that two of the salesmen leave to set up their own business. What is the best allocation of the remaining salesmen?

12.6 Examples of the dynamic programming method

In this section we shall discuss two examples illustrating the wide range of problems which are amenable to the DP method. Further examples can be found in the exercises at the end of the chapter.

Example A hiker can fill his knapsack with cans of soup, packets of biscuits, guidebooks, and so on, each kind of object having a known weight and a certain 'utility' for the journey. Let W be the total weight he can carry, and let $P_1, P_2, \ldots, P_n$, denote the various kinds of object, where each item P_i has weight w_i and utility u_i. Suppose also that x_i denotes the number of items P_i selected. The *knapsack problem* is to find $x_1, x_2, \ldots, x_n$ such that the utility

$$u_1x_1 + u_2x_2 + \cdots + u_nx_n$$

is maximized, given that the total weight is fixed by the equation

$$w_1x_1 + w_2x_2 + \cdots + w_nx_n = W.$$

(Clearly, the problem arises in many practical situations, many of them more serious than hiking.)

Formulate the general problem in DP terminology, and use the DP method to solve the problem when $n = 3$, $W = 10$, and w_i and u_i are given by the table

i	1	2	3
w_i	4	3	1
u_i	5	2	1

Solution We take the ith stage to be the allocation of P_i $(1 \leq i \leq n)$. The decision variables are simply the x_i as defined above, while the state variables and stage returns are

$$y_i = \text{weight of allocation of } P_1, P_2, \ldots, P_i,$$
$$r_i = \text{utility of allocation of } P_i \ (= u_i x_i).$$

Consequently, the stage transformations are

$$y_i = y_{i-1} + w_i x_i \quad (1 \leq i \leq n)$$

with boundary conditions $y_0 = 0$ and $y_n = W$. The recursion is

$$f_1(y_1) = u_1 x_1, \qquad f_i(y_i) = \max\{f_{i-1}(y_{i-1}) + u_i x_i\} \quad (2 \leq i \leq n),$$

where $y_1 = w_1 x_1$, and the maximum is taken over those values of x_i and y_{i-1} for which $y_{i-1} + w_i x_i = y_i$.

In the numerical example, we note first that the given values of W and w_i imply that $0 \leq x_1 \leq 2$, $0 \leq x_2 \leq 3$. Hence the calculation for the first two stages goes as follows.

Stage 1

x_1	0	1	2
y_1	0	4	8
$f_1(y_1) = u_1 x_1$	0	5	10.

Stage 2

x_2	0	0.	0	1	1	2	2	3
y_1	0	4	8	0	4	0	4	0
y_2	0	4	8	3	7	6	10	9
$f_2(y_2) = f_1(y_1) + u_2 x_2$	0	5	10	2	7	4	9	6.

Fortunately, each value of y_2 arises from only one pair (x_2, y_1), so the calculation of $f_2(y_2)$ involves no maximization. At the third stage we need only consider those pairs (x_3, y_2) for which $y_2 + \dot{x}_3 = y_3 = 10$.

Stage 3

x_3	0	1	2	3	4	5	6	7	8	9	10
y_2	10	9	8	7	6	—	4	3	—	—	0

y_3	10	10	10	10	10		10	10			10
$f_3(y_3) = f_2(y_2) + u_3x_3$	9	7	12	10	8		11	9			10.

Hence $f_3(10)$ is the maximum of the values on the last line, that is, $f_3(10) = 12$. Checking back through the working we find that this arises from the allocation $x_1 = 2$, $x_2 = 0$, $x_3 = 2$. ☐

Example Find the shortest directed path from s to t in the network illustrated in Fig. 12.2.

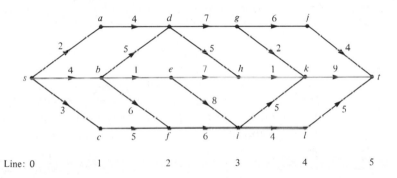

Line: 0   1   2   3   4   5

Fig. 12.2
Find the shortest path
from s to t.

Solution The diagram of the network suggests that we should arrange the vertices in six 'lines', so that line 0 contains s, line 1 contains a, b, c, and so on. Then we can regard the process of going from line $i-1$ to line i as Stage i of a dynamic program. The decision and state variables are

$$x_i = \text{arc chosen to lead from line } i-1 \text{ to line } i,$$

$$y_i = \text{vertex in line } i.$$

The stage return r_i is $l(x_i)$, the length of x_i, as given in the diagram, and the stage transformation is the rule which, given a vertex y_{i-1} in line $i-1$ and an arc $x_i = (y_{i-1}, y_i)$, defines the vertex y_i in line i. The boundary conditions are $y_0 = s$ and $y_5 = t$.

The recursion is

$$f_1(y_1) = l(x_1) \quad \text{where} \quad x_1 = (s, y_1),$$
$$f_i(y_i) = \min\{f_{i-1}(y_{i-1}) + l(x_i)\} \quad (2 \leqslant i \leqslant 5),$$

where the minimum is taken over those pairs x_i and y_{i-1} for which x_i is an arc from y_{i-1} to y_i.

The calculation proceeds as follows.

Stage 1	y_1	a	b	c
	$f_1(y_1)$	2	4	3

Stage 2	y_2	d	e	f
	$f_2(y_2)$	6	5	8

Stage 3	y_3	g	h	i
	$f_3(y_3)$	13	11	13

Stage 4	y_4	j	k	l
	$f_4(y_4)$	19	12	17

Stage 5	y_5	t
	$f_5(y_5)$	21

Thus the shortest directed path has length 21, and it is s, a, d, h, k, t. (It should be noted that the values $f_i(y_i)$ assigned to the vertices here are obtained in exactly the same way as the labels used in the more general algorithm discussed in Section 9.6.) $\qquad\square$

Exercises 12.6

1 Find the shortest directed path from s to t in the network illustrated in Fig. 12.3.

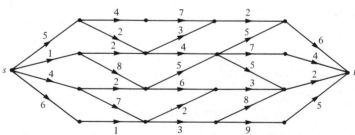

Fig. 12.3
Find the shortest directed path from s to t.

2 Solve the knapsack problem with $n = 3$ and $W = 19$ when w_i and u_i are given by the table

i	1	2	3
w_i	6	4	3
u_i	11	7	5 .

12.7 Miscellaneous Exercises

1 Find explicit formulae for the terms of the sequences defined by

(i) $u_0 = 0$, $u_1 = 1$, $u_{n+2} + u_{n-1} - 2u_n = 0$ $(n \geqslant 0)$;

(ii) $u_0 = 1$, $u_1 = 0$, $u_{n+2} - 6u_{n+1} + 8u_n = 0$ $(n \geqslant 0)$.

2 Show that the equation

$$n(n+1)u_{n+2} - 5n(n+2)u_{n+1} + 4(n+1)(n+2)u_n = 0$$

is satisfied by $u_n = n$. Use the substitution $u_n = nv_n$ to show that the solution for which $u_1 = 12$ and $u_2 = 60$ is

$$u_n = 3n2^{2n-1} + 6n.$$

3 Find a formula for the nth term of the sequence (u_n) defined by

$$u_0 = X, \qquad u_1 = Y, \qquad u_{n+2} = u_n + n \quad (n \geqslant 0).$$

4 An ordered triple (a, b, c) of integers is said to be *linear* if $a < b < c$ and $b - a = c - b$. Let L_n denote the number of linear triples whose elements belong to $\mathbb{N}_n$. Show that

$$L_{2n+1} = L_{2n} + n$$

and derive a similar equation for L_{2n}. Deduce that L_n satisfies the recursion in Ex. 3 and find a formula for L_n.

5 Let C_n denote the cycle graph with n vertices, as defined in Section 8.3, and let $f_n(k)$ be the number of vertex-colourings of C_n when there are k colours available. By splitting the set of colourings into two parts, according as vertices 0 and 2 do or do not have the same colour, show that

$$f_n(k) = (k-1)f_{n-2}(k) + (k-2)f_{n-1}(k) \quad (n \geqslant 5).$$

Deduce that

$$f_n(k) = (k-1)[(k-1)^{n-1} + (-1)^n] \quad (n \geqslant 3).$$

6 Show that the number of vertex-colourings of any tree with n vertices, when there are k colours available, is $k(k-1)^{n-1}$.

7   Let us say that permutation π of $\mathbb{N}_n$ is *equable* if

either $\pi(i) \leqslant \pi(i-1)+2$ or $\pi(i) \leqslant \pi(i+1)+2$ $(2 \leqslant i \leqslant n-1)$

and

$$\pi(1) \leqslant \pi(2)+2, \qquad \pi(n) \leqslant \pi(n-1)+2.$$

Let x_n be the number of equable permutations for which $n = \pi(k)$ and $n-1 = \pi(k-1)$ or $\pi(k+1)$, for some k in $\mathbb{N}_n$, and let y_n be the number of equable permutations which do not have this property. Show that

$$x_{n+1} = 2x_n + 2y_n$$
$$y_{n+1} = x_n + 2y_n,$$

and find an explicit formula for the total number of equable permutations.

8   Show that if a_n satisfies a divide-and-conquer recursion of the form

$$a_{2n} = Aa_n + Bn$$

then a_n is $O(n \log n)$ if $A = 2$ and a_n is $O(n^{\log_2 A})$ if $A > 2$.

9   Let $t(n)$ be the time needed to multiply two n-digit integers by the divide-and-conquer method indicated in Ex. 7.9.5. Assuming that all the operations apart from the single-digit multiplications can be done in time cn (where c is a constant), show that

$$t(2n) = 3t(n) + 2cn$$

and deduce that $t(n)$ is $O(n^{\log_2 3})$.

10   Let (F_n) be the sequence of Fibonacci numbers as defined in Ex. 12.2.2. Show that

(i) $F_2 + F_4 + F_6 + \cdots + F_{2n} = F_{2n+1} - 1$,
(ii) $F_{n+1}^3 + F_n^3 + F_{n-1}^3 = F_{3n}$.

11   Let $\lambda(n, k)$ be the number of k-subsets of $\mathbb{N}_n$ which do not contain two consecutive integers. Show that

$$\lambda(n, k) = \lambda(n-2, k-1) + \lambda(n-1, k)$$

and hence verify that

$$\lambda(n, k) = \binom{n-k+1}{k}.$$

12   Let $\mu(n, k)$ be the number of ways of selecting k objects from n objects arranged in a circle, in such a way that no two are adjacent. Show that if $\lambda(n, k)$ is as in the previous exercise then

$$\mu(n, k) = \lambda(n-1, k) + \lambda(n-3, k-1)$$

and deduce that

$$\mu(n, k) = \frac{n}{n-k}\binom{n-k}{k}.$$

13 Find the solution to the investment manager's problem (as discussed in Section 12.4) when the amount available is £6 m and there are three companies giving returns as follows.

	$x = 6$	$x = 5$	$x = 4$	$x = 3$	$x = 2$	$x = 1$	$x = 0$
C_1	9	8	8	6	4	2	0
C_2	12	12	12	9	7	3	0
C_3	11	10	10	7	2	0	0

14 The production of Desk Diaries for Absentminded Academics takes place throughout the year, and the diaries are stored in the factory warehouse ready for despatch at the end of September. The total number of diaries needed is 19,000, and the maximum production per quarter is 6000, with costs as shown in Table 12.7.1.

Table 12.7.1

Number (1000)	Cost (£100)			
	Oct–Dec	Jan–Mar	Apr–Jun	Jul–Sep
0	2	6	5	4
1	4	7	8	5
2	8	9	11	9
3	9	11	15	13
4	11	15	16	15
5	12	19	17	17
6	14	20	20	20

It costs £100 per thousand per complete quarter to store the diaries (there is no cost for part of a quarter). Draw up a table giving the overall cost of production and storage, and find the cheapest production schedule.

15 A smuggler operates in an area where three countries X, Y, Z have common boundaries in pairs. Each night he crosses from one country (i) to another (j) and makes a profit r_{ij} as in Table 12.7.2.

Table 12.7.2

	j		
i	X	Y	Z
X	—	10	7
Y	1	—	4
Z	8	2	—

Today he is in X. Show that his best profit over the next ten nights is 77 and determine the optimal route.

16   Suppose we have an investment problem which can be expressed in the DP framework with N stages, and such that each decision variable x_i $(1 \leqslant i \leqslant N)$ can take n integer values. Show that if we regard N as being fixed then

(i) the number of comparisons required to find the optimal solution by the DP method is $O(n^2)$;
(ii) the number of comparisons required to find the optimal solution by evaluating all the possible solutions explicitly is $O(n^N)$.

17   Formulate the following problem as an N-stage dynamic program. Find positive integers $x_1, x_2, \ldots, x_k$ whose sum is N and such that the product $x_1 x_2 \cdots x_k$ is as large as possible.

18   Solve the special cases of the problem stated in the previous exercise when $N = 13$ and $k = 3$, and when $N = 15$ and $k = 4$.

19   A burglar can carry not more than 100 kg of stolen objects. He has the opportunity to steal six objects whose weight and value is as given in the table. Which objects should he steal? (This is a *zero-one* knapsack problem, since the decision variables take only the values 0 and 1.)

Object	1	2	3	4	5	6
Weight (kg)	20	50	10	35	48	21
Value (£K)	2	8	2	3	5	4

Part III
Algebraic methods

The third part of the book is the last, and the largest. In it we shall study the application of algebraic techniques to problems of Discrete Mathematics. It will be necessary to discuss parts of what is usually called 'abstract' algebra, although we shall keep our feet firmly on the ground.

The prerequisites for this part have been covered in Part I and Part II, except that we assume a basic knowledge of matrix algebra in Chapter 13, 15, and 17.

13
Groups

13.1 The axioms for a group

In order to make a serious study of Discrete Mathematics it is essential to be familiar with modern algebraic techniques. The theories of permutations, designs, and latin squares (for example) are inextricably linked with certain aspects of the algebraic theories of groups, rings, and fields. In this book we shall study these algebraic structures from a utilitarian point of view, and hope that the reader will thereby be inspired to proceed to the study of algebra for its own sake.

The basic idea underlying the definition of an algebraic structure is that of a set with a 'binary operation'. Suppose we have a set X of objects, with the property that any pair of them, x and y, can be combined in some way to form an object z. This rule of combination can be expressed by the equation

$$x * y = z,$$

where the symbol $*$ indicates a **binary operation**, the word 'binary' signifying here that two objects are involved. The most familiar examples are the arithmetical operations like $+$ and $\times$, defined on the set of integers $\mathbb{Z}$. Another important example is the rule of composition defined on the set S_n of permutations of $\{1, 2, \ldots, n\}$.

Any given binary operation has certain algebraic properties. At the very beginning of this book we set out the properties of the arithmetical operations on $\mathbb{Z}$ and, clearly, those properties are of fundamental importance. In Theorem 3.6 we obtained four properties of the composition operation in S_n, and remarked that the full significance of those properties would emerge later. That time is nigh, for we shall see that the four properties are shared by many other systems, and that they have far-reaching consequences. For these reasons, mathematicians use the special name *group* to describe such a system.

Definition A **group** consists of a set G, together with a binary operation $*$ defined on G which satisfies the following axioms.

G1 (*Closure*).   For all x and y in G

$$x * y \text{ is in } G.$$

G2 (*Associativity*).   For all x, y, and z in G

$$(x * y) * z = x * (y * z).$$

G3 (*Identity*).   There is an element e in G such that

$$e * x = x * e = x$$

for all x in G.

G4 (*Inverse*).   For all x in G there is an x' in G such that

$$x * x' = x' * x = e.$$

The axioms are known by the names indicated. Also, an element e with the property stated in **G3** is said to be an **identity** element for the $*$ operation in G, and an element x' in **G4** is said to be an **inverse** element for x.

If G is a group and $|G|$ is finite, then $|G|$ is known as the **order** of G; a group with infinitely many elements is said to have **infinite order**.

Exercises 13.1

1   Let $G = \mathbb{Z}$. Complete the following table, where $+$, $-$, and $\times$ represent the usual operations of arithmetic.

$*$	Closure	Associativity	Identity	Inverse
$+$	$\checkmark$			
$-$	$\checkmark$			
$\times$	$\checkmark$			

2   Repeat Ex.1 taking $G = \mathbb{N}$, with the same operations.

13.2 Examples of groups

We have already met two examples of groups. The set of permutations of $\{1, 2, \ldots, n\}$ is a group with respect to the composition operation. It is known as the **symmetric group** (which explains the notation S_n) and its order is $n!$. The set $\mathbb{Z}$ of integers is a group with respect to the addition operation, and it has infinite order. If these were the only examples, the study of groups would be useful for that reason alone. But in fact the concept is so widely applicable that it pervades the whole of higher mathematics. We give two more examples here, typical of the many that can be cited in support of this claim.

Our first example is geometrical. Let $\triangle$ be an equilateral triangle—it may help to think of $\triangle$ as a flat piece of card with its corners labelled ABC. There are six different transformations of $\triangle$ which have the property that $\triangle$ occupies the same position in space before and after the transformation. These transformations are known as **symmetries** of $\triangle$, and they are indicated in Fig. 13.1 by their effect on the initial position of $\triangle$ in which A is at the top and A, B, C are in clockwise order.

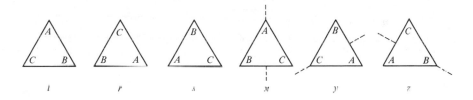

Fig. 13.1
Symmetries of an equilateral triangle.

In the figure, i is the trivial symmetry in which nothing happens, r and s are rotations through $120°$ and $240°$ about the centroid of $\triangle$, and x, y, z are reflections in the axes indicated. Alternatively, x, y, z can be regarded as the operations of turning $\triangle$ over about the relevant axis. (Note that the axes are fixed in space, and do not move when $\triangle$ is transformed.)

Example 1 Show that $G_\triangle = \{i, r, s, x, y, z\}$ is a group, with respect to the operation $*$ representing the successive implementation of symmetry transformations.

Solution It follows from the geometrical interpretation that the successive implementation of two symmetries is another symmetry. Thus, if we consider y and s, for example, the result of doing s first and then y is denoted by $y * s$ (note the order), and to find the value of $y * s$ we can trace the effect of the transformations as in Fig. 13.2. Comparing the final effect of $y * s$ with Fig. 13.1 we see that $y * s = z$.

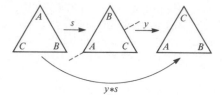

Fig. 13.2
$y*s$ has the same effect
as z.

The results of all such calculations can be conveniently arranged
in the **group table** for G (Table 13.2.1).

Table 13.2.1

	i	r	s	x	y	z
i	i	r	s	x	y	z
r	r	s	i	y	z	x
s	s	i	r	z	x	y
x	x	z	y	i	s	r
y	y	x	z	r	i	s
z	z	y	x	s	r	i

In this table the symbol in row y and column s is z, corresponding
to the fact that $y * s = z$. The other entries in the table are obtained
similarly.

Now it is easy to check that all the group axioms are satisfied. The
closure and associativity properties are immediate consequences of
the nature of symmetry transformations. The identity for $G_\triangle$ is
clearly the trivial symmetry i, and the inverse of each element can
be found from Table 13.2.1, as follows.

Element: i r s x y z
Inverse: i s r x y z

Hence $G_\triangle$ is a group.   □

Example 2 Let G_M denote the set of 2×2 matrices of the form

$$\begin{bmatrix} \alpha & \beta \\ 0 & 1 \end{bmatrix},$$

where α and β are elements of $\mathbb{Z}_3$ and $\alpha \neq 0$. Show that G_M is a
group with respect to the usual rule for multiplying matrices.

Solution We shall verify the axioms **G1** – **G4** in turn.

(**G1**) The product of two matrices in G_M is

$$\begin{bmatrix} \alpha & \beta \\ 0 & 1 \end{bmatrix}\begin{bmatrix} \gamma & \delta \\ 0 & 1 \end{bmatrix} = \begin{bmatrix} \alpha\gamma & \alpha\delta + \beta \\ 0 & 1 \end{bmatrix},$$

and since the right-hand side has the correct form it is an element of G_M. (Note that $\alpha\gamma$ cannot be zero since $\alpha \neq 0$ and $\gamma \neq 0$.)

(**G2**) The multiplication of matrices is always associative—in the present example this could be verified explicitly if necessary.

(**G3**) Taking $\alpha = 1$ and $\beta = 0$ we obtain the identity matrix, which therefore is the identity element of G_M.

(**G4**) In order to find the inverse of a typical element of G_M we note that

$$\begin{bmatrix} \alpha & \beta \\ 0 & 1 \end{bmatrix}\begin{bmatrix} \gamma & \delta \\ 0 & 1 \end{bmatrix} = \begin{bmatrix} 1 & 0 \\ 0 & 1 \end{bmatrix}$$

if and only if

$$\alpha\gamma = 1 \quad \text{and} \quad \alpha\delta + \beta = 0.$$

When α and β are given we can solve these equations for γ and δ (remembering that $\alpha \neq 0$ so α^{-1} exists in $\mathbb{Z}_3$). Explicitly

$$\gamma = \alpha^{-1} \quad \text{and} \quad \delta = -(\alpha^{-1}\beta).$$

Thus each member of G_M has an inverse determined by these equations. □

Exercises 13.2

1 Write down explicitly the matrices belonging to the group in *Example 2*. (There are just six of them.) Find the inverse of each one.

2 There are eight symmetry transformations of a square. List them, and draw up the group table as in *Example 1*.

3 For which values of $n \geqslant 2$ is it true that the following are groups?

(i) $\mathbb{Z}_n$ with the + operation.
(ii) $\mathbb{Z}_n$ with the × operation.
(iii) $\mathbb{Z}_n \backslash \{0\}$ with the × operation.

4 Draw up the group table for the set of complex numbers $\{1, -1, i, -i\}$ with respect to multiplication, verifying that it is indeed a group.

13.3 Basic algebra in groups

In this section we shall begin to study the consequences of the four axioms for a group. It will be necessary to avoid making any assumptions about the symbols being manipulated, except for the fact that they satisfy the group axioms. Students sometimes find this difficult because they have had many years of training in algebraic manipulation, and have learned many rules, only some of which apply here. For this reason it is helpful to remember that the present study is part of *abstract algebra*, where the symbols have no properties beyond what is stated in the axioms.

From the outset we shall drop the cumbersome $*$ notation for the group operation and use what is usually called the *multiplicative* notation, in which

$$
\begin{array}{lll}
x * y & \text{becomes} & xy, \\
e & \text{becomes} & 1, \\
x' & \text{becomes} & x^{-1}.
\end{array}
$$

This notation has the advantage of economy, but its disadvantage is that it can be confused with the ordinary multiplication of numbers. It must be stressed again that the only properties which can be assumed are the group properties, that is

$$
x(yz) = (xy)z, \qquad x1 = 1x = x, \qquad xx^{-1} = x^{-1}x = 1.
$$

In particular, it must *not* be assumed that $xy = yx$. If a given pair of elements do satisfy $xy = yx$ then we say that x and y **commute**; a group in which every pair commutes is said to be **commutative**. (An alternative term is **abelian**, commemorating the mathematician N. H. Abel (1802–1829).)

Theorem 13.3.1

(i) Let x, y, z, a, b, be any elements of a group G. Then

$$
xy = xz \quad \Rightarrow \quad y = z \qquad \text{(left cancellation)},
$$

and

$$
ax = bx \quad \Rightarrow \quad a = b \qquad \text{(right cancellation)}.
$$

Proof

Since G is a group, x has an inverse x^{-1}. Multiplying the equation $xy = xz$ on the left by x^{-1} and using the axioms as indicated, we obtain

$$
\begin{array}{ll}
x^{-1}(xy) = x^{-1}(xz) & \\
(x^{-1}x)y = (x^{-1}x)z & \text{(by \textbf{G2})} \\
1y = 1z & \text{(by \textbf{G4})} \\
y = z. & \text{(by \textbf{G3})}.
\end{array}
$$

The right cancellation rule is proved similarly. □

Let $\{g_1, g_2, \ldots, g_n\}$ be the elements of a finite group. The row of the group table corresponding to an element g_i contains the entries

$$g_i g_1, \ g_i g_2, \ldots, \ g_i g_n.$$

These are all different, since if $g_i g_r = g_i g_s$ then the left cancellation rule implies that $g_r = g_s$. Similarly, the right cancellation rule implies that the entries in every column of the group table are all different. In other words

a group table is a latin square.

The simplest example of a latin square of order m is the group table of the group $\mathbb{Z}_m$ with respect to the $+$ operation. We now know that *any* finite group will yield a latin square, but as we might expect, not every latin square is the group table for some group (see Ex. 13.3.5).

Theorem 13.3.2

If a and b are any elements of a group G, then the equation

$$ax = b$$

has a unique solution in G.

Proof

If x and $\bar{x}$ are both solutions of the equation, then

$$ax = a\bar{x} \quad (=b),$$

and so, by the left cancellation rule, $x = \bar{x}$. Hence there is at most one solution. Clearly $x = a^{-1}b$ is a solution, since

$$a(a^{-1}b) = (aa^{-1})b = 1b = b,$$

and so there is exactly one solution. □

In the case $a = b$ the theorem implies that there is a unique solution of the equation $ax = a$. Since any identity element of G satisfies this equation we conclude that

G has a unique identity element.

Similarly, in the case $b = 1$ the theorem implies that there is a unique solution of the equation $ax = 1$. Since any inverse of a satisfies this equation, we conclude that

each element of G has a unique inverse.

As a consequence of these observations we are justified in speaking of *the* identity element of G and *the* inverse of a in G.

Exercises 13.3

1   Show that the inverse of ab is $b^{-1}a^{-1}$.

2   Establish the following implications, where x and y are any elements of a group:

(i) $xy = 1 \Rightarrow yx = 1$;
(ii) $(xy)^2 = x^2y^2 \Rightarrow xy = yx$.

(In part (ii) x^2 stands for xx.)

3   Suppose that G is a group with the property that $g^2 = 1$ for all g in G. Prove that G is a commutative group.

4   (i) Let $G = \{1, a, b, c\}$ be a group, where 1 is the identity and $a^2 = b^2 = c^2 = 1$. Using the latin square property, write out the complete group table for G.

(ii) Give a reason why the latin square obtained from the group table for $\mathbb{Z}_4$ (with respect to $+$) is essentially different from the one obtained in part (i).

5   Show that the following latin square of order 5 is not a group table.

1	a	b	c	d
a	b	1	d	c
b	c	d	a	1
c	d	a	1	b
d	1	c	b	a

13.4   The order of a group element

Given any element x of a group G we may define the positive and negative powers of x recursively, as follows:

$$x^1 = x, \qquad x^r = xx^{r-1} \qquad (r \geqslant 2),$$
$$x^{-1} = x^{-1}, \qquad x^{-s} = x^{-1}x^{-(s-1)} \quad (s \geqslant 2).$$

If we make the convention that $x^0 = 1$ (the identity) then x^n is defined for all integers n, and the familiar rules for manipulating powers hold. That is

$$x^m x^n = x^{m+n}, \qquad (x^m)^n = x^{mn} \quad (m, n \in \mathbb{Z}).$$

Since G is a group, x^n is a member of G for each n, but this does not mean that all the powers x^n represent different elements of G. Indeed, when G is a *finite* group some of the powers must be equal, since there are infinitely many of them and only finitely many elements of G. Suppose that $x^a = x^b$, where $a > b$. Multiplying both sides of the equation by x^{-b} we obtain $x^{a-b} = 1$, where $a - b > 0$. So we can be sure that there is some positive integer n for which $x^n = 1$ and, by the well-ordering axiom, there is a least integer with this property.

Definition If x is an element of a finite group G then the least positive integer m such that $x^m = 1$ is called the **order** of x in G. When G is an infinite group the order of x is defined in the same way provided that m exists; otherwise x is said to have **infinite order**.

In practice we compute the order of a group element by calculating its positive powers until the identity is obtained. For example, in the group of the triangle $G_\triangle$, the powers of the element r are

$$r^1 = r, \qquad r^2 = s, \qquad r^3 = rs = i.$$

Since i is the identity in $G_\triangle$ it follows that the order of r is 3.

The most useful fact about the order of a group element is contained in the next theorem.

Theorem 13.4 Let x be an element of order m in a finite group G. Then

$$x^s = 1 \quad \text{in} \quad G$$

if and only if s is a multiple of m.

Proof If s is a multiple of m, say $s = mk$, then

$$x^s = x^{mk} = (x^m)^k = (1)^k = 1,$$

where we use the rules for manipulating powers and the fact that $x^m - 1$.

Conversely, suppose $x^s = 1$. By Theorem 1.5 we can write

$$s = mq + r, \qquad 0 \leqslant r < m.$$

Then

$$1 = x^s = x^{mq+r} = (x^m)^q x^r = x^r$$

since $x^m = 1$. Now if $r > 0$ the equation $x^r = 1$ contradicts the definition of m as the least positive integer for which $x^m = 1$. Consequently we must have $r = 0$ and $s = mq$, so s is a multiple of m, as required. $\square$

1 Let α and β denote the permutations of $\mathbb{N}_7$ whose representations in cycle notation are

$$\alpha = (15)(27436), \qquad \beta = (1372)(46)(5).$$

Calculate the orders of α and β, considered as elements of the symmetric group S_7. What are the orders of $\alpha\beta$ and $\beta\alpha$?

2 Let x and y be elements of a finite group G. Show that the orders of x and yxy^{-1} are the same.

3 Let M denote the set of matrices of the form

$$\begin{bmatrix} a & b \\ 0 & 1 \end{bmatrix}$$

where the entries are members of $\mathbb{Z}_7$ and $a \neq 0$. Show that M is a group with respect to matrix multiplication, and find the orders of the elements

$$A = \begin{bmatrix} 1 & 1 \\ 0 & 1 \end{bmatrix}, \qquad B = \begin{bmatrix} 3 & 0 \\ 0 & 1 \end{bmatrix}.$$

4 Let G be the group defined as in Ex.3, except that the entries of the matrices are now real numbers rather than elements of $\mathbb{Z}_7$. Show that G contains infinitely many elements of order 2.

5 Let u and v be elements of a commutative group, and suppose that their orders are r and s respectively. Show that if r and s are coprime then the order of uv is rs.

13.5 Isomorphism of groups

Recall the two examples considered in Section 13.2. In *Example 1* we discussed the group $G_\triangle$ whose elements are the six symmetries of an equilateral triangle, $G_\triangle = \{i, r, s, x, y, z\}$. In *Example 2* we discussed another group G_M which turns out to have six elements; they are the matrices which we shall denote by

$$I = \begin{bmatrix} 1 & 0 \\ 0 & 1 \end{bmatrix}, \qquad R = \begin{bmatrix} 1 & 1 \\ 0 & 1 \end{bmatrix}, \qquad S = \begin{bmatrix} 1 & 2 \\ 0 & 1 \end{bmatrix},$$

$$X = \begin{bmatrix} 2 & 0 \\ 0 & 1 \end{bmatrix}, \qquad Y = \begin{bmatrix} 2 & 1 \\ 0 & 1 \end{bmatrix}, \qquad Z = \begin{bmatrix} 2 & 2 \\ 0 & 1 \end{bmatrix},$$

where the symbols 0, 1, and 2 are elements of $\mathbb{Z}_3$. For the moment let us forget the definitions of $G_\triangle$ and G_M and concentrate on their group tables (Tables 13.5.1a,b).

Table 13.5.1(a)

	i	r	s	x	y	z
i	i	r	s	x	y	z
r	r	s	i	y	z	x
s	s	i	r	z	x	y
x	x	z	y	i	s	r
y	y	x	z	r	i	s
z	z	y	x	s	r	i

Table 13.5.1(b)

	I	R	S	X	Y	Z
I	I	R	S	X	Y	Z
R	R	S	I	Y	Z	X
S	S	I	R	Z	X	Y
X	X	Z	Y	I	S	R
Y	Y	X	Z	R	I	S
Z	Z	Y	X	S	R	I

Clearly, the group tables are essentially the same—the notation has been carefully chosen to emphasize this. Thus, insofar as group properties are concerned, $G_\triangle$ and G_M are identical; they differ only in the names given to their elements. More formally, we have a bijection $\beta: G_\triangle \to G_M$ which takes i to I, r to R, and so on, and which preserves the group operation. For example, $rx = y$ in $G_\triangle$ and $RX = Y$ in G_M, which implies that

$$\beta(rx) = \beta(y) = Y = RX = \beta(r)\beta(x).$$

Definition If G_1 and G_2 are groups (both written in the multiplicative notation), then a bijection $\beta: G_1 \to G_2$ is said to be an **isomorphism** if, for all g and g' in G_1

$$\beta(gg') = \beta(g)\beta(g').$$

When there is such an isomorphism, G_1 and G_2 are said to be **isomorphic**, and we write $G_1 \approx G_2$.

Exercises 13.5

1 Describe the four symmetries of a rectangle, and construct the group table. By writing down a suitable bijection show that your group is isomorphic to the one whose group table is Table 13.5.2.

Table 13.5.2

	1	a	b	c
1	1	a	b	c
a	a	1	c	b
b	b	c	1	a
c	c	b	a	1

2 By analysing the possible group tables show that, if isomorphic groups are regarded as the same, then

(i) there is just one group of order 2;
(ii) there is just one group of order 3;
(iii) there are just two groups of order 4.

3 Show that the isomorphism relation $\approx$ is an equivalence relation.

4 Suppose that G_1 and G_2 are finite groups and $\beta: G_1 \to G_2$ is an isomorphism. If $x_2 = \beta(x_1)$ for a given element x_1 in G_1, prove that x_1 and x_2 have the same order.

13.6 Cyclic groups

From the abstract point of view, two isomorphic groups are the same. So the notion of isomorphism allows us to classify groups in an obvious way. In practice, if we encounter some group G in a particular context, we usually set out to show that it is isomorphic to a 'standard example' H, whose properties are already worked out. Then the group-theoretical properties of G are precisely those of H, and we do not have to study them afresh.

In order to carry out the programme outlined above we must, naturally, have a stock of standard examples. We now begin to establish this stock.

Definition   A group G is said to be **cyclic** if it contains an element x such that every member of G is a power of x. The element x is said to **generate** G, and we write $G = \langle x \rangle$.

If x generates G, and all the powers of x are *distinct*, then

$$G = \{\ldots, x^{-3}, x^{-2}, x^{-1}, 1, x, x^2, x^3, \ldots\}.$$

In this case we say that G is an **infinite cyclic group**, and we use the symbol C_∞ for a typical group of this kind. Groups isomorphic to C_∞ are very common and they occur in many disguises.

Example   Show that the set $\mathbb{Z}$ of integers, with the ordinary addition operation, is an infinite cyclic group.

Solution   We have to construct a bijection β from $\mathbb{Z}$ to C_∞ such that $\beta(n_1 + n_2) = \beta(n_1)\beta(n_2)$. Note that the $+$ sign occurs on the left-hand side because addition is the specified operation in $\mathbb{Z}$. Taking x to be a generator of C_∞ define β by the rule

$$\beta(n) = x^n \quad (n \in \mathbb{Z}).$$

There is a power x^n in C_∞ for each n in $\mathbb{Z}$, and these powers are all distinct, so β is a bijection. Furthermore

$$\beta(n_1 + n_2) = x^{n_1 + n_2} = x^{n_1} x^{n_2} = \beta(n_1)\beta(n_2),$$

so that β is an isomorphism, as required. $\qquad\qquad\qquad\square$

We now turn to the case of a cyclic group G with generator x such that the powers of x are not all distinct. In this case x is an element of finite order m and we can show that

$$G = \{1, x, x^2, \ldots, x^{m-1}\}.$$

For if k is any integer, we can write $k = mq + r$ with $0 \le r < m$, and so

$$x^k = x^{mq+r} = (x^m)^q x^r = 1^q x^r = x^r.$$

Consequently, any power of x is equal to one of the m elements listed above. Furthermore, these elements are all different, since if $x^i = x^j$ $(0 \le i < j \le m-1)$ then $x^{j-i} = 1$, where $0 < j - i < m$, contradicting the definition of m. In this case G is said to be a **cyclic group of order m**, denoted by C_m. The most familiar concrete instance of a group isomorphic to C_m is the set $\mathbb{Z}_m$ of integers modulo m with respect to the $+$ operation. As in the infinite case it is easy to show that the function β from $\mathbb{Z}_m$ to C_m defined by $\beta(n) = x^n$ is an isomorphism.

One way of extending the list of standard examples of groups is to combine known groups in some way. For example, if we are given two groups A and B (both written in the multiplicative notation) we can form their **direct product** $A \times B$ as follows. The elements of $A \times B$ are the ordered pairs

$$(a, b) \quad (a \in A, \, b \in B),$$

and these elements are combined under the group operation defined by

$$(a_1, b_1)(a_2, b_2) = (a_1 a_2, b_1 b_2).$$

Note that on the right-hand side $a_1 a_2$ denotes the result of combining a_1 and a_2 under the group operation in A, and $b_1 b_2$ denotes the result of combining b_1 and b_2 under the group operation in B.

It is easy to check that $A \times B$ is indeed a group under the operation defined above (Ex. 13.6.3).

Example   List the elements of $C_2 \times C_3$ and $C_2 \times C_4$. Show that $C_2 \times C_3$ is isomorphic to C_6, but $C_2 \times C_4$ is not isomorphic to C_8.

Solution   Suppose C_2 and C_3 are generated by x and y respectively, so that

$$C_2 = \langle x \rangle = \{1, x\}, \qquad C_3 = \langle y \rangle = \{1, y, y^2\}.$$

According to the definition of the direct product, the elements of $C_2 \times C_3$ are:

$$(1, 1), (1, y), (1, y^2), (x, 1), (x, y), (x, y^2).$$

If we let $z = (x, y)$, then calculating according to the rule given above we find that

$$z^2 = (1, y^2), \qquad z^3 = (x, 1), \qquad z^4 = (1, y),$$
$$z^5 = (x, y^2), \qquad z^6 = (1, 1).$$

Since $(1, 1)$ is the identity in $C_2 \times C_3$, we have shown that the elements of $C_2 \times C_3$ can be written as $1, z, z^2, z^3, z^4, z^5$, and consequently the group is a cyclic group of order 6.

Suppose that C_4 is generated by u, so that

$$C_4 = \langle u \rangle = \{1, u, u^2, u^3\}.$$

The elements of $C_2 \times C_4$ are:

$$(1, 1), (1, u), (1, u^2), (1, u^3), (x, 1), (x, u), (x, u^2), (x, u^3),$$

and their orders are, respectively,

$$1, \quad 4, \quad 2, \quad 4, \quad 2, \quad 4, \quad 2, \quad 4.$$

Thus there is no element of order 8 in $C_2 \times C_4$ and consequently $C_2 \times C_4$ cannot be isomorphic to C_8. $\square$

The fact that $C_2 \times C_3 \approx C_6$ is a special case of the following theorem.

Theorem 13.6   If m and n are coprime positive integers then

$$C_m \times C_n \approx C_{mn}.$$

Proof   Suppose C_m, C_n are generated by x, y respectively, and let z denote the element (x, y) of $C_m \times C_n$. Let r be the order of z in $C_m \times C_n$. We shall show that $r = mn$.

Since $z^r = (x^r, y^r)$, and the identity of $C_m \times C_n$ is $(1, 1)$, the equation $z^r = 1$ implies that $x^r = 1$ in C_m and $y^r = 1$ in C_n. Consequently, by Theorem 13.4, r is a multiple of m and a multiple of n.

Since r is the least positive integer for which $z^r = 1$, it must be the *least* common multiple of m and n. Hence, by Ex. 1.9.8,

$$r = \text{lcm}\,(m, n) = \frac{mn}{\gcd\,(m, n)} = mn,$$

since m and n are coprime.

Since $C_m \times C_n$ has mn elements, and it contains an element z of order mn, it must be a cyclic group. $\qquad\square$

Exercises 13.6

1   Let U be the subset of $\mathbb{Z}_7$ which contains all the elements of $\mathbb{Z}_7$ except 0. Show that multiplication in $\mathbb{Z}_7$ defines a group operation for U, and that $U \approx C_6$.

2   Let M denote the set of 2×2 matrices of the form

$$\begin{bmatrix} 1 & m \\ 0 & 1 \end{bmatrix},$$

where the entries are integers. Show that M, with respect to matrix multiplication, is an infinite cyclic group.

3   Write out the details of the proof that the direct product $A \times B$ of two groups is a group.

4   Show that the two distinct groups of order 4 which you obtained in Ex. 13.5.2 are isomorphic to C_4 and $C_2 \times C_2$.

13.7 Subgroups

A subset H of a group G is said to be a **subgroup** of G if the members of H form a group with respect to the group operation in G.

Example   Let $G_\triangle = \{i, r, s, x, y, z\}$ be the group of symmetries of the triangle, as in *Example 1*, Section 13.2. Which of the following subsets of $G_\triangle$ are subgroups?

$$H_1 = \{r, s, z\}, \qquad H_2 = \{i, r, s\}, \qquad H_3 = \{i, r, s, x\}.$$

Solution H_1 is not a subgroup, for several reasons. It is not closed, because although r and z are H_1 the product rz is equal to x, which is not in H_1. Also H_1 has no identity element.

On the other hand, H_2 *is* a subgroup of $G_\triangle$. Constructing the 'multiplication table' for the elements of H_2 we obtain

	i	r	s
i	i	r	s
r	r	s	i
s	s	i	r

which shows that H_2 is closed. The associativity property holds in H_2 since it holds in $G_\triangle$, and clearly i is the identity in H_2, as in $G_\triangle$. Finally $i^{-1} = i$, $r^{-1} = s$, and $s^{-1} = r$.

The subset H_3 is not a subgroup since it is not closed; for example, $rx = y$ and y is not in H_3. ☐

Although we can always verify the subgroup property by working out the group table, this is not often a very practical method. Indeed, it might be said that the main purpose of group theory is to study groups without recourse to group tables. For this reason we shall formulate a theorem giving sufficient conditions for a subset to be a subgroup.

Theorem 13.7

Let G be a group, and suppose that H is a non-empty subset of G satisfying the conditions

S1. $x, y \in H \ \Rightarrow \ xy \in H$;

S2. $x \in H \ \Rightarrow \ x^{-1} \in H$.

Then H is a subgroup of G. If G is finite, then **S1** alone is sufficient to ensure that H is a subgroup.

Proof

The stated conditions assert that H is closed, and that every element of H has an inverse in G. The associativity property for elements of H follows from the corresponding property in G, since the elements of H all belong to G. Finally, to show that the identity 1 of G is in H (and consequently that it is the identity in H), we may argue as follows. If x is any member of H then x^{-1} is in H (by **S2**), whence xx^{-1} is in H (by **S1**). But $xx^{-1} = 1$, so 1 is in H, as required.

It remains to show that, in the finite case, **S1** implies **S2**. If H contains only the element 1 it is clearly a subgroup. Otherwise, let x be any element of H except 1, and let m be the order of x. Multiplying the equation $x^m = 1$ by x^{-1} we obtain $x^{m-1} = x^{-1}$ and, since $m > 1$, x^{-1} is equal to a positive power of x. But using **S1** repeatedly we see that any positive power of x is in H, and so x^{-1} is in H. Thus **S2** is a consequence of **S1**. ☐

Example   Given any group G, let $Z(G)$ denote the subset consisting of those elements which commute with every element of G, that is

$$Z(G) = \{z \in G \mid zg = gz \text{ for all } g \text{ in } G\}.$$

Show that $Z(G)$ is a subgroup of G (it is called the **centre** of G).

Solution   We verify the conditions **S1** and **S2**. Suppose x and y are in $Z(G)$ so that

$$xg = gx \quad \text{and} \quad yg = gy$$

for all g in G. We have

$$(xy)g = x(yg) = x(gy) = (xg)y = (gx)y = g(xy),$$

so that xy is in G, and **S1** is verified. Also, multiplying the equation $xg = gx$ by x^{-1} on both sides we get

$$x^{-1}(xg)x^{-1} = x^{-1}(gx)x^{-1},$$

and using the associative law we obtain $gx^{-1} = x^{-1}g$. Thus x^{-1} is in $Z(G)$ and **S2** is verified.   □

If x is an element of order m in a group G then the elements

$$1, x, x^2, \ldots, x^{m-1}$$

of G form the **cyclic subgroup** $\langle x \rangle$ generated by x. It is very easy to verify that $\langle x \rangle$ is indeed a subgroup (Ex. 13.7.6), and clearly

the order of x is equal to the order of the subgroup $\langle x \rangle$.

For example, let U be the group of non-zero elements of $\mathbb{Z}_7$, with respect to multiplication (Ex. 13.6.1). Computing the powers of 2 in U we find

$$2^1 = 2, \qquad 2^2 = 4, \qquad 2^3 = 1,$$

so 2 has order 3 and the cyclic subgroup $\langle 2 \rangle$ contains the three elements 1, 2, 4. On the other hand the cyclic subgroup $\langle 3 \rangle$ is the whole of U since

$$3^1 = 3, \quad 3^2 = 2, \quad 3^3 = 6, \quad 3^4 = 4, \quad 3^5 = 5, \quad 3^6 = 1.$$

Exercises 13.7

1   Which of the following are subgroups of $G_\triangle$?

$$K_1 = \{i, x\}, \qquad K_2 = \{i, x, y\}, \qquad K_3 = \{i, r, s, x, y\}.$$

2   Use the group $G_\triangle$ to provide an example of the fact that if H and K are subgroups then $H \cup K$ need not be a subgroup.

3   Show that if H and K are subgroups of G then so is $H \cap K$.

4   Let g be a given element of a group G and let $C(g)$ denote the set of elements of G which commute with g, that is

$$C(g) = \{x \in G \mid xg = gx\}.$$

Show that $C(g)$ is a subgroup of G. What is the relationship between these subgroups and the centre $Z(G)$?

5   If G is the group of the square (Ex. 13.2.2), find $C(g)$ for each g in G, and thence find $Z(G)$.

6   Use Theorem 13.7 to verify that if x is an element of order m in a group G then $\langle x \rangle = \{1, x, \ldots, x^{m-1}\}$ is a subgroup of G.

13.8   Cosets and Lagrange's theorem

In this section we shall prove the first really interesting theorem about groups. Anyone who understands the significance of this theorem will have no difficulty in believing that the theory of groups is a subject rich in elegant and fascinating results.

The theorem asserts that if H is a subgroup of a finite group G then $|H|$ is a divisor of $|G|$. So, for example, a group of order 20 can have subgroups only of the orders 1, 2, 4, 5, 10, 20. The idea of the proof is to partition G in such a way that each of the parts has the same size as H. If there are k such parts, then we must have $|G| = k |H|$, and the result follows. The parts are known as cosets.

Definition   Let H be a subgroup of a (not necessarily finite) group G. The **left coset** gH of H with respect to an element g in G is defined to be the set obtained by multiplying each element of H on the left by g, that is

$$gH = \{x \in G \mid x = gh \text{ for some } h \in H\}.$$

The **right coset** of H with respect to g is defined analogously:

$$Hg = \{x \in G \mid x = hg \text{ for some } h \in H\}.$$

If H is a finite subgroup, say $H = \{h_1, h_2, \ldots, h_m\}$, then the elements which belong to the left coset gH are

$$gh_1, gh_2, \ldots, gh_m.$$

It is clear that they are all distinct, since if $gh_i = gh_j$ the left cancellation rule implies that $h_i = h_j$. Thus we have a fundamental property of cosets:

$$|gH| = |H| \quad (g \in G).$$

For example, let $G_\triangle = \{i, r, s, x, y, z\}$ be the group of the triangle, and let H be the subgroup $\{i, x\}$. The left cosets of H in $G_\triangle$ are

$$iH = \{ii, ix\} = \{i, x\}, \qquad rH = \{ri, rx\} = \{r, y\},$$
$$sH = \{si, sx\} = \{s, z\}, \qquad xH = \{xi, xx\} = \{x, i\},$$
$$yH = \{yi, yx\} = \{y, r\}, \qquad zH = \{zi, zx\} = \{z, s\}.$$

We remark that only three distinct subsets of $G_\triangle$ occur as left cosets H, and they are disjoint. Consequently, we have the partition

$$G_\triangle = \{i, x\} \cup \{r, y\} \cup \{s, z\},$$

as displayed in Fig. 13.3, where each part is a left coset of H.

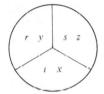

Fig. 13.3
The left cosets of $\{i, x\}$ in $G_\triangle$.

     The point which often confuses beginners is that the parts have alternative names when they are regarded as cosets, so that $\{r, y\}$ (for example) may be denoted by rH or yH. Consequently there are several different ways of writing $G_\triangle$ as a disjoint union of left cosets, such as

$$G_\triangle = iH \cup rH \cup sH \quad \text{or} \quad G_\triangle = xH \cup yH \cup zH,$$

although the parts themselves are the same in each case.

    The next theorem establishes the general property observed in the example.

Theorem 13.8.1   Let H be a subgroup of a group G. If g_1 and g_2 are any elements of G, the left cosets g_1H and g_2H are either identical or they have no elements in common.

Proof   We shall show that if g_1H and g_2H have one common element, then they are identical. Suppose that x belongs to g_1H and g_2H, that is

$$x = g_1h_1 \text{ for some } h_1 \in H, \qquad x = g_2h_2 \text{ for some } h_2 \in H.$$

In order to show that $g_1H \subseteq g_2H$, let y be any element of g_1H, so that $y = g_1 h$ for some $h \in H$. Then

$$
\begin{aligned}
y = g_1 h = (x h_1^{-1})h &= x(h_1^{-1} h) \\
&= (g_2 h_2)h_1^{-1} h \\
&= g_2(h_2 h_1^{-1} h).
\end{aligned}
$$

Since H is a subgroup, $h_2 h_1^{-1} h$ is in H, and it follows that y is in $g_2 H$. Thus $g_1 H \subseteq g_2 H$. An analogous argument with g_1 and g_2 reversed shows that $g_2 H \subseteq g_1 H$, so $g_1 H$ and $g_2 H$ are identical, as claimed.   □

Another way of proving Theorem 13.8.1 uses the theory of equivalence relations. If we define a relation $\sim$ on G by saying that

$$
x \sim y \quad \text{means} \quad x^{-1}y \in H
$$

then $\sim$ is an equivalence relation and the equivalence classes are the left cosets (Ex. 13.8.1). If follows from the basic property of equivalence relations (Theorem 5.2) that the distinct left cosets form a partition of G. This fundamental observation leads immediately to our main theorem, which is known as **Lagrange's theorem,** after J. L. Lagrange (1736–1813).

Theorem 13.8.2   If G is a finite group of order n and H is a subgroup of order m, then m is a divisor of n.

Proof   We have seen that each left coset has the same cardinality m as H, and the distinct left cosets form a partition of G. So if there are k distinct left cosets we must have $n = km$.   □

The number of distinct left cosets is called the **index** of H in G, and is written $|G:H|$; that is

$$
|G:H| = |G|/|H|.
$$

Of course, we could use right cosets instead of left cosets, and obtain the same results. However, it should be noted that although the numbers of left and right cosets are the same, it is not true that the left cosets and the right cosets form the same partition of G. (See Ex. 13.8.2.)

Many useful theorems about groups flow directly from Lagrange's theorem. We give two examples here.

Theorem 13.8.3   Let g be an element of a finite group G and suppose $|G| = n$. Then

(i) the order of g divides n,   (ii) $g^n = 1$.

Proof (i) The order of g is the same as the order of the cyclic subgroup $\langle g \rangle$, and by Lagrange's theorem this is a divisor d of n.
(ii) If $dk = n$, then, since $x^d = 1$, we have

$$x^n = (x^d)^k = 1^k = 1. \qquad \square$$

Theorem 13.8.4 If G is a group whose order is a prime p, then G is isomorphic to the cyclic group C_p.

Proof Since $p > 1$, G has an element $x \neq 1$. The order of the cyclic subgroup $\langle x \rangle$ is greater than 1, and by Lagrange's theorem it is a divisor of p. Since p is prime, the order of $\langle x \rangle$ is p and so $\langle x \rangle$ is the whole of G. Consequently, G is a cyclic group of order p. $\square$

In practice, Lagrange's theorem is important because it restricts the order of subgroups (and of group elements) so drastically. However, it does not provide any information about the number of subgroups. In the next section we shall show that when G is cyclic there is exactly one subgroup corresponding to each divisor d of $|G|$, but in general there may be any number, including zero, of subgroups of order d. The following *Example* displays some of the more complicated features which the set of subgroups can have in the general case.

Example Show that the set A_4 of all even permutations of $\{1, 2, 3, 4\}$ is a group of order 12, and find all its subgroups.

Solution In Section 5.6 we showed that the composite of two even permutations is even and so, by Theorem 13.7, A_4 is a group (actually, a subgroup of the symmetric group S_4). According to Theorem 5.6.2 the order of A_4 is $\frac{1}{2}(4!) = 12$, and the list of permutations which belong to A_4 is as follows.

The identity permutation: id.
Three permutations of order 2: (12)(34), (13)(24), (14)(23).
Eight permutations of order 3: (123), (132), (124), (142), (134), (143), (234), (243).

By Lagrange's theorem, the possible orders of subgroups of A_4 are 1, 2, 3, 4, 6, and 12. We examine the possibilities in turn.

Order 1 The trivial subgroup $\{\text{id}\}$ is the only possibility.

Order 2 By Theorem 13.8.4 (or the simple-minded method of Ex. 13.5.2) any subgroup of order 2 must be cyclic. Since A_4 has just three elements of order 2 there are three subgroups of order 2, such as $\{\text{id}, (12)(34)\}$.

Order 3 By similar arguments, there are four (why not eight?) subgroups of order 3, such as {id, (123), (132)}.

Order 4 A subgroup of order 4 is isomorphic to C_4 or $C_2 \times C_2$ (Ex. 13.6.4). Since there are no elements of order 4, C_4 is impossible. The three elements of order 2, namely (12)(34), (13)(24), (14)(23), together with the identity, do form a subgroup isomorphic to $C_2 \times C_2$; and this is the only possibility, since the elements of order 3 cannot belong to a subgroup of order 4 (why?).

Order 6 Suppose K is a subgroup of order 6. Since there are only four elements of A_4 which do not have order 3, K must contain at least one element of order 3, say $x = (123)$, and its inverse $x^{-1} = (132)$. The elements of order 3 in K occur in pairs, like x and x^{-1}, and so there are an even number of them. But K also contains the identity, and so in order to make the total number of elements even, K must contain at least one of the elements of order 2. Now if one such element y is in K, so are xyx^{-1} and $x^{-1}yx$ (by the subgroup property). These elements are conjugate to y and so they have the same type as y, and they are distinct. Hence the elements y, xyx^{-1}, and $x^{-1}yx$ are the three elements of order 2 (that is, (12)(34), (13)(24), and (14)(23)) in some order.

But we showed that these three elements, together with the identity, form a subgroup of order 4, which must be a subgroup of K. By Lagrange's theorem, the order of K is a multiple of 4, contradicting the assumption that $|K| = 6$. Hence there are no subgroups of order six.

Order 12 Clearly the only possibility is A_4 itself. ☐

It is convenient to arrange the subgroups in a *lattice* (Fig. 13.4). In general, the lattice of subgroups can be extremely complicated, but in the next section we shall see that in the case of a cyclic group the lattice can be described in simple arithmetical terms.

Exercises 13.8

1   Let H be a subgroup of G, and define a relation $\sim$ on G by the rule that $x \sim y$ means $x^{-1}y \in H$. Show that $\sim$ is an equivalence relation and its equivalence classes are the left cosets of H.

2   Describe explicitly the partition of the triangle group by the *right* cosets of the subgroup $H = \{i, x\}$. Check that the partition is not the same as that given by the left cosets of H.

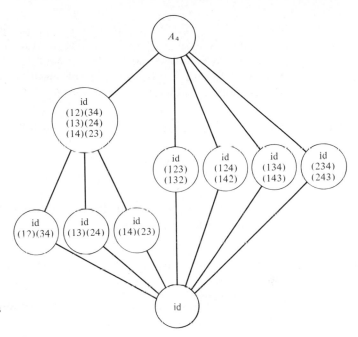

Fig. 13.4
The lattice of subgroups
of A_4.

3 The symmetry group of a regular pentagon is a group of order 10. Show that it has subgroups of each of the orders allowed by Lagrange's theorem, and sketch the lattice of subgroups.

4 Suppose that a finite group G and a prime number p are given, and G has exactly m subgroups of order p. Show that the number of elements of order p in G is $m(p-1)$.

5 Use Ex. 4 to show that a non-cyclic group of order 55 has at least one subgroup of order 5 and at least one subgroup of order 11.

6 Sketch the lattice of subgroups of the symmetric group S_4. [Hint: you will need a large sheet of paper.]

13.9 Characterization of cyclic groups

In Lagrange's theorem we begin with an algebraic hypothesis (H is a subgroup of G) and end with an arithmetical conclusion ($|H|$ divides $|G|$). Results of this kind are extremely useful because they reduce hard algebraic problems to simpler arithmetical ones. In this section we shall discuss the arithmetical properties of cyclic groups, and show that such groups can be characterized by numerical properties. We shall use some of the counting techniques developed in Chapter 3 and 4, in particular the method of Möbius inversion and its relation with Euler's function ϕ.

Theorem 13.9

If G is a finite group of order $n \geqslant 2$ the following statements are equivalent.

(i) G is a cyclic group.

(ii) For each divisor d of n the number of elements x in G which satisfy $x^d \doteq 1$ is d.

(iii) For each divisor d of n the number of elements x in G which have order d is $\phi(d)$.

Proof

We shall show that (i) $\Rightarrow$ (ii) $\Rightarrow$ (iii) $\Rightarrow$ (i), from which it follows that the statements are equivalent.

(i) $\Rightarrow$ (ii)   Suppose G is a cyclic group of order n generated by g. Given any divisor d of n, let $dk = n$. The elements

$$1, g^k, g^{2k}, \ldots, g^{(d-1)k}$$

are distinct, and each of them satisfies the equation $x^d = 1$ since

$$(g^{ik})^d = (g^{kd})^i = (g^n)^i = 1^i = 1.$$

Thus we have d elements of G satisfying $x^d = 1$. We must show that there are no other solutions. Let y be any element of G such that $y^d = 1$; since G is generated by g we have $y = g^e$ for some $e \geqslant 0$, and consequently

$$g^{ed} = (g^e)^d = y^d = 1.$$

The order of g is n and so, by Theorem 13.4, ed is a multiple of n, say ln. Thus

$$ed = ln = l(dk),$$

so $e = lk$ and $y = g^e = g^{lk}$, which is equal to one of the elements g^{ik} $(0 \leqslant i \leqslant d-1)$ already known to be solutions of $x^d = 1$. Hence these are the only solutions.

(ii) $\Rightarrow$ (iii)   Suppose statement (ii) holds. By Theorem 13.4 an element x satisfies $x^d = 1$ if and only if the order of x is a divisor c of d. Hence if there are $\alpha(c)$ elements of order c we must have

$$d = \sum_{c \mid d} \alpha(c).$$

By the Möbius inversion formula,

$$\alpha(d) = \sum_{c|d} \mu(c)\frac{n}{c}.$$

But the relation between μ and ϕ (page 77) shows that the right-hand side is equal to $\phi(d)$. Hence $\alpha(d) = \phi(d)$ as claimed.

(iii) $\Rightarrow$ (i)   If (iii) holds we know, in particular, that the number of elements of order n is $\phi(n)$. Now $\phi(n) \geqslant 1$, since 1 is always coprime to n, and hence G contains at least one element of order n. This element generates the whole of G (since $|G| = n$), and so G is a cyclic group.   □

In Chapter 16 we shall use this numerical characterization to show that an important class of groups, arising in a wider algebraic context, are cyclic groups.

For the moment, it is instructive to use the theorem to determine all the subgroups H of a cyclic group G of order n. By Lagrange's theorem, we know that $|H| = d$, where $d \mid n$, and by Theorem 13.8.3 (ii), each of the d elements of H satisfies $x^d = 1$. But we have shown that G contains *exactly* d elements satisfying $x^d = 1$: specifically, the elements $1, g^k, \ldots, g^{(d-1)k}$, where g generates G and $dk = n$. Thus H must contain precisely those elements. In summary, we have shown that

> a cyclic group of order n has just one subgroup of
> each order d dividing n, and these subgroups are cyclic.

For example, consider the cyclic group C_{12} generated by an element z. Each of the elements $1, z^1, \ldots, z^{11}$ of C_{12} generates a cyclic subgroup of C_{12}, and we now know that these are the only subgroups. Furthermore, any two subgroups having the same order are identical. By simple calculations we can verify these facts explicitly, as in Table 13.9.1. Another way of illustrating the result is to say that the lattice of subgroups of C_{12} is the same as the lattice of divisors of 12 (Fig. 13.5).

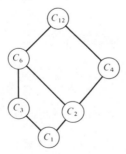

Fig. 13.5
The lattice of subgroups
of C_{12}.

Table 13.9.1

Subgroup	Elements	Isomorphism class
$\langle 1 \rangle$	1	C_1
$\langle z^6 \rangle$	$1, z^6$	C_2
$\left.\begin{array}{l}\langle z^4 \rangle \\ \langle z^8 \rangle\end{array}\right\}$	$1, z^4, z^8$	C_3
$\left.\begin{array}{l}\langle z^3 \rangle \\ \langle z^9 \rangle\end{array}\right\}$	$1, z^3, z^6, z^9$	C_4
$\left.\begin{array}{l}\langle z^2 \rangle \\ \langle z^{10} \rangle\end{array}\right\}$	$1, z^2, z^4, z^6, z^8, z^{10}$	C_6
$\left.\begin{array}{l}\langle z \rangle \\ \langle z^5 \rangle \\ \langle z^7 \rangle \\ \langle z^{11} \rangle\end{array}\right\}$	$1, z, z^2, \ldots, z^{11}$	C_{12}

Exercises 13.9

1 Sketch the lattice of subgroups of the cyclic group C_{24}. If z is a generator of C_{24}, identify the subgroups generated by z^7, z^8, and z^9.

2 How many elements of C_{60} generate the whole group?

3 Suppose that r and s are divisors of n, and the cyclic group C_n is generated by x. Write down generators for the cyclic subgroups C_r and C_s of C_n. Find the order of $C_r \cap C_s$, and write down a generator for this subgroup.

4 Explain how Theorem 13.8.4 can be deduced from Theorem 13.9.

13.10 Miscellaneous Exercises

1 Suppose x, y, and z are elements of a group G. Simplify the following expressions in G.

(i) $(x^{-1}z^{-1})(y^{-1}x)^{-1}y$,   (ii) $(xzy)^{-1}x(xyz^{-1})^{-1}$.

2   When x and y are elements of a group G the **conjugate** of x with respect to y is yxy^{-1}, which is written x^y, and the **commutator** of x and y is $xyx^{-1}y^{-1}$, which is written $[x, y]$. Show that, for any elements a, b, c in G,

(i) $[ab, c] = [b, c]^a[a, c],$ (ii) $[a, bc] = [a, b][a, c]^b.$

3   Let $t_{a,b}$ denote the function from the set $\mathbb{R}$ of real numbers to itself defined by

$$t_{a,b}(x) = ax + b,$$

where a and b are real numbers and $a \neq 0$. Show that the set of all such functions forms a group under the operation of composition of functions.

4   Show that the group defined in the previous exercise contains infinitely many elements of order 2.

5   Show that if H is a subgroup of index 2 in a group G then the left coset gH is the same as the right coset Hg for all g in G.

6   Define a binary operation $*$ on the set $\mathbb{R}$ of real numbers by $x * y = xy + x + y$, where the symbols on the right-hand side of the equation have their usual meanings. Show that, with respect to $*$, $\mathbb{R}$ satisfies three of the axioms for a group, but that not every element of $\mathbb{R}$ has an inverse.

7   Let α and β denote the functions defined (on a suitable subset of the real numbers) by

$$\alpha(x) = 1/x, \qquad \beta(x) = 1/(1 - x).$$

Let K denote the group obtained by the composition of α and β and their powers in all possible ways. Show that K is isomorphic to the triangle group $G_\triangle$ defined in Section 13.2.

8   Let J_m denote the set of complex numbers z which satisfy $z^m = 1$, where m is a positive integer. Show that J_m is a cyclic group of order m, with respect to the usual multiplication of complex numbers.

9   Show that the set of all 2×2 matrices

$$\begin{bmatrix} a & b \\ c & d \end{bmatrix} \qquad (a, b, c, d \in \mathbb{R}),$$

such that $ad \neq bc$, is a group under the usual matrix multiplication.

10   Let K and L be finite subgroups of a group G and let

$$KL = \{g \in G \mid g = kl \text{ for some } k \in K \text{ and } l \in L\}$$

Prove that $|K||L| = |KL||K \cap L|$.

11   Show that the subset KL defined in the previous exercise is a subgroup of G if and only if $KL = LK$.

12   Show that any group of order six is isomorphic to C_6 or S_3.

13 Let H and K be subgroups of a finite group G such that gcd $(|H|, |K|) = 1$. Show that $|H \cap K| = 1$.

14 Let X be the set of ordered pairs (n, f) where n is an integer and f is a rational number (fraction). Define a binary operation $\circ$ on X by the rule

$$(n_1, f_1) \circ (n_2, f_2) = (n_1 + n_2, 2^{n_2}f_1 + f_2).$$

Show that X is a group with respect to the $\circ$ operation. Which of the following are subgroups of X?
  (i) The subset of elements of the form $(n, 0)$.
  (ii) The subset of elements of the form $(0, f)$.

15 Let M be the group obtained by the construction given in Ex. 3 when the set $\mathbb{R}$ is replaced by the set $\mathbb{Z}_5$ of integers modulo 5. Show that
  (i) $|M| = 20$;
  (ii) M contains subgroups of each of the orders allowed by Lagrange's theorem.
Sketch the lattice of subgroups of this group. (It may be helpful to consider elements of M as permutations of $\mathbb{Z}_5 = \{0, 1, 2, 3, 4\}$.)

16 Write out the partition of A_4 (Section 13.8, *Example*) formed by the left cosets of the cyclic subgroup generated by (123), and verify that the partition formed by the right cosets of the same subgroup is different.

17 Show that S_5 has no subgroup of order 15.

18 Let p be a prime. The group $GL_2(\mathbb{Z}_p)$ is defined to be the group of those 2×2 matrices with entries in $\mathbb{Z}_p$ which have an inverse with respect to matrix multiplication. Show that

  (i) $GL_2(\mathbb{Z}_2)$ is isomorphic to the symmetric group S_3.
  (ii) The order of $GL_2(\mathbb{Z}_p)$ is $p(p^2 - 1)$.
  (iii) The centre of $GL_2(\mathbb{Z}_p)$ consists of the matrices αI, where I is the identity matrix and $\alpha \neq 0$ in $\mathbb{Z}_p$.

19 Let F denote the set of all functions $f : \mathbb{Z} \to \mathbb{Z}$, and let V be the set of ordered pairs (n, f) such that n is in $\mathbb{Z}$ and f is in F. Define a binary operation $*$ on V by the rule

$$(n_1, f_1) * (n_2, f_2) = (n_1 + n_2, f_3)$$

where $f_3(m) = f_1(m - n_1) + f_2(m)$. Show that V is a group with respect to the $*$ operation.

20 Let V and F be as in the previous exercise, and let E be the subset of F containing those functions f for which $f(n) = n$ except for a finite set of integers n. Show that the subset of elements of the form $(0, f)$ for which is f is in E is a subgroup of V.

21 Show that there are just five different (that is, mutually non-isomorphic) groups of order eight.

22 Let x be an element of order m in a finite group G. Show that the order of x^t in G is m/d, where $d = \gcd(m, t)$.

23   Let G be a finite group and H a subgroup of index k in G. Prove that there is a set $\{g_1, g_2, \ldots, g_k\}$ of elements of G which is simultaneously a set of representatives for the left cosets, and for the right cosets, of H in G. In other words, we can write

$$G = g_1H \cup \cdots \cup g_kH = Hg_1 \cup \cdots \cup Hg_k.$$

[Hint: Ex. 10.7.15.]

14
Groups of permutations

14.1 Definitions and examples

When we began our study of permutations we remarked that the rule of composition in the symmetric group S_n has four fundamental properties. In the previous chapter we used those properties as the axioms for a group. Now we shall return to the study of permutations, using the group-theoretical terminology to guide our investigations.

Let G be a set of permutations of a finite set X. If G is a group (with respect to the rule for combining permutations) then we say that G is a **group of permutations of X**. We often say that G **acts** on X, but it is as well to note that this latter term also covers a more general situation (Section 14.5).

If we take $X = \{1, 2, \ldots, n\}$ then a group of permutations of X is simply a subgroup of S_n. For example, here is a list of all the subgroups of S_3, each of which is also a group of permutations of $\{1, 2, 3\}$.

$$H_1 = \{\text{id}\}, \qquad H_2 = \{\text{id}, (12)\}, \qquad H_3 = \{\text{id}, (13)\},$$
$$H_4 = \{\text{id}, (23)\}, \qquad H_5 = \{\text{id}, (123), (132)\}, \qquad H_6 = S_3.$$

In order to check whether or not a given subset of S_n is a subgroup it is convenient to use Theorem 13.7, which tells us that (since S_n is finite) we need only verify the closure property.

Exercises 14.1

1   Which of the following are groups of permutations of the set $\{1, 2, 3, 4, 5\}$, that is, which of them are subgroups of S_5?

   (i) $\{(12345), (124)(35)\}$.
   (ii) $\{\text{id}, (12345), (13524), (14253), (15432)\}$.
   (iii) $\{\text{id}, (12)(34), (13)(24), (14)(23)\}$.
   (iv) $\{\text{id}, (12)(345), (135)(24), (15324), (12)(45), (134)(25), (143)(25)\}$.

An important subgroup of S_n is the subgroup consisting of all the even permutations, which is known as the **alternating group** A_n. The results obtained in Section 5.6 imply that the composite of two even permutations is even (so that A_n is indeed a subgroup of S_n) and that the order of A_n is $\frac{1}{2}n!$.

Many examples of groups of permutations arise as the symmetry groups of geometrical objects. For example, if we label the corners of a square in clockwise order 1, 2, 3, 4, then each symmetry induces a permutation of the set $\{1, 2, 3, 4\}$, and we get the eight permutations listed in Table 14.1.1.

Table 14.1.1

Identity	id
Clockwise rotation through 90°	(1234)
Clockwise rotation through 180°	(13)(24)
Clockwise rotation through 270°	(1432)
Reflection in diagonal 13	(24)
Reflection in diagonal 24	(13)
Reflection in perpendicular bisector of 12	(12)(34)
Reflection in perpendicular bisector of 14	(14)(23).

It follows from the geometrical interpretation that these eight permutations form a group; more specifically, it is a subgroup of S_4.

A similar situation arises when we study graphs rather than geometric objects. In this case the 'symmetries' are the permutations of the vertices which transform edges into edges. Such a permutation is called an **automorphism** of the graph. The permutation (15)(24) is an automorphism of the graph depicted in Fig. 14.1, whereas (12345) is not an automorphism since the edge $\{2, 4\}$ is transformed into $\{3, 5\}$, which is not an edge. Clearly the set of all automorphisms of any graph forms a group, known as the **automorphism group** of the graph.

Fig. 14.1
A graph with two automorphisms.

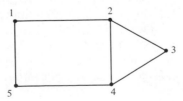

The most important fact about automorphisms is that if v and w are vertices of a graph Γ, and there is an automorphism α such that $\alpha(v) = w$, then v and w have the same properties with respect to Γ. For instance, every edge containing v is transformed by α into an edge containing w, and so the valency of w is the same as that of v. (See Fig. 14.2.) Similarly, every cycle through v is transformed into

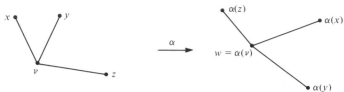

Fig. 14.2
Automorphisms preserve valency.

a cycle of the same length through w. In simple cases we can use such facts to determine the automorphism group of G completely.

Example   Find the automorphism group of the graph depicted in Fig. 14.3.

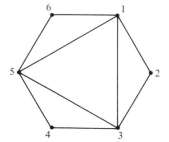

Fig. 14.3
How many automorphisms?

Solution   We notice first that the vertices fall naturally into two sets: the set $\{1, 3, 5\}$ having valency four and the set $\{2, 4, 6\}$ having valency two. For the reasons given above, no automorphism can transform a member of the first set to a member of the second set. On the other hand, it is easy to see that we can take *any* permutation of $\{1, 3, 5\}$ and extend it to an automorphism of the graph. For example, if the permutation (135) is part of an automorphism α, then α must transform 2 to 4, since 2 is the only vertex adjacent to both 1 and 3, and 4 is only vertex adjacent to both 3 and 5. Similarly, α must transform 4 to 6 and 6 to 2, and so we must have $\alpha = (135)(246)$. In the same way, each of the six permutations of $\{1, 3, 5\}$ can be extended in a unique manner to give an automorphism of the graph:

id   extends to   id,   (13) extends to   (13)(46),
(135)   extends to   (135)(246),   (15) extends to   (15)(24),
(153)   extends to   (153)(264),   (35) extends to   (35)(26).

It follows that there are just six automorphisms of the graph, and
they are the six extended permutations listed above.   □

Exercises
14.1
(continued)

2 Find the orders of the following permutations, considered as
elements of the symmetric group S_8:

(i) (1235)(48)(67);   (ii) (12)(35)(48)(67);   (iii) (13672)(458).

3 According to Theorem 13.8.3 the order of any element of S_8 is a
divisor of $|S_8| = 8!$. By considering the cycle structure of the permu-
tations in S_8 write down the orders of elements which actually occur
in S_8, and give an example of a divisor of $8!$ which is not the order
of an element of S_8.

4 List all the symmetries of a regular pentagon, regarded as
permutations of the corners 1, 2, 3, 4, 5, labelled in cyclic order.

5 Find the group of automorphisms of the graph given by the
following adjacency list. (A picture will help.)

1	2	3	4	5	6	7	8
2	1	1	1	2	3	4	4
3	3	2	7	7	7	5	5
4	5	6	8	8	8	6	6

14.2 Orbits and stabilizers

Suppose that G is a group of permutations of a set X. We shall show
that the group structure of G leads naturally to a partition of X.
  Define a relation $\sim$ on X by the rule

$$x \sim y \quad \Leftrightarrow \quad g(x) = y \text{ for some } g \in G.$$

We can check that $\sim$ is an equivalence relation in the usual way.

(Reflexivity) Since id belongs to any group, and $\text{id}(x) = x$ for all
          $x \in X$, we have $x \sim x$.

(Symmetry) Suppose $x \sim y$, so that $g(x) = y$ for some $g \in G$. Since G is a group, g^{-1} belongs to G, and since $g^{-1}(y) = x$ we have $y \sim x$, as required.

(Transitivity) If $x \sim y$ and $y \sim z$ we must have $y = g_1(x)$ and $z = g_2(y)$ for some g_1 and g_2 in G. Since G is a group, $g_2 g_1$ is in G and since $g_2 g_1(x) = z$ we have $x \sim z$, as required.

It follows that the distinct equivalence classes of $\sim$ form a partition of X; x and y are in the same part if and only if there is a permutation in G which transforms x to y. These parts (equivalence classes) are known as **orbits** of G on X. The orbit of x contains all the members of X which are of the form $g(x)$ for some $g \in G$, and it is usually written as Gx. Explicitly,

$$Gx = \{y \in X \mid y = g(x) \text{ for some } g \in G\}.$$

Intuitively, the orbit Gx contains all the objects which are indistinguishable from x under the action of G. For example, when G is the group of automorphisms of the graph shown in Fig. 14.3 the vertex-set is partitioned into two orbits $\{1, 3, 5\}$ and $\{2, 4, 6\}$, and we have

$$G1 = G3 = G5 = \{1, 3, 5\}, \qquad G2 = G4 = G6 = \{2, 4, 6\}.$$

Exercises 14.2

1   Write down all the automorphisms of the graph shown in Fig. 14.1. (There are only two of them!) Show that the group of automorphisms induces a partition of the vertex-set into three orbits.

2   Let G be the group of automorphisms of the tree shown in Fig. 14.4a, acting on the set X of vertices. Determine the orbits of G on X.

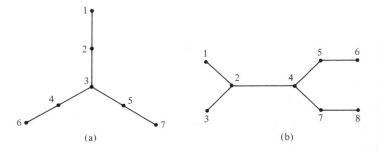

Fig. 14.4
Find the orbits.

(a)   (b)

3 Repeat Ex. 2 for the tree shown in Fig. 14.4b.

There are two obvious numerical problems concerning orbits: we need to know how to find the size of each orbit, and how to find the number of orbits. The solution of these problems involves the combination of group-theoretical ideas on the one hand and elementary counting techniques on the other.

When G is a group of permutations of a set X we shall denote by $G(x \rightarrow y)$ the set of elements of G which take x to y; that is

$$G(x \rightarrow y) = \{g \in G \mid g(x) = y\}.$$

In particular, when $x = y$ the set $G(x \rightarrow x)$ contains all the permutations γ in G which **fix** x: that is, the permutations such that $\gamma(x) = x$. The set $G(x \rightarrow x)$ is called the **stabilizer** of x, and is written G_x. We note that if γ_1 and γ_2 are in G_x then

$$\gamma_1 \gamma_2(x) = \gamma_1(x) = x,$$

and so $\gamma_1 \gamma_2$ also belongs to G_x. Thus G_x is actually a *subgroup* of G.

Theorem 14.2

Let G be a group of permutations of X and suppose that h belongs to $G(x \rightarrow y)$. Then

$$G(x \rightarrow y) = hG_x,$$

the left coset of G_x with respect to h.

Proof

If α is in the left coset hG_x we have $\alpha = h\beta$ for some β in G_x. Thus

$$\alpha(x) = h\beta(x) = h(x) = y,$$

and so α belongs to the set $G(x \rightarrow y)$. Conversely, if γ is in $G(x \rightarrow y)$ then

$$h^{-1}\gamma(x) = h^{-1}(y) = x$$

so that $h^{-1}\gamma = \delta$ where δ is in the stabilizer G_x. Thus $\gamma = h\delta$ is in hG_x. We have proved that every member of hG_x is also in $G(x \rightarrow y)$ and conversely, so it follows that the two sets are the same. $\square$

Theorem 14.2 is best regarded as the basis for a rule for finding the size of the set $G(x \rightarrow y)$. We recall that any coset of a given subgroup has the same size as the subgroup itself, and so $|hG_x| = |G_x|$. Thus whenever there is an element h of G which takes x to y (in other words, whenever y is in the orbit Gx) we have the equation

$$|G(x \to y)| = |G_x| \quad (y \in Gx).$$

This holds for any y in Gx. On the other hand, if y is not in the orbit Gx then, by definition, there are no permutations taking x to y, and so

$$|G(x \to y)| = 0 \quad (y \notin Gx).$$

Exercises 14.2 (continued)

4 In the *Example* given in Section 14.1 verify explicitly that

    (i) $\quad |G(2 \to 6)| = |G_2|$,   (ii) $\quad |G(3 \to 1)| = |G_3|$.

5 Let G be a group of permutations of a set X and let k be an element of $G(x \to y)$. Prove that $G(x \to y)$ is equal to the *right* coset $G_y k$, and deduce that if u and v are any two elements in the same orbit of G then $|G_u| = |G_v|$.

6 Let $X = \mathbb{Z}_5$ and suppose that G is the cyclic group of permutations of X generated by the permutation π defined by the rule $\pi(x) = 2x$. Write down the elements of G in cycle notation and determine the orbits of G on X.

14.3 The size of an orbit

In this section we shall establish a fundamental relationship between the size of an orbit Gx and the size of the stabilizer G_x. We shall need the results obtained in the previous section, together with the techniques for counting sets of pairs developed in Section 3.2.

Let G be a group of permutations of a set X, and let x be a chosen element of X. The set S of pairs (g, y) such that $g(x) = y$ can be described by means of a table as in Section 3.2.

$$\cdots \quad y \quad \cdots$$

$\vdots$		
g	$\sqrt{}$ means that	$r_g(S)$
$\vdots$	(g, y) is in S	

$$c_y(S)$$

The two methods of counting S, using the row totals $r_g(S)$ and the column totals $c_y(S)$, form the basis for the proof of our main theorem.

Theorem 14.3

Let G be a group of permutations of a set X, and let x be any chosen element of X. Then we have the equation

$$|Gx| \times |G_x| = |G|.$$

Proof

Let S denote the set of pairs illustrated in the table above, that is

$$S = \{(g, y) \mid g(x) = y\}.$$

Since g is a permutation there is just one y such that $g(x) = y$, for each g. In other words, each row total $r_g(S)$ is equal to 1.

The column total $c_y(S)$ is the number of g such that $g(x) = y$, that is $|G(x \to y)|$. So if y is in the orbt Gx we have

$$c_y(S) = |G(x \to y)| = |G_x|.$$

On the other hand, if y is not in Gx there are no permutations in G which take x to y, and so $c_y(S) = 0$.

The two methods for counting S give the equation

$$\sum_{y \in X} c_y(S) = \sum_{g \in G} r_g(S).$$

On the left-hand side there are $|Gx|$ terms equal to $|G_x|$ and the rest are zero, while on the right-hand side there are $|G|$ terms equal to 1. Hence we have the result. $\square$

For example, it is easy to verify the result when G is the group of symmetries of a square, regarded as permutations of the corners (as in Section 14.1). To determine the orbit of the corner 1 (say), we note that G contains permutations which transform

1 to 1:	id	1 to 2:	(1234),
1 to 3:	(13)(24),	1 to 4:	(1432).

Thus the orbit $G1$ is the whole set and $|G1| = 4$. The stabilizer of 1 is

$$G_1 = \{id, (24)\},$$

and so

$$|G1| \times |G_1| = 4 \times 2 = 8,$$

as expected, since there are eight symmetries in all.

The result can also be used to compute the order of a group of permutations, provided we can calculate the size of an orbit and the corresponding stabilizer.

Example   Let T be a regular tetrahedron in three-dimensional space (Fig. 14.5). Find the order of the group of rotational symmetries of T.

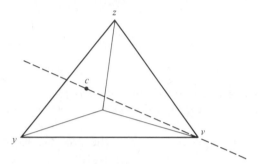

Fig. 14.5
A regular tetrahedron.

Solution   Let G be the group of permutations of the corners which correspond to rotational symmetries, and let z be any corner of T. Given any other corner y there is an edge yz of T and there are two faces of T which are bounded by yz. Let c be the centroid of one of these faces and let v be the opposite corner of T (Fig. 14.5). Then a rotation of 120° about the axis cv (in the appropriate sense) takes z to y. Hence the orbit Gz contains all four vertices, and we have $|Gz| = 4$.

The only rotational symmetries which fix z are the rotations through 0°, 120°, and 240° about the altitude passing through z, and so the stabilizer G_z has order 3. Hence

$$|G| = |Gz| \times |G_z| = 4 \times 3 = 12.$$ □

**Exercises
14.3**

1   Label the corners of a regular tetrahedron T as 1, 2, 3, 4. Write down the permutations corresponding to the twelve rotational symmetries of T and verify that the group obtained is the alternating group A_4.

2   Let X denote the set of corners of a cube and let G denote the group of permutations of X which correspond to rotations of the cube. Show that:

   (i) G has just one orbit on X;
   (ii) if z is any corner, then $|G_z| = 3$;
   (iii) $|G| = 24$.

3   Let V be the vertex-set of the graph Γ shown in Fig. 14.6, and let G be the automorphism group of Γ. Determine the orbits of G on V, and compute the orders of G_a, G_b, and G.

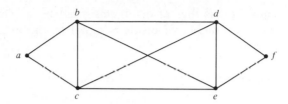

Fig. 14.6
Illustrating Ex. 14.3.3.

14.4   The number of orbits

We turn now to the problem of counting the number of orbits when we are given a group G of permutations of a set X. Each orbit is a subset of X whose members are indistinguishable under the action of G, and so the number of orbits tells us the number of distinguishably different types of object in X.

For example, suppose it is proposed to manufacture identity cards from plastic squares, marked with a 3×3 grid on both sides, and punched with two holes, as in Fig. 14.7. Since there are 9 positions

Fig. 14.7
Some identity cards.

and 2 holes, the number of ways of punching the holes is $\binom{9}{2}=36$. We shall refer to these as *configurations*. Now, not all the configurations are distinguishable, since the cards may be rotated and overturned. The first two configurations shown in Fig. 14.7 are indistinguishable, but the third one is essentially different from them.

Clearly, the group G which acts here is the familiar group of eight symmetries of a square. But we must consider its action on the set X of 36 configurations, rather than on the four corners as hitherto. Then the number of orbits of G on X is just the number of distinguishably different identity cards.

We could find the number of orbits by labelling the 36 members of X in some way, and writing down the eight permutations of these 36 configurations explicitly. However this would be rather laborious, and fortunately there is a better way. Given any group G of permutations of a set X we define, for each g in G, a set

$$F(g) = \{x \in X \mid g(x) = x\}.$$

Thus $F(g)$ is the set of objects *fixed* by g. The next theorem says that the number of orbits is equal to the average size of the sets $F(g)$.

Theorem 14.4   The number of orbits of G on X is

$$\frac{1}{|G|} \sum_{g \in G} |F(g)|.$$

Proof   Once again, we use the method of counting pairs. Let

$$E = \{(g, x) \mid g(x) = x\}.$$

Then the row total $r_g(E)$ is equal to the number of x fixed by g, that is $|F(g)|$. Also, the column total $c_x(E)$ is equal to the number of g which fix x, that is $|G_x|$. Hence the two methods for counting E lead to the equation

$$\sum_{g \in G} |F(g)| = \sum_{x \in X} |G_x|.$$

Suppose there are t orbits, and let z be a chosen element of X. According to Ex. 14.2.5 if x belongs to the orbit Gz then $|G_x| = |G_z|$. Hence on the right-hand side of the equation above there are $|Gz|$ terms equal to $|G_z|$, one for each x in Gz. The total contribution from these terms is

$$|Gz| \times |G_z| = |G|,$$

by Theorem 14.3. In other words, the total contribution from the members of each orbit is $|G|$. Since there are t orbits altogether the right-hand side is equal to $t|G|$, and rearranging the equation we obtain

$$t = \frac{1}{|G|} \sum_{g \in G} |F(g)|. \qquad \square$$

Now we can solve the identity card problem. For each of the eight symmetries g we need only compute $|F(g)|$, the number of configurations fixed by g. For example, when g is the rotation through 180°, there are four fixed configurations, as depicted in Fig. 14.8. In the same way we can verify that the number of configurations fixed

Fig. 14.8
Configurations fixed by
rotation through 180°.

by each of the eight symmetries is as listed in Table 14.4.1. Hence the number of orbits is

$$\tfrac{1}{8}(36+0+4+0+6+6+6+6)=8.$$

We conclude that just eight different identity cards can be produced in this way.

Table 14.4.1

Identity	36
Clockwise rotation through 90°	0
Clockwise rotation through 180°	4
Clockwise rotation through 270°	0
Reflection in diagonal 13	6
Reflection in diagonal 24	6
Reflection in perpendicular bisector of 12	6
Reflection of perpendicular bisector of 14	6

Of course, in this particular case it would be quite easy to list the eight cards by trial and error. However the result of Theorem 14.4 is applicable in much greater generality, and it is useful in the solution of many other problems involving symmetry.

Example   Necklaces are manufactured by arranging 13 white beads and three black beads on a loop of string. How many different necklaces can be produced in this way? (The position of the fastening may be ignored.)

Solution   We can think of the 16 beads as being placed at the corners of a regular polygon with 16 sides Each configuration is specified by the choice of the three corners which are occupied by black beads, and so there are $\binom{16}{3} = 560$ configurations in all. Two configurations give the same necklace if one can be obtained from the other by a symmetry transformation of the polygon: either a rotation or a reflection, the latter being equivalent to overturning the polygon. There are 32 symmetries in all, as listed below.

(a) The identity fixes all 560 configurations.
(b) There are 15 rotations through angles $2\pi n/16$ ($n = 1, 2, \ldots, 15$), and each of them has no fixed configurations. (Why?)
(c) There are eight reflections in axes joining the mid-points of opposite sides, and each of them has no fixed configurations.
(d) There are eight reflections in axes passing through opposite corners. The positions of the three black beads are unchanged (as a threesome) by such a reflection only if one of the beads occupies one of the two corners lying on the axis, and the other pair occupy one of the seven pairs of corners symmetrically placed with respect to this axis. Hence there are $2 \times 7 = 14$ fixed configurations for each reflection of this kind.

It follows that the number of different necklaces is

$$\tfrac{1}{32}(650 + (8 \times 14)) = 21.$$ □

Exercises 14.4

1 Show that there are just five different necklaces which can be constructed from five white beads and three black beads. Sketch them.

2 Suppose identity cards are manufactured from square cards ruled with a 4×4 grid, with two holes punched. How many different cards can be produced in this way?

3 Let V be the vertex-set of the binary tree shown in Fig. 14.9, and let G be the group of automorphisms of the tree. Write down the elements of G (as permutations of V), and verify that Theorem 14.4 holds in this case.

4 Let X denote the set of 'coloured trees' which result when each vertex of the tree in Fig. 14.9 is assigned one of the colours red or blue. How many different coloured trees of this kind are there?

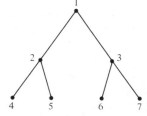

Fig. 14.9
Illustrating Ex. 14.4.3 and Ex. 14.4.4.

5 Let G be a group of permutations of X, and let $I(x)$ be an expression which is constant on each orbit of G, so that

$$I(g(x)) = I(x) \quad \text{for all} \quad g \in G, x \in X.$$

Let D be a set of representatives, one from each orbit, and let $E = \{(g, x) \mid g(x) = x\}$, as in the proof of Theorem 14.4. By evaluating the sum

$$\sum_{(g,x) \in E} I(x)$$

in two different ways, show that

$$\sum_{x \in D} I(x) = \frac{1}{|G|} \sum_{g \in G} \sum_{x \in F(g)} I(x).$$

(This is a 'weighted' version of Theorem 14.4, and reduces to it when $I(x) = 1$ for all $x \in X$. It will be used in Chapter 20.)

14.5 Representation of groups by permutations

Suppose we are given a group G, not specifically a group of permutations, and a set X. A **representation** of G by permutations of X assigns to each element g of G a permutation $\hat{g}$ of X, in such a way that the composition of group elements is compatible with the composition of the corresponding permutations. In other words,

$$\widehat{g_1 g_2} = \hat{g}_1 \hat{g}_2 \quad \text{for all } g_1 \text{ and } g_2 \text{ in } G.$$

The compatibility condition means that the set of all the permutations $\hat{g}$ is a group $\hat{G}$ of permutations of X. However it is important to notice that G and $\hat{G}$ are not necessarily isomorphic groups, because we do not insist that different group elements are represented by different permutations. If we do have the property

$$\hat{g}_1 = \hat{g}_2 \quad \Leftrightarrow \quad g_1 = g_2 \quad (g_1, g_2 \in G)$$

then we say that the representation is **faithful**; otherwise it is **unfaithful**. In the faithful case the function which takes g to $\hat{g}$ is a bijection, and the compatibility condition implies that it is an isomorphism from G to $\hat{G}$.

Example Let G be the group of symmetry transformations of a square. Show that the representation in which each symmetry is represented by the corresponding permutation of the corners of the square is faithful, whereas the representation in which each symmetry is represented by the corresponding permutation of the diagonals (considered as lines without direction) is unfaithful.

Solution   Label the corners 1, 2, 3, 4 in cyclic order, and denote the elements of G by i (the trivial symmetry), ρ_1, ρ_2, ρ_3 (rotations through 90°, 180°, 270°), μ_1, μ_2 (reflections in the diagonals), and μ_3, μ_4 (reflections in the perpendicular bisectors of the sides). We can write down the permutations of the corners corresponding to the symmetries as in Section 14.1:

$$\hat{i} = \mathrm{id}, \qquad \hat{\rho}_1 = (1234), \qquad \hat{\rho}_2 = (13)(24), \qquad \hat{\rho}_3 = (1432),$$
$$\hat{\mu}_1 = (24), \qquad \hat{\mu}_2 = (13), \qquad \hat{\mu}_3 = (12)(34), \qquad \hat{\mu}_4 = (14)(23).$$

Since the permutations are all different we have a faithful representation. (Actually we have tacitly assumed this fact in some of the examples in previous sections.)

Let a and b denote the diagonals 13 and 24 respectively. The permutations of $\{a, b\}$ corresponding to the symmetries are:

$$\hat{i} = \mathrm{id}, \qquad \hat{\rho}_1 = (ab), \qquad \hat{\rho}_2 = \mathrm{id}, \qquad \hat{\rho}_3 = (ab),$$
$$\hat{\mu}_1 = \mathrm{id}, \qquad \hat{\mu}_2 = \mathrm{id}, \qquad \hat{\mu}_3 = (ab), \qquad \hat{\mu}_4 = (ab).$$

So in this case the representation is unfaithful.   □

Whenever we have a representation of G by permutations of X we may say that G **acts** on X. This terminology has already been used in the case when the representation is faithful, but it also applies when the representation is unfaithful.

The technique of representing a group by permutations of a set is useful because it is easier to compute with permutations than with abstract group elements. Naturally the technique is most valuable when the representation is faithful. The next theorem shows that every finite group has a faithful representation.

Theorem 14.5   Let G be a finite group and let X denote the set of elements of G (in other words X is the same set as G, but the group structure is disregarded). For each g in G define a permutation $\hat{g}$ of X by the rule

$$\hat{g}(h) = gh \quad (h \in X),$$

where the symbol gh is simply the composite of g and h in G. Then this rule defines a faithful representation of G by permutations of X.

Proof   Given g_1 and g_2 in G and any h in X, we have

$$\widehat{g_1 g_2}(h) = (g_1 g_2)h = g_1(g_2 h) = \hat{g}_1(\hat{g}_2(h)),$$

so that $\widehat{g_1 g_2} = \hat{g}_1 \hat{g}_2$ and we do have a representation. Furthermore

$$\hat{g}_1 = \hat{g}_2 \quad \Leftrightarrow \quad g_1(h) = g_2(h) \quad (h \in X)$$
$$\Leftrightarrow \quad g_1 h = g_2 h$$
$$\Leftrightarrow \quad g_1 = g_2,$$

by the cancellation rule. Hence the representation is faithful. $\quad\square$

It follows from the theorem that every finite group is isomorphic to a group of permutations. But in general the number of objects permuted is unreasonably large, since it is equal to the order of the group. In practice we try to find a faithful representation by permutations of a relatively small set so that calculations can be carried out in an efficient way. An example will be given in Ex. 14.5.4.

Exercises 14.5

1 Let $G = \{g_1, g_2, g_3, g_4, g_5, g_6, g_7, g_8\}$ be the group whose table is Table 14.5.1. Write down the permutations $\hat{g}_i$ $(1 \le i \le 8)$ as defined in Theorem 14.5 and hence determine the order of each element of G. (It is convenient to use the subscript i instead of g_i when writing out the permutations: thus $\hat{g}_2 = (1234)(5678)$.)

Table 14.5.1

	g_1	g_2	g_3	g_4	g_5	g_6	g_7	g_8
g_1	g_1	g_2	g_3	g_4	g_5	g_6	g_7	g_8
g_2	g_2	g_3	g_4	g_1	g_6	g_7	g_8	g_5
g_3	g_3	g_4	g_1	g_2	g_7	g_8	g_5	g_6
g_4	g_4	g_1	g_3	g_3	g_8	g_5	g_6	g_7
g_5	g_5	g_8	g_7	g_6	g_3	g_2	g_1	g_4
g_6	g_6	g_5	g_8	g_7	g_4	g_3	g_2	g_1
g_7	g_7	g_6	g_5	g_8	g_1	g_4	g_3	g_2
g_8	g_8	g_7	g_6	g_5	g_2	g_1	g_4	g_3

2 Let G be the group of symmetries of a regular hexagon. In each of the following cases, say whether or not the representation of G by permutations of the given set X is faithful.

    (i) $X =$ corners;
    (ii) $X =$ sides (considered as unordered pairs of corners);
    (iii) $X =$ diagonals;
    (iv) $X =$ perpendicular bisectors of the sides.

3   Show that if we have an unfaithful representation of a group G then there is an element $g \neq 1$ in G for which $\hat{g} = \mathrm{id}$.

4   Let G be the group of rotational symmetries of a cube (Ex. 14.3.2), and let D denote the set of four space diagonals of the cube. (A space diagonal is a line joining opposite corners which passes through the centre of the cube.) By examining the various kinds of rotational symmetry show that the representation which assigns to each rotation the corresponding permutation of D is faithful. Deduce that $G \approx S_4$.

14.6   Applications to group theory

In this section we shall study an *unfaithful* representation of a finite group by permutations of itself, and thereby obtain some important results in the theory of groups. We begin by showing that the relationship between the sizes of an orbit and the corresponding stabilizer obtained in Theorem 14.3 also holds when G has an unfaithful representation on X.

Given any representation of G by permutations of X (faithful or not) we may define the orbit Gx to be the set

$$Gx = \{y \in X \mid y = \hat{g}(x) \text{ for some } g \text{ in } G\}.$$

Clearly this is just the same as $\hat{G}x$. On the other hand, the stabilizer G_x is defined to be the subgroup

$$G_x = \{g \in G \mid \hat{g}(x) = x\},$$

and this is not the same as $\hat{G}_x$ since there may be several group elements g which induce the same permutation $\hat{g}$. However, if we copy the proof of Theorem 14.3 using the set $\{(g, y) \mid \hat{g}(x) = y\}$, we obtain the expected relation between $|Gx|$, $|G_x|$, and $|G|$; that is

$$|Gx| \times |G_x| = |G|.$$

Let A be a finite group and for each element a of A define a permutation $\hat{a}$ of A by the rule

$$\hat{a}(x) = axa^{-1} \quad (x \in A).$$

The element axa^{-1} is known as the **conjugate** of x by a. This assignment satisfies the compatibility condition, since

$$\begin{aligned}
\widehat{a_1 a_2}(x) &= (a_1 a_2)x(a_1 a_2)^{-1} \\
&= a_1(a_2 x a_2^{-1})a_1^{-1} \\
&= \hat{a}_1 \hat{a}_2(x),
\end{aligned}$$

and so we have a representation of A by permutations of itself. However the representation is not necessarily faithful, because if a commutes with x we have

$$\hat{a}(x) = axa^{-1} = xaa^{-1} = x.$$

In Section 13.7 we defined the *centre* $Z(A)$ to be the set of those elements of A which commute with every element of A. It follows from the equation obtained above that if a belongs to $Z(A)$ then $\hat{a} = \text{id}$, and so when the centre contains elements other than the identity the representation is unfaithful. In particular if A is a commutative group then $Z(A) = A$ and we have the trivial representation in which every element of A is represented by the identity permutation.

In the general case the stabilizer of x in this representation is the subgroup containing those elements a for which $\hat{a}(x) = x$, that is $axa^{-1} = x$, or equivalently $ax = xa$. In other words it is the subgroup consisting of those elements which commute with x, which is usually called the **centralizer** $C(x)$. The orbit of x is the set of all conjugates of x, and so by the extended form of Theorem 14.3 we have the equation

$$|C(x)| \times (\text{number of conjugates of } x) = |A|.$$

In particular this implies that the size of a class of conjugates is a divisor of $|A|$. When A is the symmetric group S_n we recall (Theorem 5.5) that the conjugacy class of x consists of the permutations which have the same type as x. If x has α_i cycles of length i then the total number of permutations of this type is

$$\frac{n!}{1^{\alpha_1} 2^{\alpha_2} \cdots n^{\alpha_n} \alpha_1! \, \alpha_2! \cdots \alpha_n!}.$$

Since $|S_n| = n!$ we can deduce that the number of permutations in S_n which commute with x is

$$|C(x)| = 1^{\alpha_1} 2^{\alpha_2} \cdots n^{\alpha_n} \alpha_1! \, \alpha_2! \cdots \alpha_n!.$$

In any finite group A the orbits of the representation defined by $\hat{a}(x) = axa^{-1}$ correspond to the conjugacy classes in A and form a partition of A. This fact may be used to derive a useful formula involving the size of the centre $Z(A)$. Since each element of $Z(A)$, including the identity, is conjugate only to itself, there are $|Z(A)|$ conjugacy classes of size 1. Suppose the other classes have sizes $n_1, n_2, \ldots, n_c$ where, as we noted above, each n_i is a divisor of $|A|$. Then the total number of elements of A is

$$|A| = |Z(A)| + (n_1 + n_2 + \cdots + n_c).$$

This equation is known as the **class equation** for A. The fact that each term n_i is a divisor of $|A|$ has a remarkable consequence when the order of A is a power of a prime.

Theorem 14.6

If A is a group of order p^r (p prime, $r \geqslant 1$) then $Z(A)$ contains an element other than the identity.

Proof

In the class equation for A we have $|A| = p^r$ and each n_i ($1 \leqslant i \leqslant c$) is a divisor of p^r. Since $n_i \neq 1$, we must have $n_i = p^{e_i}$ for some $e_i \geqslant 1$. Thus p is a divisor of $|A|$ and of each n_i. It follows that p is also a divisor of $|Z(A)|$, and so $|Z(A)| > 1$, as required. $\square$

Theorem 14.6 indicates that simple counting arguments can have far-reaching consequences. We are given an abstract group with a certain number of elements, and we deduce that one of the non-identity elements must commute with all the others. Such unexpected results may explain why so many mathematicians are fascinated by the abstract theory of groups.

Exercises 14.6

1   How many permutations in S_8 commute with $(135)(24)(67)(8)$?

2   Let π be a permutation in S_n which has a single cycle of length n. Show that the centralizer $C(\pi)$ is the cyclic subgroup of S_n generated by π.

3   According to Theorem 14.6 any group G of order 8 ($= 2^3$) has a non-trivial centre. Verify this by finding a non-identity element which commutes with every other element when G is the group of the square, and when G is the group discussed in Ex. 14.5.1.

4   Find all the conjugacy classes for the group given in Ex. 14.5.1, and verify the class equation for this group.

5   Show that the order of the centralizer of $\pi = (123)(45)$ in S_7 is 12. Prove that the centralizer is isomorphic to $C_6 \times C_2$, where C_6 is generated by π and C_2 is generated by a permutation σ, which is to be found.

14.7   Miscellaneous Exercises

1   Let G be any subgroup of S_n. Prove that either all the permutations in G are even, or exactly half of them are even.

2 Show that A_n is the only subgroup of index two in S_n.

3 Let π be a permutation which has cycles of lengths $n_1, n_2, \ldots, n_r$. Show that the order of π is the least common multiple of $n_1, n_2, \ldots, n_r$.

4 Construct subgroups of S_5 which have orders 1, 2, 3, 4, 5, 6, 8, 10, 12, 20, 24, 60, 120.

5 Investigate the possibility of subgroups of S_5 whose orders are not among the numbers listed in the previous exercise.

6 Let G be the group of automorphisms of the graph $K_{3,3}$ (as defined in Ex. 10.1.2) and let v be any vertex of the graph. Calculate $|G_v|$ and $|G|$.

7 Show that the group of automorphisms of Petersen's graph has order 120.

8 Suppose that each corner of a regular tetrahedron is assigned one of the colours red, white, and blue. Find the number of distinguishable tetrahedra of this kind.

9 Show that 57 different cubes can be constructed by painting each *face* of a cube red, white, or blue.

10 Let Y be the set of partitions of an 8-set into two parts of size 4. Let A_8 act on Y in the manner corresponding to its action on the 8-set. Show that there is just one orbit in this action, and calculate the order of a stabilizer.

11 Suppose that G is a group of permutations of a set X which is abelian and has just one orbit on X. Show that $|G| = |X|$.

12 Show that the number of different ID cards with an $n \times n$ square grid and two holes (as in Section 14.4) is

$$\tfrac{1}{16}(n^4 + 6n^2 - 4n) \qquad \text{if} \quad n \text{ is even,}$$

$$\tfrac{1}{16}(n^4 + 8n^2 - 8n - 1) \quad \text{if} \quad n \text{ is odd.}$$

13 Suppose that ID cards are made in the shape of an equilateral triangle, with n equidistant lines ruled parallel to each edge (on both sides of the card) and a hole punched in one of the small triangles so formed. If n is an odd number of the form $6m + 1$ or $6m + 3$ show that the number of ID cards which can be constructed is $\tfrac{1}{6}(n+2)(n+3)$. What is the result when $n = 6m + 5$?

14 How many different necklaces can be made from seven black beads and three white beads?

15 A circular disc is divided on one side into eight equal sectors, five of which are to be painted blue and three red. How many different discs can be made in this way?

16 Let p be a prime and define T_p to be the set of permutations of $\mathbb{Z}_p$ given by functions of the form

$$t(x) = ax + b \quad (x \in \mathbb{Z}_p)$$

where a and b are in $\mathbb{Z}_p$ and $a \neq 0$. Show that T_p is a group of order $p(p-1)$. Prove also that given any two ordered pairs (x_1, y_1) and (x_2, y_2) of distinct elements of $\mathbb{Z}_p$, then there is a unique permutation t in T_p such that $t(x_1) = x_2$ and $t(y_1) = y_2$.

17 Let X be any set of transpositions in S_n, and define a graph $\Gamma(X)$ as follows. The vertices are the integers $1, 2, \dots, n$, and ij is an edge if and only if there is a transposition in X which switches i and j. Show that $\Gamma(X)$ is connected if and only if any element of S_n can be expressed as a composite of transpositions belonging to X.

18 An **inversion** in a permutation π of $\mathbb{N}_n$ is a pair (i, j) such that $i < j$ and $\pi(i) > \pi(j)$. Show that if $q(\pi)$ is the total number of inversions in σ, then $\operatorname{sgn}(\pi) = (-1)^{q(\pi)}$.

In the next three exercises we shall consider a finite group G of rotations in three-dimensional space, each rotation having an axis which passes through the origin.

19 Let S be a sphere with centre at the origin and suppose ρ is in G. The axis of ρ meets S in two points called the *poles* of ρ. Show that if x is a pole of ρ and $y = \theta(x)$ for some θ in G, then y is a pole of $\theta\rho\theta^{-1}$. Deduce that G acts on the set P of poles.

20 A pole x is said to have order m if it is the pole of a rotation of order m. Show that if x has order m_x then the size of the orbit of x is $|G|/m_x$. Deduce that

$$2(|G| - 1) = \sum \frac{|G|}{m_x}(m_x - 1),$$

where the sum is taken over the orbits of G on P.

21 Show that the formula obtained in the previous exercise can be rewritten as

$$2\left(1 - \frac{1}{|G|}\right) = \sum\left(1 - \frac{1}{m_x}\right).$$

Deduce that there are either two or three orbits, and that the possibilities are as given in the following table.

$|G|$:	n	$2n$	12	24	60
Orders of poles:	n, n	$2, 2, n$	$2, 3, 3$	$2, 3, 4$	$2, 3, 5$

15
Rings, fields, and polynomials

15.1 Rings

Although the concept of a group is a very useful one, a group is a rather restricted algebraic object, since it has only one operation. We are accustomed to dealing with structures in which there are two basic operations, like the addition and multiplication in $\mathbb{Z}$.

The most basic object of this kind is known in mathematics as a *ring*. The axioms for a ring are very similar to the arithmetical axioms **I1–I6** for the integers, but slightly weaker (and more general, of course). We shall present the axioms in a compact form, using the group concept to amalgamate several axioms into a single one.

Definition   A **ring** is a set R on which there are defined two binary operations $+$ and $\times$, satisfying the following axioms.

R1. R with the operation $+$ is a commutative group.
R2. The operation $\times$ has the closure, associativity, and identity properties.
R3. (The distributive laws.) For all a, b, and c in R,

$$a \times (b + c) = (a \times b) + (a \times c)$$
$$(a + b) \times c = (a \times c) + (b \times c).$$

Almost always we shall suppress the $\times$ sign, and write ab instead of $a \times b$.

Let us review the implications of the definition in detail. According to **R1**, the $+$ operation has the properties

$$(a + b) + c = a + (b + c),$$
$$a + 0 = 0 + a = a,$$
$$a + (-a) = (-a) + a = 0,$$
$$a + b = b + a,$$

where the existence of the element 0 and the additive inverse $-a$ is part of the definition. Similarly the $\times$ operation satisfies

$$(ab)c = a(bc),$$

$$a1 = 1a = a,$$

where the existence of the element 1 is also guaranteed. However, it must be stressed that the existence of a multiplicative inverse a^{-1} for each element a is *not* assumed. Also it is *not* assumed that the $\times$ operation is commutative. Finally, the axiom **R3** gives the rules for dealing with 'brackets' which are familiar from elementary algebra.

Clearly the prototype of a ring is the set $\mathbb{Z}$ of integers with the usual addition and multiplication. The integers also have two extra algebraic properties, not included in the general definition of a ring: the multiplication is commutative (axiom **I2**, part (ii)), and we can cancel a non-zero integer from both sides of an equation (axiom **I7**).

Another familiar example of a ring is the set $\mathbb{Z}_m$ of integers modulo m, with the operations defined in Section 6.2. The ring $\mathbb{Z}_m$ has a commutative multiplication, but the cancellation rule does not hold, since in $\mathbb{Z}_6$ (for example) we have $3 \times 1 = 3 \times 5$ but we cannot conclude that $1 = 5$.

As an example of a ring in which the commutative law for multiplication does not hold we can take the set of 2×2 matrices with integer entries, and the usual operations of matrix addition and multiplication. It is easy to check that axioms **R1**, **R2**, and **R3** hold, but multiplication is not commutative. For example, if

$$A = \begin{bmatrix} 2 & 1 \\ 0 & 1 \end{bmatrix}, \qquad B = \begin{bmatrix} 1 & 0 \\ 2 & 2 \end{bmatrix},$$

then

$$AB = \begin{bmatrix} 4 & 2 \\ 2 & 2 \end{bmatrix}, \qquad BA = \begin{bmatrix} 2 & 1 \\ 4 & 4 \end{bmatrix}.$$

Exercises 15.1

1   If R is a ring then the set $M_2(R)$ of 2×2 matrices over R is also a ring. (You need not verify this.)

(i) What is the additive identity in $M_2(R)$?
(ii) What is the additive inverse $-A$ of a matrix A in $M_2(R)$?
(iii) What is the multiplicative identity in $M_2(R)$?
(iv) What is the cardinality of $M_2(\mathbb{Z}_m)$?
(v) Show that multiplication is not commutative in $M_2(\mathbb{Z}_2)$.
(vi) Which elements of $M_2(\mathbb{Z}_2)$ have a multiplicative inverse?

2 By using axiom **I7** for $\mathbb{Z}$, prove that if x and y are integers such that $xy = 0$, then $x = 0$ or $y = 0$. Show by examples that the corresponding result does not hold when $\mathbb{Z}$ is replaced by $\mathbb{Z}_6$ or by $M_2(\mathbb{Z})$.

3 Show that if x and y are members of a ring R then $(-x)y = -xy$ and $(-x)(-y) = xy$. (At each stage of the proof, explain which property of R you are using.)

15.2 Invertible elements of a ring

An element x of a ring R is said to be **invertible** if x has a multiplicative inverse in R, that is, if there is an element u in R such that

$$ux = xu = 1.$$

A simple argument (Ex. 15.2.2) shows that if x is invertible then the element u is unique, and so we can use symbol x^{-1} for u without ambiguity. The element x^{-1} is **the inverse** of x, and the set of invertible elements of R is denoted by $U(R)$.

The only invertible elements of the ring $\mathbb{Z}$ are 1 and -1, each of which is its own inverse. In Section 6.3 we studied the invertible elements of $\mathbb{Z}_m$, and according to the results obtained there we have $U(\mathbb{Z}_8) = \{1, 3, 5, 7\}$, for example.

Theorem 15.2 The set $U(R)$ of invertible elements of a ring R is a group with respect to the multiplication in R.

Proof If x and y are invertible, let x^{-1} and y^{-1} be their inverses. Then

$$(xy)(y^{-1}x^{-1}) = (y^{-1}x^{-1})(xy) = 1$$

so that $y^{-1}x^{-1}$ is the inverse of xy. Thus $U(R)$ is closed under multiplication in R. Furthermore, multiplication is associative, and 1 is its own inverse so it certainly belongs to $U(R)$. Finally, if x is in $U(R)$ then its inverse x^{-1} is invertible (its inverse is x) and so x^{-1} is in $U(R)$. $\qquad\square$

In Section 6.3 we found that the element r is invertible in $\mathbb{Z}_m$ if and only if $\gcd(r, m) = 1$. It follows that $U(\mathbb{Z}_m)$ is a group of order $\phi(m)$. For example, $U(\mathbb{Z}_8) = \{1, 3, 5, 7\}$ is a group of order $\phi(8) = 4$; its group table is as given in Table 15.2.1.

Table 15.2.1

	1	3	5	7
1	1	3	5	7
3	3	1	7	5
5	5	7	1	3
7	7	5	3	1

In this case the group is the non-cyclic group $C_2 \times C_2$ of order 4. On the other hand,

$$U(\mathbb{Z}_7) = \{1, 2, 3, 4, 5, 6\}$$

is a cyclic group C_6, with generator 3, since

$$3^1 = 3, \qquad 3^2 = 2, \qquad 3^3 = 6, \qquad 3^4 = 4, \qquad 3^5 = 5, \qquad 3^6 = 1.$$

In the next chapter we shall prove that $U(\mathbb{Z}_p)$ is a cyclic group of order $\phi(p) = p - 1$ whenever p is a prime.

Exercises 15.2

1 Find the orders of the groups $U(\mathbb{Z}_{10})$, $U(\mathbb{Z}_{11})$, and $U(\mathbb{Z}_{12})$, and describe their structure.

2 Show that if x is an element of a ring R and u, v are elements of R such that

$$ux = xu = 1, \qquad vx = xv = 1,$$

then $u = v$.

3 A complex number of the form $m + ni$, where m and n are integers, is known as a **Gaussian integer**. Verify that the set Γ of Gaussian integers is a ring with respect to the usual addition and multiplication of complex numbers. (You need not verify explicitly the standard properties of complex numbers.) Find the invertible elements of Γ and describe the group structure of $U(\Gamma)$.

15.3 Fields

A **field** is a ring in which the multiplication is commutative and every element except 0 has a multiplicative inverse. Thus, in a field F we have

$$U(F) = F \backslash \{0\}.$$

In order to avoid difficulties about trivial cases, we insist that a field has at least two elements.

We can reorganize the definition, and make it more explicit, by saying that a set F is a field with respect to the operations $+$ and $\times$ if

(i) F is a commutative group with respect to $+$,
(ii) $F \backslash \{0\}$ is a commutative group with respect to $\times$,
(iii) the distributive laws (**R3**) hold.

The groups in (i) and (ii) are usually referred to as the **additive group** and the **multiplicative group** of the field, respectively.

Certainly, $\mathbb{Z}$ is *not* a field since only 1 and -1 have inverses in $\mathbb{Z}$. Similarly, the ring $\mathbb{Z}_m$ is not generally a field, but it follows from Theorem 6.3.1 that

when p is a prime $\mathbb{Z}_p$ is a field.

Of course, the most familiar field is the field $\mathbb{R}$ of real numbers, but although this is the field which underlies most of elementary mathematics and Calculus, its definition and properties are far from being elementary. Fortunately, in Discrete Mathematics the more amenable finite fields like $\mathbb{Z}_p$ are just as important as is the field $\mathbb{R}$ in Calculus.

In fact there are other finite fields besides the fields $\mathbb{Z}_p$ (p prime), and we shall study their construction and properties in the next chapter. The following *Example* contains a construction of one such field.

Example Let F be a field and let $S_2(F)$ denote the set of 2×2 *skew-symmetric* matrices over F; that is, matrices M of the form

$$M = \begin{bmatrix} x & y \\ -y & x \end{bmatrix}, \qquad (x, y \in F).$$

Prove that:

(i) $S_2(F)$ is a ring with respect to the usual matrix operations,
(ii) multiplication in $S_2(F)$ is commutative,
(iii) $S_2(F)$ is a field when $F = \mathbb{Z}_3$ but not when $F = \mathbb{Z}_5$.

Solution (i) The crucial fact is that $S_2(F)$ is closed under addition and multiplication. To prove this, we calculate that

$$\begin{bmatrix} x_1 & y_1 \\ -y_1 & x_1 \end{bmatrix} + \begin{bmatrix} x_2 & y_2 \\ -y_2 & x_2 \end{bmatrix} = \begin{bmatrix} x_1+x_2 & y_1+y_2 \\ -(y_1+y_2) & x_1+x_2 \end{bmatrix},$$

$$\begin{bmatrix} x_1 & y_1 \\ -y_1 & x_1 \end{bmatrix} \times \begin{bmatrix} x_2 & y_2 \\ -y_2 & x_2 \end{bmatrix} = \begin{bmatrix} x_1x_2-y_1y_2 & x_1y_2+x_2y_1 \\ -(x_1y_2+x_2y_1) & x_1x_2-y_1y_2 \end{bmatrix},$$

and remark that the matrices on the right-hand side are indeed skew-symmetric. We should remark also that the additive and multiplicative identities

$$\begin{bmatrix} 0 & 0 \\ 0 & 0 \end{bmatrix} \quad \text{and} \quad \begin{bmatrix} 1 & 0 \\ 0 & 1 \end{bmatrix}$$

are both skew-symmetric, and that if M is skew-symmetric then so is $-M$. The remaining ring axioms are consequences of the standard properties of matrix algebra. For example, matrix addition and multiplication are always associative, and the distributive laws hold.

(ii) In general, matrix multiplication is not commutative. But when the matrices are 2×2 skew-symmetric matrices, we can check that the commutative property does hold. We calculated the product of two such matrices in one order above; in the other order we have

$$\begin{bmatrix} x_2 & y_2 \\ -y_2 & x_2 \end{bmatrix} \times \begin{bmatrix} x_1 & y_1 \\ -y_1 & x_1 \end{bmatrix} = \begin{bmatrix} x_2x_1 - y_2y_1 & x_2y_1 + y_2x_1 \\ -(x_2y_1 + y_2x_1) & x_2x_1 - y_2y_1 \end{bmatrix},$$

which, by virtue of the field axioms for F, is the same as before.

(iii) Suppose that

$$\begin{bmatrix} x & y \\ -y & x \end{bmatrix} \quad \text{has inverse} \quad \begin{bmatrix} a & b \\ -b & a \end{bmatrix}.$$

Then

$$\begin{bmatrix} a & b \\ -b & a \end{bmatrix} \times \begin{bmatrix} x & y \\ -y & x \end{bmatrix} = \begin{bmatrix} 1 & 0 \\ 0 & 1 \end{bmatrix}$$

so that $ax - by = 1$, $ay + bx = 0$. Solving formally for a and b we obtain

$$a = x(x^2 + y^2)^{-1}, \qquad b = -y(x^2 + y^2)^{-1}.$$

Since F is a field, the element $x^2 + y^2$ will have a multiplicative inverse in F unless it is zero. If x and y are both zero, then we have the zero matrix, and we do not require an inverse matrix. We have to ask whether it is possible that $x^2 + y^2 = 0$ in F when x and y are not both zero.

In $\mathbb{Z}_3$ we can verify explicitly that $x^2 + y^2 \neq 0$ when $(x, y) \neq (0, 0)$:

x	0	0	1	1	1	2	2	2
y	1	2	0	1	2	0	1	2
$x^2 + y^2$	1	1	1	2	2	1	2	2.

We conclude that every non-zero matrix in $S_2(\mathbb{Z}_3)$ has an inverse, and so $S_2(\mathbb{Z}_3)$ is a field.

On the other hand, in $\mathbb{Z}_5$

$$1^2 + 2^2 = 0,$$

and so the matrix with $x = 1$, $y = 2$ has no inverse in $S_2(\mathbb{Z}_5)$, and we do not have a field. □

Exercises 15.3

1 In the *Example* we showed that $S_2(\mathbb{Z}_3)$ is a field. It has nine elements.

    (i) Denote the elements of $S_2(\mathbb{Z}_3)$ by $O, I, A_1, A_2, \ldots, A_7$, where O, I denote the additive and multiplicative identities, and $A_1, A_2, \ldots, A_7$ are the remaining elements in some order.

    (ii) Write out the group table for the additive group.

    (iii) Show that the additive group is not cyclic.

    (iv) Write out the group table for the multiplicative group.

    (v) Show that the multiplicative group is cyclic.

2 By finding a suitable generator, show that the multiplicative group of the field $\mathbb{Z}_{23}$ is cyclic.

3 Suppose x and y are elements of a field such that $xy = 0$. Prove that $x = 0$ or $y = 0$.

15.4 Polynomials

In elementary algebra the word *polynomial* is used to describe expressions like

$$x^2 + 4x + 3, \quad 7x^4 + 2x^2 + 3x + 1.$$

We do not usually trouble ourselves about the meaning of the symbol x, since the context signifies what is intended. For instance, if we are asked to solve the *equation*

$$x^2 + 4x + 3 = 0,$$

then it is understood that x is to be replaced by a suitable number so that the statement becomes a valid equality between numbers.

    When we compute with polynomials it becomes clear that the symbols $x, x^2, x^3, \ldots$ can be regarded simply as labels for the positions of the coefficients. In order to compute the sum of two polynomials, we add the corresponding coefficients of each x^i. In order to find the coefficient of x^i in the product of two polynomials, we find the product of the coefficient of x^j in the first one and the coefficient of x^{i-j} in the second one, and add these products for $j = 0, 1, \ldots, i$. Such considerations lead to the conclusion that the important thing about a polynomial is its sequence of coefficients.

It is convenient to rely on the foregoing observations when we have to make formal definitions about polynomials. Suppose R is a ring with commutative multiplication. A finite sequence

$$(a_0, a_1, a_2, \ldots, a_n)$$

of elements of R is said to be a **polynomial** with the coefficients in R. Usually we represent this polynomial in traditional form, by writing it as

$$a(x) = a_0 + a_1 x + a_2 x^2 + \cdots + a_n x^n.$$

There is no need to be more explicit about x, since it is introduced as part of the notation, rather than the definition. The set of all polynomials with coefficients in R is denoted by $R[x]$. The polynomials of the form (a_0) are known as **constant** polynomials, and they may be identified in the obvious way with the elements of the ring R.

Suppose we are given two polynomials

$$a(x) = a_0 + a_1 x + \cdots + a_n x^n, \qquad b(x) = b_0 + b_1 x + \cdots + b_m x^m.$$

There is no loss of generality in supposing that $n \geq m$, and if $n > m$ we shall set $b_{m+1} = b_{m+2} = \cdots = b_n = 0$. We define the **sum** $a(x) + b(x)$, and the **product** $a(x)b(x)$ of the polynomials in the following way:

$$a(x) + b(x) = (a_0 + b_0) + (a_1 + b_1)x + \cdots + (a_n + b_n)x^n,$$

$$a(x)b(x) = a_0 b_0 + (a_0 b_1 + a_1 b_0)x$$
$$+ (a_0 b_2 + a_1 b_1 + a_2 b_0)x^2 + \cdots + a_n b_m x^{n+m}$$

More formally, $s(x) = a(x) + b(x)$ is the polynomial $(s_0, s_1, \ldots, s_n)$ given by

$$s_i = a_i + b_i \qquad (0 \leq i \leq n),$$

and $p(x) = a(x)b(x)$ is the polynomial $(p_0, p_1, \ldots, p_{n+m})$ given by

$$p_i = a_0 b_i + a_1 b_{i-1} + \cdots + a_i b_0 \qquad (0 \leq i \leq n+m).$$

It follows from the definitions that when the coefficients of $a(x)$ and $b(x)$ belong to a ring R the coefficients of their sum and product belong to the same ring. For example, if $a(x)$ and $b(x)$ are the members of $\mathbb{Z}_3[x]$ given by

$$a(x) = 1 + 2x + 2x^2, \qquad b(x) = 2 + x,$$

then

$$a(x) + b(x) = (1 + 2) + (2 + 1)x + (2 + 0)x^2$$
$$= 2x^2;$$
$$a(x)b(x) = (1 \times 2) + (1 \times 1 + 2 \times 2)x + (2 \times 2 + 2 \times 1)x^2 + (2 \times 1)x^3$$
$$= 2 + 2x + 2x^3.$$

In general, $R[x]$ is closed under addition and multiplication. With a lot of tedious checking, one can prove that $R[x]$ is in fact a ring with commutative multiplication, provided that R is a ring with commutative multiplication. We shall accept this fact without explicit justification.

We shall also use the standard notational conventions for dealing with polynomials. Thus the coefficient 1 is usually suppressed so that, for example, we write $2 + x$ instead of $2 + 1x$ for the polynomial $b(x)$ considered above. When a coefficient is zero, we suppress both it and the corresponding power of x (as for example, with the coefficient of x^2 in $a(x)b(x)$ above). Finally, we often write a polynomial with its **leading coefficient** first, that is, in the form

$$a_n x^n + a_{n-1} x^{n-1} + \cdots + a_1 x + a_0,$$

where $a_n \neq 0$. When $a_n = 1$ we say that the polynomial is **monic**.

Exercises 15.4

1 Compute the sum and product of the following polynomials in $\mathbb{Z}_5[x]$:

(i) $x^3 + 3x^2 + 2x + 1$ and $x^2 + 4x + 2$;
(ii) $x^4 + x^2 + 2$ and $x^3 + 4x + 1$.

2 Show that the number of multiplications required to compute the product of the polynomials

$$a_n x^n + \cdots + a_0,$$

and

$$b_n x^n + \cdots + b_0,$$

by direct use of the definition, is $O(n^2)$. (Remarkably this is *not* the most efficient method.)

3   Suppose

$$a(x) = a_0 + \cdots + a_n x^n$$

and

$$b(x) = b_0 + \cdots + b_m x^m$$

are polynomials in $\mathbb{Z}[x]$. Write a program to compute the coefficients c_i $(0 \leqslant i \leqslant n + m)$ in the product $a(x)b(x)$.

4   Show that, in $\mathbb{Z}_7[x]$,

$$(x + 1)^7 = x^7 + 1.$$

For which values of m is it true that $(x + 1)^m = x^m + 1$ in $\mathbb{Z}_m[x]$?

5   Let

$$f(x) = f_0 + f_1 x + \cdots + f_k x^k$$

be an element of the ring $\mathbb{Z}_p[x]$, where p is a prime. Denote the product of n factors $f(x)$ by $f(x)^n$ and the result of replacing x by x^n in $f(x)$ by $f(x^n)$. Show that

(i) $f(x)^p = f(x^p)$,
(ii) $f(x)^{p^r} = f(x^{p^r})$ for all $r \geqslant 1$.

[Hint: for part (i) use the multinomial theorem and Ex. 5.3.6; for part (ii) use the principle of induction.]

15.5   The division algorithm for polynomials

In elementary algebra we are taught how to find the quotient and remainder when one polynomial is 'divided into' another. The working is usually displayed as in the following example.

$$
\begin{array}{r}
x^2 + x - 4 \\
x^2 + 3x + 2 \overline{)\, x^4 + 4x^3 + x^2 + 3x + 4} \\
\underline{x^4 + 3x^3 + 2x^2 } \\
x^3 - x^2 + 3x \\
\underline{x^3 + 3x^2 + 2x } \\
-4x^2 + x + 4 \\
\underline{-4x^2 - 12x - 8} \\
13x + 12
\end{array}
$$

Here $x^2 + 3x + 2$ is divided into $x^4 + 4x^3 + x^2 + 3x + 4$, and the quotient and remainder are $x^2 + x - 4$ and $13x + 12$ respectively. Explicitly we have

$$x^4 + 4x^3 + x^2 + 3x + 4 = (x^2 + 3x + 2)(x^2 + x - 4) + (13x + 12).$$

This is an equation in $\mathbb{Z}[x]$, the ring of polynomials with coefficients in $\mathbb{Z}$. It is fairly clear that essentially the same method will work more generally, and we shall now investigate the details of the general case.

The **degree** of a polynomial is the largest value of d for which the coefficient of x^d is non-zero. For example, the degree of $x^2 + 3x + 2$ is 2. We shall write $\deg f(x)$ for the degree of the polynomial $f(x)$, noting that according to our definition $\deg 0$ is not defined (where 0 denotes the zero polynomial). It is convenient for technical reasons to treat the zero polynomial as a special case, and we shall therefore be content to leave its degree undefined.

When the coefficient ring R is a ring like $\mathbb{Z}_m$ the degree does not always have the properties which our familiarity with polynomials over $\mathbb{Z}$ or $\mathbb{R}$ leads us to expect. For instance, the degree of the sum of two polynomials may be strictly less than the degree of either polynomial: in $\mathbb{Z}_3[x]$ we have the example

$$a(x) = x^2 + x + 1, \qquad b(x) = 2x^2 + x + 1,$$
$$a(x) + b(x) = 2x + 2,$$

so that the degree of $a(x) + b(x)$ is 1, whereas both $a(x)$ and $b(x)$ have degree 2.

It can also happen that the degree of the product of two polynomials is strictly less than the sum of their degrees. For if

$$f(x) = f_n x^n + f_{n-1} x^{n-1} + \cdots + f_0,$$
$$g(x) = g_m x^m + g_{m-1} x^{m-1} + \cdots + g_0$$

then the coefficient of x^{n+m} in the product $f(x)g(x)$ is $f_n g_m$, and this can be zero even if $f_n \neq 0$ and $g_m \neq 0$. For example, in $\mathbb{Z}_6$ we have $2 \times 3 = 0$, and so in $\mathbb{Z}_6[x]$ we have

$$(2x^2 + x + 4)(3x + 1) = 5x^2 + x + 4,$$

where the degree of the product is strictly less than the sum of the degrees of the factors. However, when the coefficients belong to a field F we have the important property that

$$uv = 0 \quad \Rightarrow \quad u = 0 \text{ or } v = 0$$

(Ex. 15.3.3). Clearly this implies that in $F[x]$

$$\deg f(x)g(x) = \deg f(x) + \deg g(x).$$

The next theorem is the analogue in $F[x]$ of the division theorem for integers (Theorem 1.5). The fact that the degree in $F[x]$ satisfies the equation given above is a vital part of the proof.

Theorem 15.5

Let F be a field and suppose that $a(x)$ and $b(x)$ are polynomials in $F[x]$, with $b(x) \neq 0$. Then there are unique polynomials $q(x)$ and $r(x)$ in $F[x]$ such that

$$a(x) = b(x)q(x) + r(x),$$

where either deg $r(x) <$ deg $b(x)$ or $r(x)$ is the zero polynomial.

Proof

We shall suppose that $b(x)$ is given, and use induction on the degree of $a(x)$. If deg $a(x) <$ deg $b(x)$ then the conditions are satisfied by taking $q(x) = 0$ and $r(x) = a(x)$, since

$$a(x) = b(x) \times 0 + a(x),$$

and by assumption deg $a(x) <$ deg $b(x)$.

If deg $a(x) \geqslant$ deg $b(x)$ we make the induction hypothesis that the result is true for all polynomials whose degree is strictly less than the degree of $a(x)$. Suppose

$$a(x) = a_{d+k} x^{d+k} + \cdots + a_0, \qquad b(x) = b_d x^d + \cdots + b_0,$$

where $a_{d+k} \neq 0$, $b_d \neq 0$, and $k \geqslant 0$. Let

$$\bar{a}(x) = a(x) - a_{d+k} b_d^{-1} x^k b(x).$$

The coefficient of x^{d+k} in $\bar{a}(x)$ is

$$a_{d+k} - (a_{d+k} b_d^{-1}) b_d = 0,$$

and so deg $\bar{a}(x) <$ deg $a(x)$. It follows from the induction hypothesis that there are polynomials $\bar{q}(x)$ and $r(x)$ such that

$$\bar{a}(x) = b(x)\bar{q}(x) + r(x),$$

where either deg $r(x) <$ deg $b(x)$ or $r(x) = 0$. Thus, if we put

$$q(x) = \bar{q}(x) + a_{d+k} b_k^{-1} x^k,$$

then it follows that

$$a(x) = b(x)q(x) + r(x),$$

as required. Hence the induction step is verified and the result is true for all values of deg $a(x)$.

In order to show that $q(x)$ and $r(x)$ are unique, suppose that

$$a(x) = b(x)q_1(x) + r_1(x) = b(x)q_2(x) + r_2(x),$$

where either $\deg r_1(x) < \deg b(x)$ or $r_1(x) = 0$, and either $\deg r_2(x) < \deg b(x)$ or $r_2(x) = 0$. Then

$$b(x)(q_1(x) - q_2(x)) = r_2(x) - r_1(x).$$

The left-hand side is zero if $q_1(x) = q_2(x)$; otherwise its degree is at least equal to that of $b(x)$. On the other hand, if the right-hand side is not zero then its degree is strictly less than that of $b(x)$. Hence both sides must be zero and $q_1(x) = q_2(x)$ and $r_1(x) = r_2(x)$. $\qquad\square$

It is worth remarking that the construction of $\bar{a}(x)$ in the proof is precisely how we proceed in practical 'long division'. In the example at the beginning of the section, we begin with

$$a(x) = x^4 + 4x^3 + x^2 + 3x + 4, \qquad b(x) = x^2 + 3x + 2$$

and obtain at the first step

$$\bar{a}(x) = x^3 - x^2 + (3x + 4).$$

Of course when the calculation is set out in the usual style, we do not 'bring down' the terms $3x$ and 4 until they are required in the working.

**Exercises
15.5**

1   Find the quotient and remainder when $x^3 + x^2 + 1$ is divided by $x^2 + x + 1$ in $\mathbb{Z}_2[x]$.

2   Find the quotient and remainder when $x^2 + 2x + 3$ is divided into $x^5 + x^4 + 2x^3 + x^2 + 4x + 2$ in $\mathbb{Z}_5[x]$.

3   Repeat Ex. 2 when the polynomials are regarded as members of $\mathbb{Z}[x]$ and comment on the relationship between the two results.

4   Without doing any further long division, write down the quotient and remainder when $x^2 + 2x + 3$ is divided into $x^5 + x^4 + 2x^3 + x^2 + 4x + 2$ in $\mathbb{Z}_7[x]$; and in $\mathbb{Z}_{73}[x]$.

5   Let F be a field. Show that the polynomial $p(x)$ has an inverse in the ring $F[x]$ if and only if $p(x)$ is a non-zero constant polynomial.

15.6   The Euclidean algorithm for polynomials

Now that we have a division algorithm for $F[x]$, analogous to the familiar one for $\mathbb{Z}$, we can proceed with some definitions and theorems about divisibility and factorization of polynomials.

We say that $g(x)$ is a **divisor** (or **factor**) of $f(x)$ in $F[x]$ if there is a polynomial $h(x)$ in $F[x]$ such that $f(x) = g(x)h(x)$. Given any two polynomials $a(x)$ and $b(x)$ in $F[x]$, we say that $d(x)$ is a **greatest common divisor** (**gcd**) of $a(x)$ and $b(x)$ if

(i) $d(x)$ is a divisor of $a(x)$ and $b(x)$, and
(ii) any divisor of $a(x)$ and $b(x)$ is also a divisor of $d(x)$.

According to the definition there is not, in general, a unique gcd of two given polynomials. If $d_1(x)$ and $d_2(x)$ are two gcd's, then $d_1(x) = \alpha d_2(x)$ for some constant polynomial α (Ex. 15.6.4). So there will be just one gcd which is monic (that is, which has its leading coefficient equal to 1), and we can if we wish specify this as *the* gcd. But it is often convenient not to make this restriction.

In order to calculate a gcd of $a(x)$ and $b(x)$ in $F[x]$ we imitate the method for $\mathbb{Z}$ and apply repeated division; this is the *Euclidean algorithm* for $F[x]$. Thus the method follows the familiar pattern:

$$a(x) = b(x)q_0(x) + r_0(x)$$
$$b(x) = r_0(x)q_1(x) + r_1(x)$$
$$r_0(x) = r_1(x)q_2(x) + r_2(x)$$
$$\cdots$$
$$r_{n-2}(x) = r_{n-1}(x)q_n(x) + r_n(x)$$
$$r_{n-1}(x) = r_n(x)q_{n+1}(x).$$

From the last equation we see that $r_n(x)$ is a divisor of $r_{n-1}(x)$. Then the penultimate equation shows that $r_n(x)$ is also a divisor of $r_{n-2}(x)$, and using the equations in reverse order we conclude that $r_n(x)$ is a divisor of $r_{n-3}(x)$ and so on, as far as $b(x)$ and $a(x)$. On the other hand, the first equation tells us that any divisor $c(x)$ of $a(x)$ and $b(x)$ is also a divisor of $r_0(x)$. Using the equations in the given order we see that $c(x)$ is also a divisor of $r_1(x), r_2(x), \ldots, r_{n-1}(x)$, and $r_n(x)$. It follows that $r_n(x)$ is a gcd of $a(x)$ and $b(x)$.

By rearranging the equations we can express $r_n(x)$ in the form

$$\lambda(x)a(x) + \mu(x)b(x),$$

where $\lambda(x)$ and $\mu(x)$ are polynomials in $F[x]$. Furthermore, if $d(x)$ is *any* gcd of $a(x)$ and $b(x)$ then $d(x)$ is a constant multiple of $r_n(x)$ (by Ex. 15.6.4 again). So we have the following important result.

Theorem 15.6   Let F be a field and suppose $d(x)$ is a gcd of the polynomials $a(x)$ and $b(x)$ in $F[x]$. Then there are polynomials $\lambda(x)$ and $\mu(x)$ in $F[x]$

such that
$$d(x) = \lambda(x)a(x) + \mu(x)b(x).$$ $\qquad\square$

Example   Find a gcd of
$$a(x) = x^3 + 2x^2 + x + 1$$
and
$$b(x) = x^2 + 5$$
in $\mathbb{Z}_7[x]$, and express it in the form $\lambda(x)a(x) + \mu(x)b(x)m$ with $\lambda(x)$ and $\mu(x)$ in $\mathbb{Z}_7[x]$.

Solution   Remembering that the coefficients are in $\mathbb{Z}_7$, we have for the first step

$$
\begin{array}{r}
x + 2 \\
x^2 + 5\,\overline{\smash{\big)}\,x^3 + 2x^2 + x + 1} \\
x^3 + 5x \\
\hline
2x^2 + 3x + 1 \\
2x^2 + 3 \\
\hline
3x + 5.
\end{array}
$$

That is,
$$x^3 + 2x^2 + x + 1 = (x^2 + 5)(x + 2) + (3x + 5).$$

The next step is to divide $3x + 5$ into $x^2 + 5$:

$$
\begin{array}{r}
5x + 1 \\
3x + 5\,\overline{\smash{\big)}\,x^2 + 5} \\
x^2 + 4x \\
\hline
3x + 5 \\
3x + 5 \\
\hline
0.
\end{array}
$$

Thus the remainder is zero and we have
$$x^2 + 5 = (3x + 5)(5x + 1).$$

So $3x + 5$ is a gcd. Rearranging the first equation we get
$$3x + 5 = (x^3 + 2x^2 + x + 1) - (x + 2)(x^2 + 5)$$
$$= (x^3 + 2x^2 + x + 1) + (6x + 5)(x^2 + 5),$$

which is of the required form with $\lambda(x) = 1$ and $\mu(x) = 6x + 5$. $\square$

1   Find the monic gcd of x^3+x^2+x+1 and x^2+2 in $\mathbb{Z}_3[x]$ and express the result in the form

$$\lambda(x)(x^3+x^2+x+1)+\mu(x)(x^2+2),$$

where $\lambda(x)$ and $\mu(x)$ are polynomials in $\mathbb{Z}_3[x]$.

2   Find the monic gcd of $x^4+2x^3+x^2+4x+2$ and x^2+3x+1 in $\mathbb{Z}_5[x]$.

3   In $\mathbb{Z}_2[x]$ find the monic gcd of

(i) x^4+1 and x^2+1,   (ii) x^5+1 and x^2+1,

(iii) x^9+1 and x^6+1.

Can you find a general rule for the monic gcd of x^n+1 and x^m+1 in $\mathbb{Z}_2[x]$?

4   Suppose $d_1(x)$ and $d_2(x)$ are gcd's of $a(x)$ and $b(x)$ in $F[x]$, where F is a field. Prove that

(i) $d_1(x)$ is a divisor of $d_2(x)$, and $d_2(x)$ is divisor of $d_1(x)$;
(ii) if $d_1(x)=\alpha(x)d_2(x)$ and $d_2(x)=\beta(x)d_1(x)$ then $\deg\alpha(x)=\deg\beta(x)=0$, so $\alpha(x)$ and $\beta(x)$ are constant polynomials.

15.7   Factorization of polynomials in theory

The culmination of our studies of divisibility and factorization of integers in Chapter 1 was the theorem that every integer $n\geqslant 2$ has a unique factorization into primes. In this section we investigate the analogous result for $F[x]$.

First, we notice that the existence of non-zero constant polynomials allows trivial factorizations of any polynomial. This is because a non-zero constant α has an inverse β in F, which is also its inverse in $F[x]$, so that

$$f(x)=\alpha\times(\beta f(x))$$

is a factorization of $f(x)$ in $F[x]$. Clearly, we wish to exclude such trivial factorizations from the theory. For this reason, we define a polynomial $f(x)$ in $F[x]$ to be **irreducible** if it is not a constant polynomial and if, whenever

$$f(x)=g(x)h(x)\quad\text{in }F[x],$$

then either $g(x)$ or $h(x)$ is a constant polynomial. The irreducible polynomials play the same rôle in $F[x]$ as do the primes in $\mathbb{Z}$.

The proof of the following theorem follows the same lines as the proof of Theorem 1.8.2, and it will be presented as a sequence of Exercises.

Theorem 15.7

Any non-constant polynomial $f(x)$ in $F[x]$ can be expressed as a product of irreducible polynomials. If there are two such factorizations

$$f(x) = g_1(x)g_2(x) \cdots g_r(x) = h_1(x)h_2(x) \cdots h_s(x)$$

then $r = s$ and we can rearrange the order of factors so that $g_i(x)$ is a constant multiple of $h_i(x)$ $(1 \leqslant i \leqslant r)$; that is, $g_i(x) = \alpha_i h_i(x)$ for some non-zero constant polynomial α_i. $\qquad\square$

Exercises 15.7

(All polynomials in Exercises 1–5 belong to $F[x]$, where F is a field.)

1 Show that if $r(x)$ is not a constant polynomial then

$$\deg r(x)s(x) > \deg s(x).$$

2 (Existence of factorization.) Show that, if $f(x)$ is a non-constant polynomial, then either $f(x)$ is irreducible or $f(x) = g(x)h(x)$ where $g(x)$ and $h(x)$ are non-constant polynomials whose degrees are less than that of $f(x)$. Hence prove the first assertion of Theorem 15.7 by induction on the degree of $f(x)$.

3 Show that if 1 is a gcd of $r(x)$ and $s(x)$, and $r(x)$ is a divisor of $s(x)t(x)$, then $r(x)$ is a divisor of $t(x)$.

4 Show that if $r(x)$ is irreducible and $r(x)$ is a divisor of $s_1(x)s_2(x) \cdots s_k(x)$, then $r(x)$ is a divisor of $s_i(x)$ for some i in the range $1 \leqslant i \leqslant k$.

5 (Uniqueness of factorization.) Prove the uniqueness assertion in Theorem 15.7.

6 Verify that in $\mathbb{Z}_{15}[x]$

$$(x+1)(x+14) = (x+4)(x+11).$$

What is the significance of this result in relation to the theory developed above?

15.8 Factorization of polynomials in practice

The existence of a very satisfactory theorem about factorization does not mean that it is easy to find the factors in any given case. The general problem is a difficult one, but there is simple test for finding factors of a particular kind, which we shall now describe.

A polynomial $a_1x + a_0$ with $a_1 \neq 0$ has degree 1 and is said to be a **linear** polynomial. Multiplying by the constant a_1^{-1} transforms the polynomial into the form $x - \alpha$, where $\alpha = -a_1^{-1}a_0$, and we shall take this as the standard form because there is a useful test which tells us when $x - \alpha$ is a factor of $f(x)$ in $F[x]$.

Suppose that

$$f(x) = f_n x^n + f_{n-1} x^{n-1} + \cdots + f_0,$$

and for each α in F let

$$f(\alpha) = f_n \alpha^n + f_{n-1} \alpha^{n-1} + \cdots + f_0.$$

Since F is a field the expression $f(\alpha)$ is an element of F, and we say that it is obtained by **evaluating** $f(x)$ at α. The rule which takes α to $f(\alpha)$ is a function from F to F; strictly speaking it should be called the **polynomial function** corresponding to the polynomial $f(x)$. Although we do not need to stress the point here, there are good reasons for distinguishing between the function and the polynomial (which is simply a sequence of coefficients): one such reason is that different polynomials may be associated with the same function (Ex. 15.9.8 and Ex. 15.9.16).

Theorem 15.8.1 Let F be a field and suppose $f(x)$ is a polynomial in $F[x]$. Then $x - \alpha$ is a divisor of $f(x)$ in $F[x]$ if and only if $f(\alpha) = 0$ in F.

Proof Suppose $x - \alpha$ is a divisor of $f(x)$, so that

$$f(x) = (x - \alpha)g(x) \quad \text{in } F[x].$$

Evaluating both sides at α we obtain

$$f(\alpha) = (\alpha - \alpha)g(\alpha) = 0g(\alpha) = 0.$$

Conversely, suppose $f(\alpha) = 0$ in F. By Theorem 15.5 there are polynomials $q(x)$ and $r(x)$ in $F[x]$ such that

$$f(x) = (x - \alpha)q(x) + r(x),$$

where either $\deg r(x) < \deg (x - \alpha)$ or $r(x) = 0$. Evaluating both sides of the equation at α, and remembering that $f(\alpha) = 0$ we obtain

$$0 = f(\alpha) = (\alpha - \alpha)q(\alpha) + r(\alpha) = r(\alpha).$$

Now if $\deg r(x) < \deg (x - \alpha)$ then the degree of $r(x)$ must be zero and $r(x)$ is a non-zero constant polynomial; in this case $r(\alpha)$ cannot be zero. Hence we must have $r(x) = 0$, from which it follows that $x - \alpha$ is a divisor of $f(x)$.   ☐

Theorem 15.8.1 is known as the **factor theorem**. Before illustrating its use in practice we shall record for future reference an important theoretical consequence. If $f(x)$ is a polynomial in $F[x]$ and α is an element of F then we say that α is a **root** of the equation $f(x) = 0$ whenever $f(\alpha) = 0$.

Theorem 15.8.2   If F is a field and $f(x)$ is a polynomial of degree $n \geqslant 1$ in $F[x]$ then the equation $f(x) = 0$ has at most n roots in F.

Proof   Suppose the equation has m distinct roots $\alpha_1, \alpha_2, \ldots, \alpha_m$ in F. By the factor theorem $x - \alpha_1, x - \alpha_2, \ldots, x - \alpha_m$ are divisors of $f(x)$, and furthermore they are all irreducible. Hence the factorization of $f(x)$ in $F[x]$ takes the form

$$f(x) = (x - \alpha_1)(x - \alpha_2) \cdots (x - \alpha_m)g(x)$$

for some $g(x)$ in $F[x]$. Since the coefficients belong to a field F the degree of a product is the sum of the degrees of the factors, and so it follows that the degree of $f(x)$ is at least m. Equivalently, this means that the number of roots of $f(x) = 0$ is at most n.   ☐

Now we return to the practical problem of how to find the irreducible factors of a given polynomial. The most interesting and useful cases from our point of view occur when the polynomials belong to $\mathbb{Z}_p[x]$, where $\mathbb{Z}_p$ is the field of integers modulo a prime p. In such cases we can find the linear factors in a finite number of steps, since the factor theorem tells us that we need only evaluate the polynomial at each of the p elements of $\mathbb{Z}_p$. However there is no reason why the irreducible factors of a given polynomial should be linear. If the polynomial has small degree then we may be able to make some progress by simple-minded methods, and we shall now discuss such methods. There is no loss of generality in assuming that the given polynomial is monic, since we can transform a given polynomial into a monic polynomial of the same degree by multiplying by a constant.

By definition, every linear polynomial is irreducible in $\mathbb{Z}_p[x]$, and so there are p monic irreducible linear polynomials $x + a_0$. If the *quadratic* polynomial $x^2 + a_1 x + a_0$ is reducible in $\mathbb{Z}_p[x]$ then it has two linear factors which can be found by the factor theorem. Since there are p possible linear factors, there are $\frac{1}{2}p(p-1)$ monic reducible quadratics of the form $(x-\alpha)(x-\beta)$ with $\alpha\neq\beta$, and p of the form $(x-\alpha)^2$. There are p^2 monic quadratics in all, so the number of irreducible ones is

$$p^2-(\tfrac{1}{2}p(p-1)+p)=\tfrac{1}{2}p(p-1).$$

In particular, we note that there is always at least one monic irreducible quadratic polynomial in $\mathbb{Z}_p[x]$. When $p=2$ we have

$$x^2=(x+0)^2, \qquad x^2+1=(x+1)^2, \qquad x^2+x=(x+0)(x+1),$$

but x^2+x+1 is irreducible.

If a *cubic* polynomial $x^3+a_2x^2+a_1x+a_0$ is reducible then it has either three linear factors or one linear and one quadratic factor. Since there is a linear factor in either case, we can use the factor theorem here again to test for reducibility. However, when the degree of the polynomial is four or more, other methods may be required.

Example   Find the irreducible factors of x^4+1 in $\mathbb{Z}_3[x]$.

Solution   First we use the factor theorem to look for linear factors. Let $f(x)=x^4+1$; then

$$f(0)=0^4+1=1, \qquad f(1)=1^4+1=2, \qquad f(2)=2^4+1=2.$$

So there are no linear factors. The only possibility remaining is a factorization into two quadratics:

$$x^4+1=(x^2+Ax+B)(x^2+Cx+D),$$

with A, B, C, D in $\mathbb{Z}_3$. Equating coefficients of corresponding powers of x we get the equations

(i) $A+C=0$,   (ii) $B+D+AC=0$,

(iii) $AD+BC=0$,   (iv) $BD=1$.

It follows from (i) that $A=-C$ and from (iv) that $B=D$. (Why?) Taking $B=D=1$ we get $AC=1$ in (ii), which contradicts (i). Taking $B=D=2$ we get $AC=2$; then $A=1$, $C=2$ is a solution. Hence the required factorization is

$$x^4+1=(x^2+x+2)(x^2+2x+2). \qquad \Box$$

Of course, the method given in the example is very inefficient in most cases. A great deal of work has been done on the development of efficient algorithms for finding the irreducible factors of polynomials, but many of the best methods are beyond the scope of this book.

1 Find the irreducible factors of

(i) x^2+1 in $\mathbb{Z}_5[x]$;
(ii) x^3+5x^2+5 in $\mathbb{Z}_{11}[x]$;
(iii) x^4+3x^3+x+1 in $\mathbb{Z}_5[x]$.

2 Find all the monic irreducible quadratics in $\mathbb{Z}_3[x]$.

3 Show that the number of monic irreducible cubics in $\mathbb{Z}_p[x]$ is $\frac{1}{3}p(p^2-1)$, and list them when $p=2$.

4 Let $f(x)=f_nx^n+f_{n-1}x^{n-1}+\cdots+f_0$ be a polynomial of degree n in $F[x]$, and let $\alpha \in F$. Explain how $f(\alpha)$ can be evaluated by means of the recursion

$$f_0(\alpha)=f_n, \qquad f_i(\alpha)=\alpha f_{i-1}(\alpha)+f_{n-i} \quad (1\leqslant i\leqslant n).$$

What is the number of multiplications required by this method? (It is sometimes known as *Horner's method*.)

5 Suppose that $f(x)$ and α are as in Ex. 4, and $f(\alpha)$ is evaluated by computing each term $f_i\alpha^i$ individually ($0\leqslant i\leqslant n$). Find the approximate number of multiplications required if the methods for calculating α^i given in Section 7.7 are used.

5.9 Miscellaneous Exercises

1 Investigate the structures of the groups of invertible elements of $\mathbb{Z}_{15}$ and $\mathbb{Z}_{16}$.

2 Factorize the following into irreducible polynomials in $\mathbb{Z}_5[x]$:

(i) x^4+4, (ii) x^4+3x^3+2x+4.

3 Find an irreducible polynomial of degree 4 in $\mathbb{Z}_5[x]$.

4 Show that if the integer r is a root of the equation $x^n+a_{n-1}x^{n-1}+\cdots+a_1x+a_0=0$ in $\mathbb{Z}[x]$ then r is a divisor of a_0.

5 Find the monic gcd of the polynomials x^3+1 and x^4+x^2+1 in $\mathbb{Z}_3[x]$, and in $\mathbb{Z}_7[x]$.

6   Prove that the multiplication in a ring R is commutative if and only if

$$(a+b)^2 = a^2 + 2ab + b^2$$

for all a, b in R.

7   Show by means of an example that if m is not a prime then a quadratic equation in $\mathbb{Z}_m[x]$ may have more than two roots in $\mathbb{Z}_m$.

8   Let p be a prime. Prove that the polynomials $f(x) = x^p$ and $g(x) = x$ in $\mathbb{Z}_p[x]$ define the same function from $\mathbb{Z}_p$ to itself.

9   Let ϕ denote the set of all functions f from $\mathbb{Z}$ to itself, and define the addition and multiplication of such functions by the rules $(f+g)(x) = f(x) + g(x)$, $(fg)(x) = f(x)g(x)$. Which of the following subsets of Φ are rings?

(i) The functions satisfying $f(0) = 0$.
(ii) The functions satisfying $f(0) \neq 0$.
(iii) The functions satisfying $f(0) = f(1)$.
(iv) The functions satisfying $f(x) = f(x+1)$ for all x in $\mathbb{Z}$.

10   Show that when $a \neq 0$ the polynomial $ax^2 + bx + c$ is irreducible in $\mathbb{Z}_p[x]$ (where p is a prime) if and only if $b^2 - 4ac$ is not a square in $\mathbb{Z}_p$.

11   Find the monic gcd in $\mathbb{Z}[x]$ of

$$x^6 + 6x^5 + 8x^4 + x^3 + 4x^2 + 2x + 8 \quad \text{and} \quad x^4 + 5x^3 + 6x^2 + 9x + 4.$$

12   Let ω denote the complex number $e^{2\pi i/3}$. Show that $1 + \omega + \omega^2 = 0$ and deduce that the set of complex numbers of the form $m + n\omega$ ($m, n \in \mathbb{Z}$) is a ring. (You need not check properties which follow directly from the standard properties of complex numbers.)

13   Describe the group of invertible elements of the ring constructed in the previous exercise.

14   Let F be a field with an infinite number of elements. Show that if $f(x)$ and $g(x)$ are polynomials in $F[x]$ which induce the same function from F to itself, then $f(x) = g(x)$.

15   Let $f(x) = f_0 + f_1 x + \cdots + f_n x^n$ be a polynomial with coefficients in a field F. The **derivative** of $f(x)$ is defined to be the polynomial

$$f'(x) = f_1 + 2f_2 x + \cdots + nf_n x^{n-1}.$$

Show that (i) $(f+g)'(x) = f'(x) + g'(x)$; (ii) $(fg)'(x) = f'(x)g(x) + f(x)g'(x)$.

16   Find all primes p for which the polynomials $x^2 + 3x + 1$ and $x^6 + 2x^5 + x + 1$ define the same function on $\mathbb{Z}_p$.

17   Show that the set of real numbers of the form $m + n\sqrt{2}$, where m and n are integers, is a ring with respect to the usual operations of addition and multiplication. Show also that $m + n\sqrt{2}$ is invertible in this ring if and only if $m^2 - 2n^2 = \pm 1$.

18 Let i, j, k be symbols satisfying the equations

$$i^2 = j^2 = k^2 = -1, \qquad ij = -ji = k,$$
$$jk = -kj = i, \qquad ki = -ik = j.$$

Show that the set Q of expressions of the form $a + bi + cj + dk$, where a, b, c, d are real numbers, satisfies all the axioms for a field except that multiplication is not commutative. (Q is an example of a **skew field**, and its elements are usually known as **quaternions**.)

19 An element x of a ring R is said to be **nilpotent** if $x^n = 0$ in R for some positive integer n. Show that if x and y are nilpotent elements of a ring R with commutative multiplication then $x + y$ is also nilpotent.

20 Find all the nilpotent elements of the rings $\mathbb{Z}_7$, $\mathbb{Z}_8$, and $\mathbb{Z}_9$.

21 Show that any function f from $\mathbb{Z}_2$ to itself may be represented by a polynomial $f(x)$. Does the same result hold for any field $\mathbb{Z}_p$ (p prime)?

16
Finite fields and some applications

16.1 A field with nine elements

In Section 6.5 we showed that when p is a prime it is possible to use the arithmetical properties of $\mathbb{Z}_p$ to construct a set of $p-1$ mutually orthogonal latin squares. With hindsight it is clear that the key to the construction is the fact that $\mathbb{Z}_p$ is a field. Given any field with n elements we could use the same method to construct $n-1$ mutually orthogonal latin squares. For this reason alone it is natural to ask whether it is possible to construct finite fields other than $\mathbb{Z}_p$.

In fact we have already given one example of a field which does not have a prime number of elements: the field whose elements are the nine skew-symmetric 2×2 matrices over $\mathbb{Z}_3$, discussed in the *Example* of Section 15.3. We shall begin our discussion of the general question by giving another construction of a field of order nine, which we shall denote by F_9.

The elements of F_9 will be represented by 0 and the eight polynomials of degree 0 and 1 with coefficients in the field $\mathbb{Z}_3$, that is

$$F_9 = \{0, 1, 2, x, x+1, x+2, 2x, 2x+1, 2x+2\}.$$

This set is closed under the usual rule for adding polynomials, since, for example

$$(x+1)+(2x+1) = 2 \quad \text{in } \mathbb{Z}_3[x].$$

However, the set is not closed under ordinary multiplication of polynomials, since

$$(x+1) \times (2x+1) = 2x^2+1 \quad \text{in } \mathbb{Z}_3[x]$$

and $2x^2+1$ is not one of the designated elements of F_9. For this reason we shall define a modified rule for multiplication, whereby we first calculate the ordinary product in $\mathbb{Z}_3[x]$ and then *reduce modulo x^2+1*. For example,

$$\begin{aligned}
(x+1) \times (2x+1) &= 2x^2+1 && (\text{in } \mathbb{Z}_3[x]) \\
&= 2+2(x^2+1) \\
&= 2 && (\text{in } F_9).
\end{aligned}$$

Similarly

$$(2x + 1) \times (x) = 2x^2 + x$$
$$= (x + 1) + 2(x^2 + 1)$$
$$= x + 1.$$

It is fairly clear that these $+$ and $\times$ operations make F_9 into a ring. The constant polynomials 0 and 1 are indeed the 0 and 1 of the ring, and the remaining axioms can be verified quite easily. What is rather less obvious is that F_9 is actually a field, because every element except 0 has a multiplicative inverse. The sceptical reader is invited to check the following table.

Element:	1	2	x	$x+1$	$x+2$	$2x$	$2x+1$	$2x+2$
Inverse:	1	2	$2x$	$x+2$	$x+1$	x	$2x+2$	$2x+1$

The general construction of which F_9 is a special case will be explained in Section 16.3. For the moment, we shall emphasize one further property of F_9, concerning its multiplicative group. If we compute the powers of the polynomial $2x + 1$ according to the rules in F_9 we obtain the following results:

$$(2x + 1)^1 = 2x + 1, \quad (2x + 1)^2 = x, \quad (2x + 1)^3 = x + 1,$$
$$(2x + 1)^4 = 2, \quad (2x + 1)^5 = x + 2, \quad (2x + 1)^6 = 2x,$$
$$(2x + 1)^7 = 2x + 2, \quad (2x + 1)^8 = 1.$$

Thus we can express all the non-zero elements of F_9 as powers of $2x + 1$. In other words, the multiplicative group of F_9 is a cyclic group C_8 generated by $2x + 1$:

$$F_9 \backslash \{0\} = U(F_9) = \langle 2x + 1 \rangle \approx C_8.$$

In Section 16.4 we shall prove the unexpected result that the multiplicative group of *any* finite field is cyclic.

Exercises 16.1

1 Which members of F_9 can be chosen as the generators of its multiplicative group?

2 Which members of F_9 have square roots in F_9?

3 Without using explicit calculations, prove that the product of all the non-zero elements of F_9 is 2.

4 Show that the additive group of F_9 is not cyclic.

16.2   The order of a finite field

Let F be any field, finite or infinite. Since F is closed under addition, and it contains a multiplicative identity 1, the elements

$$2 = 1+1, \qquad 3 = 1+1+1, \qquad 4 = 1+1+1+1,$$

and so on, all belong to F. (We do not claim that they are all distinct.) For any positive integer n the sum of n 1's is an element of F; in more formal language we may say that these are the elements of the cyclic subgroup $\langle 1 \rangle$ of the additive group of F.

When F is finite Lagrange's theorem tells us that the order of $\langle 1 \rangle$ is a divisor of $|F|$. This number is called the **characteristic** of F. For example, in the field $\mathbb{Z}_p$ the subgroup $\langle 1 \rangle$ is the whole field, and so the characteristic of $\mathbb{Z}_p$ is p. In general, the characteristic of F is the smallest positive integer m such that $m = 0$ in F, where the latter m denotes the sum of m 1's in F. (When F is infinite the characteristic may or may not be defined.)

It is easy to see that the characteristic of any finite field must be a prime number. For a number which is not prime can be written as a product $m_1 m_2$, and if we have $m_1 m_2 = 0$ in a field then either $m_1 = 0$ or $m_2 = 0$. Thus a number which is not a prime cannot be the least number for which $m = 0$. In the next theorem we use the fact that the characteristic is prime to determine the structure of the additive group of a finite field, and to show that the order of a finite field must take a very special form.

Theorem 16.2

If F is a finite field of characteristic p, then the additive group of F is isomorphic to $(C_p)^r$, the direct product of r copies of C_p. Consequently $|F| = p^r$ for some $r \geqslant 1$.

Proof

Given any element $f \neq 0$ in F and any positive integer n there is an element

$$nf = (1+1+\cdots+1)f = f+f+\cdots+f$$

in F, where there are n terms in the sum. These elements form the cyclic subgroup $\langle f \rangle$ of the additive group of F. Since $p = 0$ in F the only relevant values of n are $0, 1, \ldots, p-1$, and we can regard n as an element of $\mathbb{Z}_p$.

Let us say that the subset $\{f_1, f_2, \ldots, f_k\}$ of F *spans* F if any element of F can be written in the form

$$f = n_1 f_1 + n_2 f_2 + \cdots + n_k f_k \quad (n_1, n_2, \ldots, n_k \in \mathbb{Z}_p).$$

Such sets certainly exist, since the whole of F is one of them.

Suppose that $\{f_1, f_2, \ldots, f_r\}$ spans F and no proper subset of it does so. Then each one of the p^r expressions

$$n_1 f_1 + n_2 f_2 + \cdots + n_r f_r \quad (n_1, n_2, \ldots, n_r \in \mathbb{Z}_p)$$

is a member of F, and each member of F is equal to such an expression. If two different expressions represent the same member of F, say

$$n_1 f_1 + n_2 f_2 + \cdots + n_r f_r = m_1 f_1 + m_2 f_2 + \cdots + m_r f_r,$$

then we can pick the first subscript i such that $n_i \neq m_i$ and rewrite the equation as

$$(n_i - m_i) f_i = (m_{i+1} - n_{i+1}) f_{i+1} + \cdots + (m_r - n_r) f_r.$$

Since $n_i - m_i \neq 0$ it has an inverse in $\mathbb{Z}_p$. Multiplying the equation by $(n_i - m_i)^{-1}$ we obtain an expression for f_i in terms of $f_{i+1}, \ldots, f_r$. Thus f_i could be eliminated from the spanning set, contrary to the assumption that no proper subset of $\{f_1, f_2, \ldots, f_r\}$ spans F.

We conclude that there is a bijection in which the elements of F correspond to the r-tuples $(n_1, n_2, \ldots, n_r)$ of elements of $\mathbb{Z}_p$. Since addition in F corresponds to addition of r-tuples, the bijection is an isomorphism of the additive group of F with the direct product of r copies of $\mathbb{Z}_p$ (regarded as a cyclic group): that is, $(C_p)^r$. $\quad\square$

Of course, the most immediate result of the theorem is that if a finite field exists then its order must be a prime power. We are familiar with the fields $\mathbb{Z}_p$ (of order p) and we have met two examples of a field of order 9 (that is, 3^2). Our task now must be to show that fields of order p^r do exist for any prime p and any $r \geqslant 1$.

Exercises 16.2

1 Tables 16.2.1(a, b) define $+$ and $\times$ operations on the set $\{w, y, z, t\}$, and the resulting structure is a field F_4.

Table 16.2.1(a)

$+$	w	y	z	t
w	w	y	z	t
y	y	w	t	z
z	z	t	w	y
t	t	z	y	w

Table 16.2.1(b)

$\times$	w	y	z	t
w	w	w	w	w
y	w	y	z	t
z	w	z	t	y
t	w	t	y	z

(i) Identify the elements 0 and 1 of F_4.

(ii) Show that the additive and multiplicative groups of F_4 are isomorphic to $C_2 \times C_2$ and C_3 respectively.

(iii) What is the characteristic of F_4?

2   Prove that the subset of a field F consisting of all the elements of the form $1 + 1 + \cdots + 1$ is itself a field F_0. (F_0 is called the **prime field** of F.)

3   Let F be a field of characteristic 3. Establish the following identities in F:

$$\text{(i)} \quad x^3 + y^3 = (x + y)^3,$$
$$\text{(ii)} \quad (x + y)^4 + x^4 + (x - y)^4 + y^4 = 0.$$

4   Show that in any field of characteristic p

$$(x + y)^p = x^p + y^p.$$

[Hint: Ex. 4.3.5.]

16.3   Construction of finite fields

The field of order nine constructed in Section 16.1 can be viewed in a slightly more abstract way. The relation $\sim$ on $\mathbb{Z}_3[x]$ defined by

$$a(x) \sim b(x) \quad \Leftrightarrow \quad a(x) - b(x) \text{ is divisible by } x^2 + 1$$

is an equivalence relation. In fact, the polynomials

$$0, 1, 2, x, x + 1, x + 2, 2x, 2x + 1, 2x + 2$$

form a complete set of representatives for the equivalence classes, and we can take the classes, rather than their representatives, as the elements of F_9. Denoting the equivalence class of $a(x)$ by $[a(x)]$, we can define the addition and multiplication of equivalence classes in the obvious way:

$$[a(x)] + [b(x)] = [a(x) + b(x)], \qquad [a(x)][b(x)] = [a(x)b(x)],$$

and clearly these definitions correspond exactly to the naive calculations in Section 16.1.

This more general viewpoint opens the way for a more general construction of finite fields. The critical point is that the polynomial used to define the equivalence classes must be *irreducible*.

Theorem 16.3

Let $k(x)$ be an irreducible polynomial of degree r in $\mathbb{Z}_p[x]$, and let $\sim$ be the equivalence relation on $\mathbb{Z}_p[x]$ defined by

$$a(x) \sim b(x) \quad \Leftrightarrow \quad a(x) - b(x) \text{ is divisible by } k(x).$$

Then the set of equivalence classes of $\sim$ in $\mathbb{Z}_p[x]$ is a field of order p^r.

Proof

It is clear that the polynomials of degree $0, 1, \ldots, r-1$ form a complete set of representatives for the classes, so there are p^r classes in all. Furthermore, it is entirely a matter of routine checking to verify that the classes (like the polynomials themselves) form a ring, and that multiplication is commutative.

The important thing is to show that every class except $[0]$ has a multiplicative inverse, and this is where the irreducibility of $k(x)$ is vital. Given any polynomial $a(x)$ in $\mathbb{Z}_p[x]$, the fact that $k(x)$ is irreducible means that the monic gcd of $a(x)$ and $k(x)$ is 1. So by Theorem 15.6 there are polynomials $f(x)$ and $g(x)$ in $\mathbb{Z}_p[x]$ such that

$$a(x)f(x) + k(x)g(x) = 1.$$

Taking equivalence classes, and remembering that $[k(x)] = [0]$, we obtain

$$[a(x)][f(x)] = [1].$$

In other words, $[a(x)]$ has the inverse $[f(x)]$. $\qquad\square$

The theorem tells us that in order to construct a field of order p^r it is only necessary to find an irreducible polynomial of degree r in $\mathbb{Z}_p[x]$. That sounds easy. Indeed, in a very down-to-earth sense it is easy, since there are tables of irreducible polynomials covering any values of p and r which are ever likely to arise in practice. But in mathematics we like to prove such things, and unfortunately the general proof that there is at least one irreducible polynomial for *every* value of p and r is rather difficult. For that reason we shall proceed to derive some general properties of finite fields and indicate some of their applications, deferring the proof of their existence in all cases until Section 16.9.

Exercises 16.3

1 Prove that $x^3 + x^2 + 1$ is irreducible in $\mathbb{Z}_2[x]$, and hence construct a field of order 8. What is the order of its multiplicative group? Describe the group explicitly.

2 Prove that the field of order 4 constructed by using the irreducible polynomial $x^2 + x + 1$ in $\mathbb{Z}_2[x]$ is essentially the same as the field F_4 constructed in Ex. 16.2.1.

3 For which of the following primes p can we construct a field of order p^2 by using the polynomial $x^2 + 1$?

$$p = 3, 5, 7, 11, 13, 19, 23.$$

Describe the multiplicative group for the first two cases in which the field can be constructed.

4 Show that for every prime p there are fields of order p^2 and p^3. [Hint: see Section 15.8.]

16.4 The primitive element theorem

In Section 16.2 we showed that the *additive* group of a finite field of order $q = p^r$ is isomorphic to $(C_p)^r$, the direct product of r cyclic groups C_p. The *multiplicative* group of the field is a group of order $q - 1$ (since 0 is excluded), and its structure is surprisingly straightforward.

Theorem 16.4

The multiplicative group of any finite field is cyclic.

Proof

Let F be a field of order q and let F^* denote its multiplicative group $F \backslash \{0\}$. If f is any element of F^* then it follows from Theorem 13.8.3 that $f^{q-1} = 1$. In other words, the equation $x^{q-1} - 1 = 0$ has $q - 1$ roots in F.

We shall show that F^* satisfies the numerical characterization of cyclic groups obtained in Theorem 13.9. Specifically, we shall prove that for each divisor d of $q - 1$ there are d elements f of F^* for which $f^d = 1$.

Suppose $dk = q - 1$; then the following equation in $F[x]$ may be verified by elementary algebra:

$$x^{q-1} - 1 = (x^d - 1)(x^{d(k-1)} + x^{d(k-2)} + \cdots + x^d + 1).$$

Let us denote the second factor by $g(x)$. Since $g(x)$ is a polynomial of degree $d(k-1)$ the equation $g(x) = 0$ has at most $d(k-1)$ roots in F, by Theorem 15.8.2. Similarly, the equation $x^d - 1 = 0$ has at most d roots in F. But we have established that the equation

$x^{q-1}-1=0$ has exactly $q-1$ roots in F, and since $d(k-1)+d = q-1$ it follows that each of the subsidiary equations must have the maximum possible number of roots. In particular, $x^d - 1 = 0$ has d roots, and so there are d elements of F^* for which $f^d = 1$. Hence the result follows from Theorem 13.9. $\qquad\square$

Recall that a group is cyclic if and only if all its elements can be expressed as powers of a single element, which is called a generator for the group. Thus if z is a generator for F^* we have

$$F^* = \{1, z, z^2, \ldots, z^{q-2}\}, \quad \text{where } z^{q-1} = 1.$$

For example, the multiplicative group of the field $\mathbb{Z}_{23}$ is a cyclic group of order 22 with generator 5, since the powers of 5 give all the non-zero elements of the field:

$$5^1 = 5, \qquad 5^2 = 2, \qquad 5^3 = 10, \qquad 5^4 = 4,$$
$$\ldots$$
$$5^{19} = 7, \qquad 5^{20} = 12, \qquad 5^{21} = 14, \qquad 5^{22} = 1.$$

In general, a generator for the multiplicative group F^* is known as a **primitive element** in the field F. Using this terminology Theorem 16.4 can be restated in the form:

every finite field has a primitive element.

Despite its elegance and simplicity, the theorem has one inevitable defect: it does not tell us how to find a primitive element in any given case. Since the number of elements of order $q-1$ in C_{q-1} is $\phi(q-1)$, it follows that there are $\phi(q-1)$ primitive elements in any field of order q. If we have to find one of them 'by hand' the best method is a refined form of trial-and-error.

Example   Find a primitive element in the field $\mathbb{Z}_{41}$.

Solution   The smallest positive integer which could conceivably represent a primitive element of $\mathbb{Z}_{41}$ is 2, so we begin by computing the powers of 2 in $\mathbb{Z}_{41}$. If 2 is a primitive element then we shall obtain all the non-zero elements; if not, we shall have some useful information. We get the following table:

n:	1	2	3	4	5	6	7	8	9	10
2^n:	2	4	8	16	32	23	5	10	20	40

n:	11	12	13	14	15	16	17	18	19	20
2^n:	39	37	33	25	9	18	36	31	21	1.

Hence the order of 2 is 20, rather than 40, and we conclude that 2 is not a primitive element. We could try 3, but we note from the table that 9 (that is, 3^2) is equal to 2^{15}, so

$$3^8 = 9^4 = 2^{60} = (2^{20})^3 = 1,$$

and the order of 3 is only 8. The elements 4 and 5 are both powers of 2, and so their orders must be divisors of 20. However the element 6 is not a power of 2, and

$$6^2 = 36 = 2^{17}.$$

The order of 2^{17} is 20 (why?), so the order of 6 is 40 and 6 is the required primitive element.   □

In practice we can resort to tables giving the least positive integer which is a primitive element of the field $\mathbb{Z}_p$. These tables are available in several computer systems, and they cover an extensive range of values of the prime p. There are also tables which help us when the order of the field is a prime power q, rather than a prime, and we shall now discuss some aspects of this case.

The field F of order $q = p^r$ is constructed by choosing an irreducible polynomial $k(x)$ of degree r in $\mathbb{Z}_p[x]$. If we are lucky, then it will turn out that the polynomial x itself is a primitive element of F, and when this is so we say that $k(x)$ is a **primitive** irreducible polynomial.

Example   Show that $x^2 + 2x + 2$ is a primitive irreducible polynomial in $\mathbb{Z}_3[x]$.

Solution   First we must show that $x^2 + 2x + 2$ is irreducible. Since it is a quadratic it can only have linear factors, and we may test for these by using the factor theorem. We find that

$$0^2 + (2 \times 0) + 2 = 2, \qquad 1^2 + (2 \times 1) + 2 = 2, \qquad 2^2 + (2 \times 2) + 2 = 1,$$

so there are no linear factors and the polynomial is irreducible. Hence the polynomials

$$0, \ 1, \ 2, \ x, \ x+1, \ x+2, \ 2x, \ 2x+1, \ 2x+2,$$

represent the elements of a field, with respect to the operations of addition and multiplication modulo $x^2 + 2x + 2$. Calculating the powers of x in this field we obtain

$$x^1 = x, \qquad x^2 = x+1, \qquad x^3 = 2x+1, \qquad x^4 = 2,$$
$$x^5 = 2x, \qquad x^6 = 2x+2, \qquad x^7 = x+2, \qquad x^8 = 1.$$

Thus x generates the multiplicative group of the field, and $x^2 + 2x + 2$ is a primitive irreducible polynomial. □

The fact that the irreducible polynomial $x^2 + 2x + 2$ is primitive means that there are advantages in using this polynomial to construct a field of order 9. For example, we can use the table of powers of x to facilitate multiplication, in much the same way as we use logarithm tables in elementary arithmetic. Thus, to multiply $x + 1$ and $2x + 1$ we proceed as follows:

$$(x + 1) \times (2x + 1) = x^2 \times x^3 = x^5 = 2x.$$

On the other hand, the polynomial $x^2 + 1$ in $\mathbb{Z}_3[x]$, which we used to construct F_9 in Section 16.1, is not really a good choice since it is not primitive. Computing powers of x in F_9 we find that $x^4 = 1$, so the polynomial x is not a primitive element in F_9.

These remarks lead naturally to the final theoretical question about finite fields. We know that if we are given a prime power $q = p^r$ we can use any irreducible polynomial of degree r in $\mathbb{Z}_p[x]$ to construct a field of order q. What is the connection between the fields constructed by using different polynomials?

At this point the reader who has faith in the beauty of mathematics will surely guess: they are all the same. And indeed it is so. We have shown that any two fields of order q have isomorphic additive groups and isomorphic multiplicative groups; so it is hardly surprising that the fields themselves are isomorphic, in the sense that there is a bijection from one to the other which is simultaneously an isomorphism of the additive and multiplicative groups. Of course, from a constructive point of view it is more important to know that a field with q elements exists than it is to know that it is unique. For this reason we shall be content with giving a formal proof of the existence only (in Section 16.9).

In summary, the theory of finite fields turns out to be remarkably simple.

> *Any finite field has prime power order $q = p^r$.*
> *There is essentially just one field of order q.*
> *The additive group of the field is $(C_p)^r$.*
> *The multiplicative group of the field is C_{q-1}.*

We shall use the notation $\mathbb{F}_q$ for the unique field of order q. These fields are often known as *Galois fields*, after Évariste Galois (1811–1832), and sometimes they are denoted by the symbol $GF(q)$. Of course when q is itself a prime p the field $\mathbb{F}_p$ (or $GF(p)$) is simply the familiar field $\mathbb{Z}_p$, whose elements are the integers modulo p.

1 Find the least positive integer which represents a primitive element

(i) in $\mathbb{Z}_7$, (ii) in $\mathbb{Z}_{11}$, (iii) in $\mathbb{Z}_{47}$.

2 Show that when $\mathbb{F}_8$ is constructed by using the irreducible polynomial $x^3 + x^2 + 1$ (as in Ex. 16.3.1), then the polynomial x is a primitive element.

3 Which of the following polynomials in $\mathbb{Z}_2[x]$ are irreducible, and which of the irreducible ones are primitive?

(i) $x^4 + 1$, (ii) $x^4 + x + 1$, (iii) $x^4 + x^2 + 1$.

4 What is the number of primitive elements in the field $\mathbb{F}_{32}$? Deduce from your answer that the polynomial $x^5 + x + 1$ is a primitive irreducible polynomial in $\mathbb{Z}_2[x]$.

5 Find all values of m for which $x^2 + mx + 2$ is a primitive irreducible polynomial in $\mathbb{Z}_{11}[x]$.

16.5 Finite fields and latin squares

In this section and the three which follow it we shall describe some of the constructive uses of finite fields.

We began this chapter by recalling the problem of constructing sets of mutually orthogonal latin squares, and it is natural to ask how large such a set can be. We know that for each prime p there is a set of $p - 1$ mutually orthogonal latin squares of order p, and soon we shall prove that the same result holds when p is replaced by a prime power $q = p^r$. But in order to put the result into perspective, we shall begin by showing that the maximum possible number of mutually orthogonal latin squares of order n is $n - 1$, for any value of n.

We remark first that the symbols in each square may be relabelled independently so that the first row of each one is the same, say

$$1 \quad 2 \quad 3 \quad \ldots \quad n.$$

This relabelling does not affect the orthogonality property of the set of squares. Consider the possibilities for the symbol in position $(2, 1)$—that is, the second row and first column—of any square. This symbol can never be 1 since 1 already appears in the first column of every square, and so there are at most $n-1$ possibilities. If two different squares have the same symbol in position $(2, 1)$, say k, then both squares have the symbol k in position $(1, k)$ also, which is contrary to the orthogonality hypothesis. So the maximum number of mutually orthogonal latin squares cannot exceed the number of possibilities for the symbol in position $(2, 1)$, which is $n-1$.

The preceding remarks indicate that the next theorem is a best possible result about sets of mutually orthogonal latin squares. The proof of the theorem is simply a transcription of the proof of Theorem 6.5.2, using the finite field $\mathbb{F}_q$ in place of $\mathbb{Z}_p$.

Theorem 16.5

Whenever q is a prime power it is possible to construct $q-1$ mutually orthogonal latin squares of order q.

Proof

For each of the $q-1$ non-zero elements t of $\mathbb{F}_q$ define a $q \times q$ array by the rule

$$L_t(i, j) = ti + j \quad (i, j \in \mathbb{F}_q).$$

L_t is a latin square, since if $L_t(i, j) = L_t(i, j')$ then we have $ti + j = ti + j'$, which implies $j = j'$. Similarly if $L_t(i, j) = L_t(i', j)$ then, using the facts that $t \neq 0$ and $\mathbb{F}_q$ is a field, we can deduce that $i = i'$.

Consider the latin squares L_t and L_u. If they have the same pair of symbols (k, k') in positions (i_1, j_1) and (i_2, j_2) then we have

$$ti_1 + j_1 = k, \qquad ui_1 + j_1 = k',$$
$$ti_2 + j_2 = k, \qquad ui_2 + j_2 = k'.$$

It follows that

$$t(i_1 - i_2) = j_1 - j_2, \qquad u(i_1 - i_2) = j_1 - j_2.$$

If $i_1 - i_2 = 0$ then $j_1 - j_2 = 0$ and the two positions are the same. If not, then $i_1 - i_2$ has an inverse in $\mathbb{F}_q$ and

$$t = u = (i_1 - i_2)^{-1}(j_1 - j_2),$$

so $L_t = L_u$. Hence if $t \neq u$ L_t and L_u are orthogonal, and we have a set of $q-1$ mutually orthogonal latin squares of order q. $\qquad \square$

Is it possible to construct $n-1$ mutually orthogonal latin squares of order n when n is not a prime power? This is one of the most famous unsolved problems in Discrete Mathematics, and its solution would be an important step forward with many applications. When $n = 6$ it is known that there cannot be a set of five mutually orthogonal latin squares: indeed, there cannot even be two such squares. (This means that it is impossible to solve Euler's problem of the 36 officers, mentioned in Section 6.5.) For many years mathematicians thought that there could be no orthogonal pair in the case $n = 10$ also, but this conjecture was disproved in 1960 when Bose, Parker, and Shrikhande succeeded in constructing a pair of orthogonal 10×10 latin squares. As yet, no one has been able to construct a set of more than two such squares, but on the other hand, no one has been able to prove that it is impossible to do so.

Exercises 16.5

1 You are required to arrange the integers from 0 to 99 in a 10×10 square in such a way that no two in the same row or column have the same first digit or the same second digit. (The integers 0 to 9 are represented by 00 to 09 for the sake of uniformity.) Complete the partial arrangement given in Table 16.5.1, and explain the relevance of this arrangement to the remarks in the last paragraph above.

Table 16.5.1

00	47	18	76	29	93	85	34	61	52
86	11	57	28	70	39	94	45	02	63
95	80	22	67	38	71	49	56	13	04
59	96	81	33	07	48	72	60	24	15
73	69	90	82	44	17	58	01	35	26
68	74	09	91	83	55	27	12	46	30
37	08	75	19	92	84				
14	25	36	40	51	62				
21	32	43	54	65	06				
42	53	64	05	16	20				

2 Construct an explicit representation of F_4 and use it to write down a set of three mutually orthogonal latin squares of order four.

3 Relabel the symbols in each of the three squares constructed in Ex. 2 so that the first row of each one is 1 2 3 4.

4 How would you go about constructing 63 mutually orthogonal latin squares of order 64? (No credit will be given for frivolous answers.)

16.6 Finite geometry and designs

Another important application of finite fields is in the construction of designs. The simplest way to introduce this application is by means of elementary coordinate geometry.

The reader will be familiar with the representation of the points of an ordinary plane by coordinate pairs (x, y), where x and y belong to the field $\mathbb{R}$ of real numbers. If we take x and y to be elements of a *finite* field $\mathbb{F}_q$, then all the algebraic manipulations of elementary coordinate geometry remain valid. We obtain a 'plane' with a finite number of 'points' and 'lines', and it turns out that the lines are the blocks of a 2-design on the set of points.

So, we begin by defining a *point* to be an ordered pair (x, y), where x and y are elements of $\mathbb{F}_q$. A *line* is a set of points (x, y) satisfying an equation of the form

$$ax + by = c$$

where a, b, c are themselves elements of $\mathbb{F}_q$, and a and b are not both zero. We note that the line determined by αa, αb, αc is the same as that determined by a, b, c, provided $\alpha \neq 0$ in $\mathbb{F}_q$.

Theorem 16.6

Suppose that a finite field $\mathbb{F}_q$ is given, and points and lines are defined as above. Then the lines are the blocks of a 2-design on the set of points, with parameters

$$v = q^2, \qquad k = q, \qquad r_2 = 1.$$

Proof

Clearly there are q^2 points, since x and y can be any of the q elements of $\mathbb{F}_q$. We must show that every line has exactly q points and that any two points belong to exactly one line.

Consider a line with equation $ax + by = c$. If $b \neq 0$ each of the q possible values of x determines a unique value $y = b^{-1}(c - ax)$ such that (x, y) is on the line, and so the line has q points. On the other hand, if $b = 0$ then $a \neq 0$ (by hypothesis) and the equation becomes $ax = c$, that is,

$$x = a^{-1}c.$$

In this case the q possible values of y in $\mathbb{F}_q$ give us q points of the form $(a^{-1}c, y)$, and these points all lie on the line.

Suppose we are given two distinct points (x_1, y_1) and (x_2, y_2). Since $x_2 - x_1$ and $y_1 - y_2$ are not both zero, the equation

$$(y_1 - y_2)x + (x_2 - x_1)y = x_2 y_1 - x_1 y_2$$

is the equation of a line, and it contains the two given points. If $ax + by = c$ is the equation of another line containing the two points, then we have

$$ax_1 + by_1 = c, \qquad ax_2 + by_2 = c.$$

Hence

$$a(x_2 - x_1) = b(y_1 - y_2).$$

If $x_1 \neq x_2$ then $(x_2 - x_1)^{-1}$ exists in $\mathbb{F}_q$, and we can write

$$a = \alpha(y_1 - y_2), \quad \text{where } \alpha = b(x_2 - x_1)^{-1}.$$

Thus $b = \alpha(x_2 - x_1)$ and substituting for c we have

$$c = ax_1 + by_1 = \alpha(x_2 y_1 - x_1 y_2).$$

Hence the line is the same as the one given above. (If $x_1 = x_2$, then $y_1 \neq y_2$ and we can use a similar argument.) Thus there is exactly one line containing the two points. $\square$

The 2-design constructed in Theorem 16.6 is usually known as the **affine plane** over $\mathbb{F}_q$. We can calculate the parameters r_1 (the number of lines containing a given point) and r_0 ($= b$, the total number of lines) by means of the general theory developed in Section 4.6, and we obtain

$$r_1 = \left(\frac{v-1}{k-1} \right) r_2 = \frac{q^2 - 1}{q - 1} = q + 1,$$

$$b = r_0 = \left(\frac{v}{k} \right) r_1 = \left(\frac{q^2}{q} \right)(q + 1) = q^2 + q.$$

Of course, the arguments used to prove the theorem apply equally well in the case of the ordinary plane, and they work because the coordinates belong to a field. If we wish we can think of the ordinary plane as 2-design with an infinite value of q, and sometimes the familiar properties of this plane help us to visualize the properties of the finite affine planes. For example, let us take $q = 3$ and consider the affine plane over $\mathbb{F}_3$, that is, over $\mathbb{Z}_3$. There are nine points

$$A = (0, 0), \quad B = (0, 1), \quad C = (0, 2),$$
$$D = (1, 0), \quad E = (1, 1), \quad F = (1, 2),$$
$$G = (2, 0), \quad H = (2, 1), \quad I = (2, 2),$$

and 12 lines, as in Table 16.6.1.

Table 16.6.1

Line	Equation	Points		
L_1	$x = 0$	A	B	C
L_2	$x = 1$	D	E	F
L_3	$x = 2$	G	H	I
L_4	$y = 0$	A	D	G
L_5	$y = 1$	B	E	H
L_6	$y = 2$	C	F	I
L_7	$x + y = 0$	A	F	H
L_8	$x + y = 1$	B	D	I
L_9	$x + y = 2$	C	E	G
L_{10}	$2x + y = 0$	A	E	I
L_{11}	$2x + y = 1$	B	F	G
L_{12}	$2x + y = 2$	C	D	H

As in the ordinary plane the lines fall naturally into classes of parallel lines; in this case there are four classes with three lines in each:

$$\{L_1, L_2, L_3\}, \{L_4, L_5, L_6\}, \{L_7, L_8, L_9\}, \{L_{10}, L_{11}, L_{12}\}.$$

What is more, all the familiar properties of parallel lines carry over from the ordinary plane to the finite plane. Each class of parallel lines contains exactly one line through each point and, whereas two parallel lines have no common point, any two lines which are not parallel have just one common point.

It is instructive to attempt to 'draw' the affine plane over $\mathbb{F}_3$ in such a way as to illustrate the notion of parallelism. Unfortunately, this can only be done by bending some of the lines (Fig. 16.1).

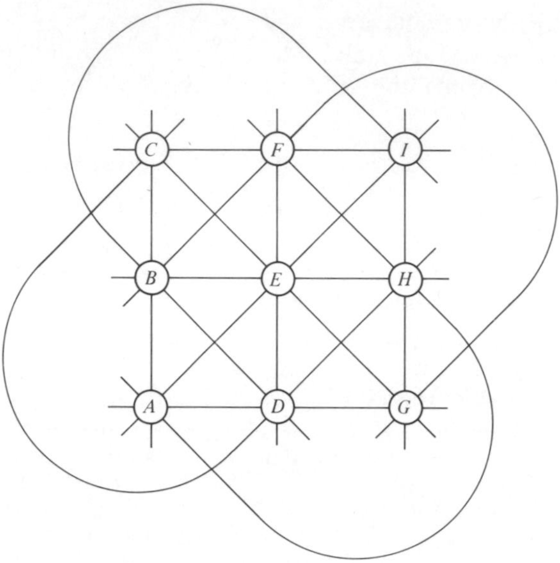

Fig. 16.1
The affine plane over $\mathbb{F}_3$.

**Exercises
16.6**

1 Make a table of the points and lines in the affine plane over $\mathbb{F}_2$, and draw the corresponding picture. Explain why this particular plane is not very interesting when considered as a 2-design.

2 When q is an odd prime power we may define the 'mid-point' of (x_1, y_1) and (x_2, y_2) to be the point whose coordinates are $(\frac{1}{2}(x_1 + x_2), \frac{1}{2}(y_1 + y_2))$, where $\frac{1}{2}$ denotes the inverse of 2 in the field $\mathbb{F}_q$. Suppose $q = 5$ and let A, B, C denote the points $(1, 2)$, $(3, 1)$, $(4, 2)$, respectively.

(i) Find the mid-points of the sides of the triangle ABC.
(ii) Find the equations of the 'medians' of the triangle ABC.
(iii) Show that the medians have a common point G, and find its coordinates.

3 How many points belong to the following 'conics' in the affine plane over $\mathbb{F}_3$?

$$\text{(i)} \quad y = x^2, \qquad \text{(ii)} \quad xy = 1,$$
$$\text{(iii)} \quad x^2 + y^2 = 1, \qquad \text{(iv)} \quad x^2 + 2y^2 = 1.$$

4 Repeat Ex. 3, taking the field to be $\mathbb{F}_5$, and generalize your results to the case when the field is $\mathbb{F}_p$ (p prime). [Hint: you may wish to read Section 16.8 before attacking the last part.]

16.7 Projective planes

From a geometrical viewpoint the affine plane has a certain lack of uniformity. Two non-parallel lines have one common point, but any two parallel lines have no common point. In elementary geometry we are taught to accept this situation as being intrinsically correct, but for many years mathematicians have studied alternative systems in which *every* two lines have exactly one common point. This kind of geometry is known as *projective geometry*. We shall study the finite version of projective geometry, and its relationship with the theory of designs.

Consider once again the affine plane over $\mathbb{F}_3$. In order to make the parallel lines meet we invent four new points W, X, Y, Z, often referred to as *points at infinity*. The point W is added to the lines L_1, L_2, L_3; X is added to L_4, L_5, L_6; Y is added to L_7, L_8, L_9; and Z is added to L_{10}, L_{11}, L_{12}. We also introduce one new line L_∞, *the line at infinity*, which contains the points W, X, Y, Z. So now we have

$\qquad$ 13 points (the original nine, and four at infinity),

and

$\qquad$ 13 lines (the original twelve, and one at infinity).

Explicitly, the lines and the points which lie on them are as follows.

L_1: ABCW $\quad$ L_2: DEFW $\quad$ L_3: GHIW $\quad$ L_4: ADGX $\quad$ L_5: BEHX

L_6: CFIX $\quad$ L_7: AFHY $\quad$ L_8: BDIY $\quad$ L_9: CEGY $\quad$ L_{10}: AEIZ

L_{11}: BFGZ $\quad$ L_{12}: CDHZ $\quad$ L_∞: WXYZ.

In particular, we notice that each line contains four points and each pair of points belongs to exactly one line. Consequently, we have a 2-design with parameters $(13, 4, 1)$.

In the next theorem we shall see how to perform the analogous construction over any finite field $\mathbb{F}_q$ and thence obtain a 2-design with parameters $(q^2 + q + 1, q + 1, 1)$. The resulting design is known as the **projective plane** over $\mathbb{F}_q$.

Theorem 16.7

For any prime power q there is a 2-design with parameters

$$v = q^2 + q + 1, \qquad k = q + 1, \qquad r_2 = 1.$$

This design has the additional property that any two blocks have just one member in common.

Proof

We must first define the general notion of parallelism in the affine plane over $\mathbb{F}_q$, by saying that the lines

$$ax + by = c \quad \text{and} \quad a'x + b'y = c'$$

are *parallel* if $ab' = a'b$ in $\mathbb{F}_q$. It is clear that this definition is independent of the representative equation chosen for each line, and that it defines an equivalence relation on the set of lines. There are $q + 1$ equivalence classes of parallel lines, represented by the q lines

$$x + fy = 0 \quad (f \in \mathbb{F}_q),$$

and the line $y = 0$. Any point of the affine plane belongs to just one line in each class.

Introduce $q + 1$ new points Ω_f $(f \in \mathbb{F}_q)$ and Ω_∞, adding Ω_f to each line parallel to $x + fy = 0$, and Ω_∞ to each line parallel to $y = 0$. Introduce also one new line L_∞ containing all the new points.

We now check that the lines are the blocks of a design with the required parameters. Clearly, there are $q^2 + q + 1$ points, and each line contains $q + 1$ points, so we have only to show that any two distinct points belong to just one line. Suppose that P and Q are two given points; then there are three cases to consider.

(i) If P and Q are both old points (that is, points of the affine plane) then they belong to a unique line in that plane, which corresponds uniquely to a line in the extended plane.

(ii) If P is an old point and Q is a new point, suppose $Q = \Omega_f$. Then P belongs to precisely one old line in the parallel class represented by Ω_f, and the corresponding new line is the unique line containing P and Q. The same is true if $Q = \Omega_\infty$.

(iii) If P and Q are both new points, then L_∞ is the unique line containing them both.

Thus any two points belong to just one line.

The fact that any two lines have just one common point is an immediate consequence of the construction. $\qquad \qquad \square$

We may calculate the parameters r_1 and r_0 $(= b)$ for the projective planes in the usual way:

$$r = \left(\frac{v-1}{k-1}\right)r_2 = \left(\frac{q^2+q}{q}\right) = q+1,$$

$$b = r_0 = \left(\frac{v}{k}\right)r_1 = \left(\frac{q^2+q+1}{q+1}\right)(q+1) = q^2+q+1.$$

(Of course, the fact that $b = q^2 + q + 1$ is obvious from the construction, since there is one new line in addition to the $q^2 + q$ lines of the affine plane.) Summarizing our results, we see that there is a reciprocal relationship between the points and lines of a projective plane, expressed by the following facts:

there are q^2+q+1 points;
there are q^2+q+1 lines;

each line contains $q+1$ points;
each point belongs to $q+1$ lines;

any two points belong to one common line;
any two lines have one common point.

Example   Use the method given in Theorem 16.7 to construct the projective plane over $\mathbb{F}_2$.

Solution   Let the points of the affine plane over $\mathbb{F}_2$ be denoted by $A = (0,0)$, $B = (1,0)$, $C = (0,1)$, $D = (1,1)$. The six lines of the affine plane and their equations are as follows:

$$AC: x \quad = 0, \qquad BD: x \quad = 1,$$
$$AD: x+y = 0, \qquad BC: x+y = 1,$$
$$AB: \quad y = 0, \qquad CD: \quad y = 1.$$

(Note that there are three classes, each containing two parallel lines.) The new points Ω_0, Ω_1, and Ω_∞ form the line at infinity, and they are added to the lines of the respective parallel classes in the order given. Thus the seven lines of the projective plane are

$$AC\Omega_0, \ BD\Omega_0, \ AD\Omega_1, \ BC\Omega_1, \ AB\Omega_\infty, \ CD\Omega_\infty, \ \Omega_0\Omega_1\Omega_\infty.$$

The plane is depicted in Fig. 16.2, with the line at infinity as the 'base'.

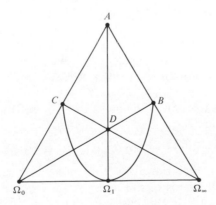

Fig. 16.2
The projective plane over $\mathbb{F}_2$.

Exercises 16.7

1 The thirteen members of the University of Folornia Satanic Rites and Bridge Club wish to organize a scheme so that each week four of them can raise the devil, or play Bridge if that fails. Because they are temperamental people no two of them can be together in more than one foursome. For how many weeks can the scheme be sustained? Draw up a suitable scheme for the maximum number of weeks.

2 A *quadrangle* in a projective plane is a set of four points, no three of which are collinear. Show that there are just seven quadrangles in the projective plane over $\mathbb{F}_2$. What is the relationship between the seven quadrangles and the seven lines?

3 If X, Y, Z, T are the points of a quadrangle in a projective plane, define the three points A, B, C to be the intersections of XY and ZT, XZ and YT, XT and YZ respectively. The points ABC are known as the *diagonal points* of the quadrangle. Show that in the projective plane over $\mathbb{F}_2$ the diagonal points of any quadrangle are collinear.

4 Give an example to show that the result of Ex. 3 does not hold in the projective plane over $\mathbb{F}_3$.

5 An *oval* in a projective plane over $\mathbb{F}_q$ is a set of $q+2$ points, no three of which are collinear. Show that the number of points common to a line and an oval is either zero or two.

16.8 Squares in finite fields

In this section we shall use the primitive element theorem to determine which elements of $\mathbb{F}_q$ have square roots. The result yields an interesting family of difference sets and an associated family of designs.

If α is a primitive element of $\mathbb{F}_q$ then the non-zero elements of the field are

$$\alpha, \alpha^2, \alpha^3, \ldots, \alpha^{q-1} = 1.$$

It is clear that the even powers have square roots, since $\alpha^{2m} = (\alpha^m)^2$. The status of the odd powers depends on whether q is even (a power of 2 in this case) or odd.

Suppose that an odd power of α has a square root, say $\alpha^{2m+1} = \beta^2$. Since β itself is a power of α, we have $\beta = \alpha^k$ and

$$\alpha^{2(m-k)+1} = \alpha^{2m+1}\alpha^{-2k} = \alpha^{2m+1}(\beta^2)^{-1} = 1.$$

The order of α in the multiplicative group of $\mathbb{F}_q$ is $q-1$ and so, by Theorem 13.4, $2(m-k)+1$ must be a multiple of $q-1$. But when q is odd, $q-1$ is even and the odd integer $2(m-k)+1$ cannot be a multiple of the even integer $q-1$. Then when q is an odd prime power a non-zero element of $\mathbb{F}_q$ has a square root in $\mathbb{F}_q$ if and only if it is an even power of a primitive element. (When q is a power of 2, every element of $\mathbb{F}_q$ has a square root—Ex. 16.8.3.)

Henceforth we shall be concerned specifically with the case when q is odd. We shall denote the set of non-zero squares in $\mathbb{F}_q$ by $\square$; equivalently $\square$ is the set of non-zero elements which have square roots, or the set of even powers of a primitive element α:

$$\square = \{\alpha^2, \alpha^4, \ldots, \alpha^{q-1}\}.$$

Thus $|\square| = \frac{1}{2}(q-1)$.

In elementary mathematics it is usual to devote some time to the discussion of the square root of -1. In $\mathbb{F}_q$ the question of the square root of -1 is also important, and the following theorem gives the answer.

Theorem 16.8.1

Let q be an odd prime power. If $q \equiv 1 \pmod 4$ then -1 has a square root in $\mathbb{F}_q$, but if $q \equiv 3 \pmod 4$ then -1 does not have a square root in $\mathbb{F}_q$.

Proof

We shall show first that

$$\alpha^{\frac{1}{2}(q-1)} = -1,$$

where α is a primitive element of $\mathbb{F}_q$. First we note that the equation $x^2 = 1$ has two solutions in $\mathbb{F}_q$ since the multiplicative group is cyclic and has even order. (This follows from the characterization of cyclic groups given in Theorem 13.9.) Obviously the two solutions are 1 and -1. On the other hand, since α has order $q-1$ both α^{q-1} and $\alpha^{\frac{1}{2}(q-1)}$ satisfy $x^2 = 1$, and so they are equal to 1 and -1 respectively.

According to the remarks preceding the theorem, $\alpha^{\frac{1}{2}(q-1)}$ is a square if and only if $\frac{1}{2}(q-1)$ is even, that is

$$\tfrac{1}{2}(q-1) = 2m,$$

or $q = 4m+1$. Thus we have the required result. □

In the next theorem we shall prove that when $q \equiv 3 \pmod 4$, in which case -1 is *not* a square, the set $\square$ of squares in $\mathbb{F}_q$ is a difference set: that is, the difference $x - y$ $(x, y \in \square)$ takes each non-zero value the same number of times.

Theorem 16.8.2   If q is a prime power of the form $4n + 3$ then the set $\square$ of non-zero squares in $\mathbb{F}_q$ is a difference set with parameters $(4n + 3, 2n + 1, n)$. In other words, each non-zero element of $\mathbb{F}_q$ can be expressed as a difference of squares in n ways.

Proof   Suppose that z is a square in $\mathbb{F}_q$, say $z = \zeta^2$, with $\zeta \neq 0$. There will be a certain number μ of ways of expressing z as a difference of squares, a typical expression being

$$z = u^2 - v^2.$$

We shall show that any other non-zero element w of $\mathbb{F}_q$ can be expressed as a difference of squares in the same number of ways. If w is a square, say $w = \omega^2$ then

$$w = (\omega\zeta^{-1})^2\zeta^2 = \beta^2 z \quad (\beta = \omega\zeta^{-1} \in \mathbb{F}_q).$$

Thus for each expression giving z as a difference of squares there is a corresponding expression for w:

$$w = (\beta u)^2 - (\beta v)^2.$$

On the other hand, if w is not a square then it is an odd power of a primitive element. Since -1 is not a square it too is an odd power of the same primitive element, and $-w = (-1)w$ is an even power of this element. Hence $-w$ is a square, say $-w = \theta^2$, and

$$w = -\theta^2 = -(\theta\zeta^{-1})^2\zeta^2 = \gamma^2(-z) \quad (\gamma = \theta\zeta^{-1} \in \mathbb{F}_q).$$

Thus for each expression giving z as a difference of squares there is a corresponding expression for w:

$$w = (\gamma v)^2 - (\gamma u)^2.$$

It follows that every non-zero element of $\mathbb{F}_q$ has the same number μ of representations as a difference of squares.

Now $|\square| = \frac{1}{2}(q - 1) = 2n + 1$, and so the total number of differences of distinct squares is $2n(2n + 1)$. Since there are $q - 1 = 4n + 2$ non-zero values, and each one occurs μ times, we have

$$2n(2n + 1) = \mu(4n + 2),$$

whence $\mu = n$ as required. $\square$

When p is a prime of the form $4n + 3$ the theorem tells us that the set $\square$ of non-zero squares in $\mathbb{Z}_p$ is a difference set with parameters $(4n+3, 2n+1, n)$, and thus we have a generalization of the result established for the case $p = 23$ in Section 6.4. It follows from Theorem 6.4 that the sets $\square + i$ $(i \in \mathbb{Z}_p)$ are the blocks of a 2-design with the same parameters.

The same method works if we use the sets $\square + i$ in any finite field $\mathbb{F}_q$, where q is congruent to 3 modulo 4. But we should note that in the general case we do not have a cyclic construction in the sense of Section 6.4, since the additive group of $\mathbb{F}_q$ is not a cyclic group. However, the important thing is that we can always construct a 2-design with parameters $(q, \frac{1}{2}(q-1), \frac{1}{4}(q-3))$ whenever q is a prime power congruent to 3 modulo 4.

One specially useful property of these designs is that they can always be extended in a simple way to yield 3-designs. Let us look at the case $q = 7$, when the 2-design arises from the set $\square = \{1, 2, 4\}$ in $\mathbb{Z}_7$. The blocks of the 2-design are

$$124, \quad 235, \quad 346, \quad 450, \quad 561, \quad 602, \quad 013.$$

The extended design has two kinds of blocks. The first kind is obtained by adding one new object ∞ to each of the old blocks, and the second kind is obtained by taking the complement (in $\mathbb{Z}_7$) of each of the old blocks. We get 14 blocks:

$$\infty124, \quad \infty235, \quad \infty346, \quad \infty450, \quad \infty561, \quad \infty602, \quad \infty013,$$
$$0356, \quad 0146, \quad 0125, \quad 1236, \quad 0234, \quad 1345, \quad 2456.$$

Now it can be verified by careful checking that we have a 3 design with parameters $(8, 4, 1)$.

It is slightly surprising, at first sight, that the same procedure works for any 2-design with parameters $(4n+3, 2n+1, n)$. But it does, and consequently we have an infinite family of 3-designs with parameters $(4n+4, 2n+2, n)$.

Exercises 16.8

1   Compute the set $\square$ of non-zero squares in $\mathbb{Z}_{19}$, and use it to construct a 2-design with parameters $(19, 9, 4)$.

2   Construct a 3-design with parameters $(20, 10, 4)$.

3   Show that if q is a power of 2 then every element of $\mathbb{F}_q$ has a square root. [Hint: an integer i in the range $1 \leq i \leq q-1$ is odd if and only if $(q-1)+i$ is even.]

4   Let F^* be the multiplicative group of the finite field $\mathbb{F}_q$. Show

that $\square$ is a subgroup of F^*, and that its index in F^* is 2 if q is odd, and 1 if q is even.

5 Explain how you would go about constructing a 3-design with parameters $(28, 14, 6)$.

16.9 Existence of finite fields

In this section we shall prove that for every prime p and every integer $n \geq 1$ there is an irreducible polynomial of degree n in $\mathbb{Z}_p[x]$. From this it follows (by Theorem 16.3) that there is a finite field of order p^n.

The method of proof is conceptually very simple: we obtain a general formula for the number $N_n(p)$ of monic irreducible polynomials of degree n in $\mathbb{Z}_p[x]$, and then we prove that $N_n(p)$ is not zero. Unfortunately, there is a lot of work involved in establishing the formula, most of which is concerned with the proof of the following algebraic result.

> In $\mathbb{Z}_p[x]$ every monic irreducible polynomial whose degree is a divisor of n occurs once and only once in the factorization of the polynomial $x^{p^n} - x$, and there are no other factors.

Before we prove this, let us see how it leads to the main result.

Theorem 16.9 For every prime p and every integer $n \geq 1$ there is a monic irreducible polynomial of degree n in $\mathbb{Z}_p[x]$.

Proof Suppose that there are $N_n(p)$ monic irreducible polynomials of degree n in $\mathbb{Z}_p[x]$. According to the assertion above, the factorization of $x^{p^n} - x$ in $\mathbb{Z}_p[x]$ contains $N_d(p)$ factors of degree d for each divisor d of n. These factors contribute $dN_d(p)$ to the total degree, which is p^n, so we have the equation

$$p^n = \sum_{d \mid n} dN_d(p).$$

Applying the Möbius inversion formula (Theorem 4.5.2) with $f(n) = p^n$ and $g(d) = dN_d(p)$ we obtain

$$nN_n(p) = \sum_{d \mid n} \mu(d) p^{n/d},$$

that is,

$$N_n(p) = \frac{1}{n} \sum_{d \mid n} \mu(d) p^{n/d}.$$

In the sum on the right-hand side the term with $d = 1$ is p^n and the other terms have the form $\pm p^i$, where i is strictly less than n. Hence a lower bound for $N_n(p)$ is given by

$$N_n(p) \geqslant \frac{1}{n} (p^n - (p^{n-1} + p^{n-2} + \cdots + p + 1))$$

$$= \frac{1}{n} \left(p^n - \frac{p^n - 1}{p - 1} \right),$$

which is strictly greater than zero. Since $N_n(p)$ is an integer it follows that $N_n(p) \geqslant 1$, as required. $\qquad\square$

Note that when $n = 1$ the formula gives $N_1(p) = p$, in agreement with the fact that every monic irreducible polynomial $x - \alpha$ with α in $\mathbb{Z}_p$ is irreducible. Also, the formula confirms the results for $N_2(p)$ and $N_3(p)$ obtained by direct arguments in Sections 15.8, that is,

$$N_2(p) = \tfrac{1}{2}(\mu(1)p^2 + \mu(2)p) = \tfrac{1}{2}(p^2 - p),$$

$$N_3(p) = \tfrac{1}{3}(\mu(1)p^3 + \mu(3)p) = \tfrac{1}{3}(p^3 - p).$$

We now embark upon the proof of the assertion about the factorization of $x^{p^n} - x$ in $\mathbb{Z}_p[x]$. We shall present the proof in three stages.

(I) If d is a divisor of n then any monic irreducible polynomial of degree d is a factor of $x^{p^n} - x$.

(II) The degree of every monic irreducible factor of $x^{p^n} - x$ is a divisor of n.

(III) The irreducible factors of $x^{p^n} - x$ occur once only.

We shall need a consequence of the formula which says that when $n = mk$

$$z^n - 1 = (z^m - 1)(z^{(k-1)m} + z^{(k-2)m} + \cdots + z^m + 1).$$

Taking z to be a prime p we deduce that

$$m \mid n \quad \Rightarrow \quad p^m - 1 \mid p^n - 1,$$

while regarding z as a polynomial indeterminate x we deduce that

$$m \mid n \quad \Rightarrow \quad x^m - 1 \mid x^n - 1.$$

Putting the two results together we conclude that

$$m \mid n \Rightarrow x^{p^m-1}-1 \mid x^{p^n-1}-1.$$

We shall regard this as a property of the polynomial ring $\mathbb{Z}_p[x]$, where the prime p will be fixed throughout the ensuing discussion.

Proof of (I)

It is clear that the polynomial x is a factor of $x^{p^n}-x$, since

$$x^{p^n}-x = x(x^{p^n-1}-1).$$

We shall show that every other monic irreducible polynomial whose degree divides n is a factor of $x^{p^n-1}-1$. Consider first a linear polynomial $x-\alpha$, where $\alpha \neq 0$. By Fermat's theorem $\alpha^{p-1}=1$, and so it follows from the factor theorem that $x-\alpha$ is a factor of $x^{p-1}-1$, which is a factor of $x^{p^n-1}-1$. Consequently $x-\alpha$ is a factor of $x^{p^n-1}-1$ also.

Suppose that $f(x)$ is a monic irreducible polynomial with degree $d \geq 2$. We know that the equivalence classes $[h(x)]$ of polynomials with respect to addition and multiplication modulo $f(x)$ form a field of order p^d. Denote the elements of this field by $[0]$ and

$$[h_1(x)], [h_2(x)], \ldots, [h_t(x)] \quad (t = p^d - 1).$$

The classes

$$[xh_1(x)], [xh_2(x)], \ldots, [xh_t(x)]$$

also represent the non-zero elements of the field in some order, and so

$$[xh_1(x)][xh_2(x)] \ldots [xh_t(x)] = [h_1(x)][h_2(x)] \ldots [h_t(x)].$$

Rearranging this equation we obtain

$$[x^t][h_1(x) \cdots h_t(x)] = [h_1(x) \cdots h_t(x)],$$

so that $[x^t] = [1]$ or

$$[x^{p^d-1}-1] = [0].$$

The class $[0]$ contains the multiples of $f(x)$, so $x^{p^d-1}-1$ is a multiple of $f(x)$. Since $d \mid n$, $x^{p^n-1}-1$ is a multiple of $x^{p^d-1}-1$ and consequently it is a multiple of $f(x)$, as required.

Proof of (II)

Suppose that the monic irreducible polynomial $g(x)$ is a factor of $x^{p^n}-x$ and that its degree m is at least 2. Let $n = ms + r$, where $0 \leq r < m$. We shall show that the case $r > 0$ is impossible.

Let F be the field of order p^m whose elements are the classes of polynomials modulo $g(x)$. By (I), $x^{p^m} - x$ is a multiple of $g(x)$, so x^{p^m} and x belong to the same class in F. Thus the identity

$$x^{p^n} = x^{p^{ms+r}} = (x^{p^m})^{p^m \cdots p^m p^r}$$

implies that x^{p^r} is in the same class as x^{p^n}. Since we are assuming that $x^{p^n} - x$ is a multiple of $g(x)$, it is also true that x belongs to the same class as x^{p^n}. Consequently x and x^{p^r} belong to the same class in F.

Given any polynomial $h(x)$, the polynomial $h(x^{p^r})$ obtained by substituting x^{p^r} for x is in the same class as $h(x)$, because of the result derived in the previous paragraph. But by Ex. 15.4.5 we have $h(x^{p^r}) = h(x)^{p^r}$, and so

$$[h(x)] = [h(x^{p^r})] = [h(x)^{p^r}] = [h(x)]^{p^r}.$$

In other words, every one of the p^m elements of F is a root of the equation $x^{p^r} - x = 0$. If $r > 0$ this is an equation of degree p^r and it cannot have as many as p^m roots, since $m > r$. The only possibility is that $r = 0$ (when the equation becomes $0 = 0$). Thus m is a divisor of n, as claimed.

Proof of (III)

It remains only to show that any irreducible polynomial $k(x)$ cannot be a repeated factor of $x^{p^n} - x$. If

$$k(x) = k_0 + k_1 x + \cdots + k_u x^u$$

then we may define its *formal derivative* $k'(x)$ to be the polynomial

$$k'(x) = k_1 + 2k_2 x + \cdots + uk_u x^{u-1}.$$

This definition is quite self-contained, and independent of any limiting processes, but it agrees with the rules obtained in Calculus. In particular the rule for finding the derivative of a product is the familiar one, and it can be proved by direct calculation (Ex. 15.9.15).

Suppose that $k(x)$ occurs more than once as a factor of $x^{p^n} - x$, so that

$$x^{p^n} - x = k(x)^2 l(x)$$

for some $l(x)$ in $\mathbb{Z}_p[x]$. Taking the formal derivative of both sides and remembering that $p^n = 0$ in $\mathbb{Z}_p$, we get

$$-1 = 2k(x)k'(x)l(x) + k(x)^2 l'(x).$$

But this implies that $k(x)$ is a factor of -1, which is impossible unless $k(x)$ is a constant. Hence we have the result. $\qquad \square$

Exercises 16.9

1 Work out the formulae for $N_4(p)$, $N_5(p)$, and $N_6(p)$.

2 What is the total number of monic irreducible polynomials of degree 4 in $\mathbb{Z}_2[x]$? Find them.

3 Find the irreducible factors of $x^{16} - x$ in $\mathbb{Z}_2[x]$.

4 Make a table of the values of $N_n(2)$ for $1 \leqslant n \leqslant 10$. For each n in this range calculate approximately the percentage of the total number of monic polynomials of degree n in $\mathbb{Z}_2[x]$ which are irreducible.

5 Let F be a finite field. A subset S of F is said to be a **subfield** if the elements of S form a field under the operations of F.

(i) Show that the characteristic of S is equal to that of F.
(ii) Show that if $|S| = p^s$ and $|F| = p^r$ then s is a divisor of r. [Hint: the multiplicative group of S is a subgroup of the multiplicative group of F.]
(iii) Write down the possible orders of subfields of $\mathbb{F}_{64}$.

16.10 Miscellaneous Exercises

1 Find the least positive integer which represents a primitive element in $\mathbb{Z}_{31}$.

2 Show that $x^4 + x^3 + x^2 + x + 1$ is irreducible in $\mathbb{Z}_2[x]$ but that it is not primitive.

3 Find all values of b for which $x^2 + x + b$ is a primitive polynomial in $\mathbb{Z}_{17}[x]$.

4 How many points lie on the following 'curves' in the affine plane over $\mathbb{F}_3$?

$$\text{(i)} \quad x^3 + y^3 = 1; \qquad \text{(ii)} \quad x^2 + xy + y^2 = 0.$$

5 A *triangle* in a projective plane is a set of three points which do not belong to a common line. Show that the number of triangles in the projective plane over $\mathbb{F}_q$ is $\frac{1}{6}q^3(q+1)(q^2+q+1)$.

6 Show that $x^2 + 1$ is irreducible in $\mathbb{Z}_p[x]$ if and only if $p \equiv 3 \pmod 4$.

7 Let X denote the set of words of length n in the alphabet $\mathbb{Z}_2$, with the exception of the zero word $00 \ldots 0$. Define the sum $\mathbf{x} + \mathbf{y}$ of two words $\mathbf{x}$ and $\mathbf{y}$ to the word whose ith digit is the sum (in $\mathbb{Z}_2$) of the ith digits of $\mathbf{x}$ and

y. Show that the set **B** of 3-subsets of X of the form $\{\mathbf{x}, \mathbf{y}, \mathbf{x} + \mathbf{y}\}$ is a 2-design with parameters $(2^n - 1, 3, 1)$.

8 Let b be a primitive element in $\mathbb{F}_q$, and for each positive integer n such that $b^n \neq -1$ define $J(n)$ by the equation $b^n + 1 = b^{J(n)}$. Construct a table of $J(n)$ for the fields $\mathbb{F}_9$ and $\mathbb{F}_{13}$ (with respect to suitable primitive elements).

9 Show that if $J(n)$ is as given in the previous exercise then, whenever $J(n - m)$ is defined, we have

$$b^m + b^n = b^{m + J(n-m)}.$$

($J(n)$ is the **Jacobi logarithm**. It is a useful device for computing in finite fields.)

10 The **incidence graph** of a projective plane is the bipartite graph $G = (V \cup W, E)$, where V is the set of points, W is the set of lines, and vw is an edge whenever the point v lies on the line w. Show that the incidence graph is a regular graph with girth 6 having the minimum number of vertices compatible with these properties. (Compare Ex. 8.8.20.)

11 Let $p_1, p_2, \ldots, p_k$ be primes congruent to 1 modulo 4 and let q be any prime divisor of $4(p_1 p_2 \cdots p_k)^2 + 1$. Show that -1 is a square in $\mathbb{Z}_q$ and deduce that there are infinitely many primes which are congruent to 1 modulo 4.

12 A polynomial $a_n x^n + a_{n-1} x^{n-1} + \cdots + a_1 x + a_0$, of even degree is said to be *reciprocal* if $a_i = a_{n-i}$ $(0 \leq i \leq \frac{1}{2}n)$. Show that a reciprocal polynomial cannot be primitive.

13 Show that if 10 is a primitive element of the field $\mathbb{Z}_p$ then the decimal expansions for the fractions r/p $(1 \leq r \leq p - 1)$ recur with period $p - 1$ and they differ only by cyclic permutations.

14 For each prime power q let $\boxtimes$ denote the complement of $\square$ in $\mathbb{F}_q$. Show that if $q = 4n + 3$ then $\boxtimes$ is a difference set with parameters $(4n + 3, 2n + 2, n + 1)$.

15 Show that any 2-design with parameters $(4n + 3, 2n + 1, n)$ can be extended to a 3-design with parameters $(4n + 4, 2n + 2, n)$ by the method outlined at the end of Section 16.8.

16 R. A. Fisher, who pioneered the use of designs and latin squares in agricultural research, posed the following problem.

In a set of sixteen people, four are English, four are Scots, four are Irish, and four Welsh. There are four each of the ages 35, 45, 55, and 65. Four are lawyers, four are doctors, four are soldiers, and four are clergymen. Four are single, four are married, four widowed, and four divorced. Finally, four are conservatives, four socialists, four liberals, and four fascists. No two of the same kind in one category are the same in another category. Three of the fascists are: a single English lawyer of 65, a married Scottish soldier of 55, and a widowed Irish doctor of 45. Furthermore, the Irish socialist is 35, the Scottish conservative is 45, and the English clergyman is 55. What can you say about the Welsh lawyer?

17 Suppose we are given a 3-design with parameters $(10, 4, 1)$. Show that if we remove the blocks which do not contain a given object x, and delete x from the remaining blocks, then the result is a 2-design with parameters $(9, 3, 1)$. Use this construction in reverse to obtain a 3-design from the affine plane over $\mathbb{F}_3$.

18 Show that if F is a field whose multiplicative group is cyclic, then $|F|$ must be finite.

19 Show that any quadratic polynomial in $\mathbb{Z}_p[x]$ can be written as the product of two linear polynomials with coefficients in $\mathbb{F}_{p^2}$.

20 Construct an explicit representation of $\mathbb{F}_{16}$ and determine all its subfields.

17
Error-correcting codes

17.1 Words, codes, and errors

When a telex message is transmitted over a long distance there may be some interference, and the message may not be received exactly as it is sent. In these circumstances we need to be able to detect and, if possible, correct errors. Ordinary language provides some facilities for this, because some sequences of letters do not make sense. Thus, if the message received is I LPVE YOU we know for sure that an error has been made, and we can be fairly confident about how to correct it. But the error-detection and error-correction properties of ordinary language are not uniform: if the received message is I LOVE LOU then we cannot be sure that we know what was intended.

A more systematic approach is to represent messages by selected words in the binary alphabet {0, 1}, for then every error is simply the result of confusing 0 and 1. The chosen binary words correspond to actual messages according to an agreed set of rules, known to both the sender and receiver. In the ideal situation, the process will operate as in Fig. 17.1.

The sender encodes the message by selecting the binary word which has been assigned to represent it. Then the coded version is

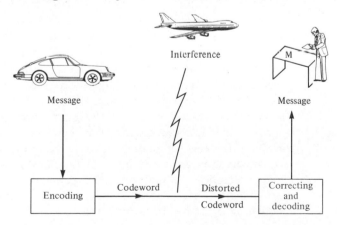

Fig. 17.1
The ideal coding operation.

Interference

Message

Message

M

Encoding — Codeword — Distorted Codeword — Correcting and decoding

transmitted over a channel, where it is subject to interference, and it becomes distorted. On receipt, the errors in the distorted message are corrected, and the receiver decodes the corrected version by using the agreed rules to translate it into the original message. It must be stressed that secrecy is not the intention of this kind of coding procedure; on the contrary, the aim is to make the received message plain, even though it may have been distorted in transmission.

It is convenient to use words of the same length in each particular case. There are 2^n binary words of length n, and we shall denote the complete set of them by V^n; for example

$$V^3 = \{000, 100, 010, 001, 110, 101, 011, 111\}.$$

Each symbol in such a word is called a **bit** (an abbreviation for binary digit). A binary **code** of **length** n is simply a subset C of V^n, and the members of C are called **codewords**. Any subset will do, but in practice we try to choose C so that there is a good prospect of being able to detect and, if possible, correct some errors.

Suppose, for example, that we wish to be able to send the messages *up, down, left, right*. Here are three codes which we might use.

Code	Length	*up*	*down*	*left*	*right*
C_1	2	00	10	01	11
C_2	3	000	110	011	101
C_3	6	000000	111000	001110	110011

The code C_1 is very economical, but it has no facility for dealing with errors. If the codeword 00 is sent, and an error in the first bit occurs in transmission, then the received word is 10. This is also a codeword, so the receiver will be unable to detect the error.

The code C_2 is better, since it is capable of *detecting* any one error. If just one bit of any codeword is altered (0 becomes 1, or 1 becomes 0) then the resulting word is not a codeword, and the receiver will know that an error has been made. But there is no reasonable way of *correcting* the error. For if 000 is sent and an error is made in the first bit, the received word 100 could equally well be the result of an error in the second bit of the codeword 110, or an error in the third bit of 101 (Fig. 17.2). In practical terms, error-detection without error-correction is useful only when it is possible for the receiver to ask for a repetition of a message in which an error has been detected.

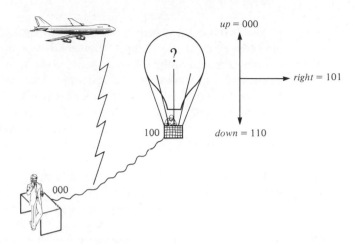

Fig. 17.2
Error detection is not
enough.

The code C_3 obviously requires more effort than the others, but it
can detect *and* correct some errors. Specifically, if we assume that
just one error has been made in transmitting any of the codewords,
then it is possible to decide which codeword was sent. For example,
if 110000 is received, the only codeword which can be obtained by
altering one bit is 111000. Of course, it is possible that two or more
bits are in error, but the likelihood of this is, in practice, much less
than the likelihood of a single error.

Now it is time to formalize the ideas introduced in the preceding
discussion. We shall use vector symbols $\mathbf{a}, \mathbf{b}, \mathbf{c}, \ldots,$ for the words in
V'', and define the **distance** $\partial(\mathbf{a}, \mathbf{b})$ to be the number of bits in which
$\mathbf{a}$ differs from $\mathbf{b}$. For example,

$$\partial(1101, 1000) = 2,$$

$$\partial(1010101, 1100100) = 3.$$

The distance ∂ has three rather obvious properties:

(i) $\partial(\mathbf{x}, \mathbf{y}) = 0 \iff \mathbf{x} = \mathbf{y},$
(ii) $\partial(\mathbf{x}, \mathbf{y}) = \partial(\mathbf{y}, \mathbf{x}),$
(iii) $\partial(\mathbf{x}, \mathbf{y}) \leqslant \partial(\mathbf{x}, \mathbf{z}) + \partial(\mathbf{z}, \mathbf{y}).$

Only the third property requires a little thought (Ex. 17.1.4).

Given any code C, we denote by δ the **minimum distance** between
pairs of distinct codewords in C:

$$\delta = \min \{\partial(\mathbf{a}, \mathbf{b}) \mid \mathbf{a}, \mathbf{b} \in C, \mathbf{a} \neq \mathbf{b}\}.$$

For the codes C_1, C_2, C_3 discussed above, δ is 1, 2, 3 respectively.

The distance between two words measures the number of errors which must be made in order that one is transformed into the other. Thus if the minimum distance of a code is δ, and not more than $\delta - 1$ errors are made in transmitting a codeword, the receiver will be able to recognize that the received word is not a codeword. If δ errors are made, the intended codeword may be transformed into another codeword. Consequently, we can say that a code with minimum distance δ is guaranteed to detect up to $\delta - 1$ errors (and in some cases, it may detect more).

In the present context, we are really more interested in correcting errors. The ideas sketched in the discussion of the code C_3 above suggest that, if an erroneous word is received, we should assume that the nearest codeword (in the sense of the distance ∂) was sent. This is known as the **nearest-neighbour** decoding principle: it amounts to saying that if $r < s$ then it is more likely that r errors have occurred than that s errors have occurred. The next theorem establishes the relationship between δ and the number of errors which can be corrected if we use this principle.

Theorem 17.1   A code C will correct e errors by the nearest-neighbour principle provided that its minimum distance δ satisfies

$$\delta \geq 2e + 1.$$

Proof   Suppose that a codeword $\mathbf{c}$ is sent and e errors are made in transmission, so that the word $\mathbf{z}$ is received. By definition of ∂ we have $\partial(\mathbf{c}, \mathbf{z}) = e$.

Let $\mathbf{c}'$ be any other codeword. Then

$$\partial(\mathbf{c}, \mathbf{z}) + \partial(\mathbf{z}, \mathbf{c}') \geq \partial(\mathbf{c}, \mathbf{c}')$$
$$\geq \delta$$
$$\geq 2e + 1,$$

by property (iii) of ∂, the definition of δ and the hypothesis. It follows that $e + \partial(\mathbf{z}, \mathbf{c}') \geq 2e + 1$, that is,

$$\partial(\mathbf{z}, \mathbf{c}') \geq e + 1.$$

Thus $\mathbf{c}$ is the only codeword within distance e of $\mathbf{z}$, and the nearest-neighbour principle will yield the correct answer. $\square$

Exercises 17.1   1   Find the minimum distance δ for each of the following codes.

   (i) $\{0000, 1100, 1010, 1001, 0110, 0101, 0011, 1111\}$ in V^4;

(ii) $\{10000, 01010, 00001\}$ in V^5;

(iii) $\{000000, 101010, 010101\}$ in V^6.

In each case, state the number of errors which can be detected and corrected.

2 Which of the codes in Ex. 1 can be extended by the inclusion of an extra codeword without altering δ?

3 Construct a code C in V^6 which will encode five messages and correct one error.

4 Prove the 'triangle inequality'

$$\partial(\mathbf{x}, \mathbf{y}) \le \partial(\mathbf{x}, \mathbf{z}) + \partial(\mathbf{z}, \mathbf{y}),$$

where $\mathbf{x}$, $\mathbf{y}$, $\mathbf{z}$ are any words in V^n.

17.2 Linear codes

The set V^n of all binary words of length n can be endowed with an algebraic structure in several ways. The simplest way is to make V^n into a group by defining $\mathbf{x} + \mathbf{y}$ to be the word obtained by adding the corresponding bits of $\mathbf{x}$ and $\mathbf{y}$ modulo 2. For example

$$1011001 + 1000111 = 0011110.$$

It is easy to check that, with this definition of the operation $+$, V^n becomes a group. As a group it is simply $(\mathbb{Z}_2)^n$, the direct product of n copies of $\mathbb{Z}_2$.

Definition A code C in V^n is **linear** if, whenever $\mathbf{a}$ and $\mathbf{b}$ are in C, then $\mathbf{a} + \mathbf{b}$ is in C. Equivalently, C is linear if and only if it is a subgroup of V^n.

The codes C_1 and C_2 discussed in the previous section are linear, but the code C_3 is not linear, since 111000 and 110011 belong to C_3 and

$$111000 + 110011 = 001011,$$

where 001011 is not in C_3.

It follows from Lagrange's theorem that, since a linear code is a subgroup of V^n, its size $|C|$ is a divisor of $|V^n| = 2^n$. Hence $|C|$ is an integer of the form 2^k, $0 \le k \le n$, and k is called the **dimension** of C.

We have now defined the three parameters (see Table 17.2.1) which determine the practical usefulness of a linear code.

Table 17.2.1

Name	Symbol	Significance
Length	n	Each codeword requires n bits.
Dimension	k	2^k different codewords are available.
Minimum distance	δ	Up to e errors can be corrected, provided $\delta \geqslant 2e + 1$.

The relationship between these parameters is what makes coding theory a challenge for mathematicians. We should like k to be reasonably large compared with n, in order that a good number of different messages can be sent without too much effort. But if there are lots of codewords, then the distance between them will be small, and few errors can be corrected.

Example Suppose that C is a linear code of length n and dimension k. Show that if e is the maximum number of errors which C will correct, then

$$2^{n-k} \geqslant 1 + \binom{n}{1} + \binom{n}{2} + \cdots + \binom{n}{e}.$$

Deduce that no linear code of length 17 and dimension 10 can correct more than one error.

Solution If $\mathbf{c}$ is a codeword of length n, then the number of words which can be obtained by altering r bits in $\mathbf{c}$ is $\binom{n}{r}$, since we may choose any r of the n bits. Let $S_e(\mathbf{c})$ denote the set of words which can be obtained by altering at most e bits in $\mathbf{c}$, so that

$$|S_e(\mathbf{c})| = 1 + \binom{n}{1} + \binom{n}{2} + \cdots + \binom{n}{e}.$$

If C corrects e errors, then the sets $S_e(\mathbf{c})$ and $S_e(\mathbf{c}')$ must be disjoint whenever $\mathbf{c}$ and $\mathbf{c}'$ are distinct members of C. Thus V^n contains $|C| = 2^k$ mutually disjoint subsets of size $|S_e(\mathbf{c})|$, so that

$$2^n \geqslant 2^k \times |S_e(\mathbf{c})|,$$

whence the result.

When $n = 17$ and $k = 10$ we have

$$1 + \binom{n}{1} + \binom{n}{2} = 1 + 17 + 136 = 154,$$

which is greater than $2^{17-10} = 128$. Hence it is impossible to correct two (or more) errors. □

One major advantage of linear codes is that the minimum distance of such a code can be calculated in a relatively simple way. This is a consequence of the fact that the distance between two words is unaltered if we add the same word to both of them:

$$\partial(\mathbf{x}+\mathbf{a}, \mathbf{y}+\mathbf{a}) = \partial(\mathbf{x}, \mathbf{y}) \quad (\mathbf{x}, \mathbf{y}, \mathbf{a} \in V^n).$$

In particular, if we define the **weight** $w(\mathbf{z})$ of a word $\mathbf{z}$ to be the number of 1's in $\mathbf{z}$, we have $w(\mathbf{z}) = \partial(\mathbf{z}, \mathbf{0})$ and

$$\partial(\mathbf{x}, \mathbf{y}) = \partial(\mathbf{x}-\mathbf{y}, \mathbf{y}-\mathbf{y}) = \partial(\mathbf{x}-\mathbf{y}, \mathbf{0})$$
$$= w(\mathbf{x}-\mathbf{y}).$$

Note that, since addition is modulo 2, $\mathbf{x}-\mathbf{y}$ is the same as $\mathbf{x}+\mathbf{y}$.

Theorem 17.2

Let C be a linear code, and let w_{min} denote the minimum weight of any codeword, except $\mathbf{0}$, in C. Then the minimum distance of C is given by

$$\delta = w_{min}.$$

Proof

Let $\mathbf{c}^*$ be a codeword for which $w(\mathbf{c}^*) = w_{min}$. Since $\mathbf{c}^*$ and $\mathbf{0}$ are both codewords, we have

$$\delta \leqslant \partial(\mathbf{c}^*, \mathbf{0}) = w(\mathbf{c}^*) = w_{min}.$$

On the other hand, if $\mathbf{c}^1$ and $\mathbf{c}^2$ are two codewords at minimum distance, $\mathbf{c}^1 - \mathbf{c}^2$ is also a codeword (because C is linear), and

$$\delta = \partial(\mathbf{c}^1, \mathbf{c}^2) = w(\mathbf{c}^1 - \mathbf{c}^2) \geqslant w_{min}. \qquad \square$$

The calculation of δ directly from the definition would involve finding the distance between each pair of codewords, whereas Theorem 17.2 tells us that it is sufficient to calculate the weight of each individual codeword—a considerable improvement (Ex. 17.2.5).

Exercises 17.2

1 For any $n \geqslant 1$ the code containing just the two words $000 \ldots 0$ and $111 \ldots 1$ of length n is a linear code. What are the values of k and δ?

2 Given any word $\mathbf{x}$ in V^n let $S_2(\mathbf{x})$ denote the set of words which can be obtained by making not more than two errors in $\mathbf{x}$. Show that

$$|S_2(\mathbf{x})| = \tfrac{1}{2}(n^2 + n + 2).$$

Deduce that if E is any code (not necessarily linear) of length 8 which will correct two errors, then $|E| \leqslant 6$.

3 What is the maximum dimension of a *linear* code of length 8 which will correct two errors? Construct such a code.

4 Let C be a linear code. Show that the subset of C containing those codewords which have even weight is also a linear code. Deduce that either all codewords in C have even weight, or exactly half of them have even weight.

5 Let C be a code with m codewords, and define an 'operation' to be the calculation of the weight of a word. Show that the number of operations required to find the distance between each pair of codewords is $O(m^2)$. If C is linear, explain how the minimum distance of C can be found by a method requiring only $O(m)$ operations.

17.3 Construction of linear codes

Let H be a binary matrix with n columns, and let $\mathbf{x}'$ denote the word $\mathbf{x}$ in V^n considered as a *column vector*. In particular, $\mathbf{0}'$ denotes the all-zero column vector. If $\mathbf{a}$ and $\mathbf{b}$ are words satisfying

$$H\mathbf{a}' = H\mathbf{b}' = \mathbf{0}',$$

then $\mathbf{a} + \mathbf{b}$ has the same property since

$$H(\mathbf{a} + \mathbf{b})' = H\mathbf{a}' + H\mathbf{b}' = \mathbf{0}'.$$

In other words, the set C defined by

$$C = \{\mathbf{x} \in V^n \mid H\mathbf{x}' = \mathbf{0}'\}$$

is a linear code. The matrix H is conventionally known as a **parity-check matrix** for C, but we shall usually abbreviate this to **check matrix**.

Example   Find all the codewords of the code determined by the check matrix

$$H = \begin{bmatrix} 1 & 0 & 1 & 0 \\ 0 & 1 & 1 & 1 \end{bmatrix}.$$

Solution   The code contains each binary word $x_1 x_2 x_3 x_4$ satisfying

$$\begin{bmatrix} 1 & 0 & 1 & 0 \\ 0 & 1 & 1 & 1 \end{bmatrix} \begin{bmatrix} x_1 \\ x_2 \\ x_3 \\ x_4 \end{bmatrix} = \begin{bmatrix} 0 \\ 0 \end{bmatrix},$$

that is,

$$x_1 + x_3 = 0, \qquad x_2 + x_3 + x_4 = 0.$$

Remembering that the arithmetic here is in $\mathbb{Z}_2$, we can rewrite the equations in the form

$$x_1 = x_3, \qquad x_2 = x_3 + x_4.$$

Thus if the values of x_3 and x_4 are assigned, x_1 and x_2 are determined. There are four ways of choosing x_3 and x_4 $(00, 10, 01, \text{and } 11)$, leading to the codewords

$$0000, \qquad 1110, \qquad 0101, \qquad 1011. \qquad \square$$

The method described in the *Example* enables us to construct codes with a given dimension in V^n. We take H to be a binary matrix of the form

$$\begin{bmatrix} 1 & & & b_{11} & b_{12} & \cdots & b_{1\,n-r} \\ & 1 & \mathbf{0} & b_{21} & b_{22} & \cdots & b_{2\,n-r} \\ & & \ddots & & & \\ \mathbf{0} & & & 1 & b_{r1} & b_{r2} & \cdots & b_{r\,n-r} \end{bmatrix}$$

where there are r rows and n columns. Then a codeword $x_1 x_2 \ldots x_n$ defined by the check matrix H satisfies the equations

$$x_1 = b_{11} x_{r+1} + b_{12} x_{r+2} + \cdots + b_{1\,n-r} x_n,$$
$$x_2 = b_{21} x_{r+1} + b_{22} x_{r+2} + \cdots + b_{2\,n-r} x_n,$$
$$\vdots$$
$$x_r = b_{r1} x_{r+1} + b_{r2} x_{r+2} + \cdots + b_{r\,n-r} x_n,$$

and these equations determine $x_1, x_2, \ldots, x_r$ when the values of $x_{r+1}, x_{r+2}, \ldots, x_n$ are given. There are 2^{n-r} ways of choosing the latter values, and so we obtain a linear code with dimension $k = n - r$.

Theoretically speaking, the order of the columns of H is not important, since a rearrangement of columns corresponds to a rearrangement of the order of the bits in each codeword. Consequently H may be taken to be any matrix which can be put in the form given above by rearranging the columns. We shall refer to this as the *standard form* for H.

**Exercises
17.3**

1   Write down all the codewords belonging to the linear code associated with the check matrix

$$\begin{bmatrix} 1 & 1 & 0 & 1 & 0 & 0 & 1 \\ 0 & 0 & 0 & 1 & 1 & 0 & 1 \\ 1 & 0 & 1 & 1 & 0 & 0 & 1 \\ 0 & 0 & 0 & 0 & 0 & 1 & 1 \end{bmatrix}.$$

2   What are the parameters n, k, δ for the code constructed in Ex. 1?

3   Suppose we wish to be able to send 128 different messages, and each message is to be represented by a binary codeword of length 11. Explain how to construct a linear code for this purpose. Is it possible to construct such a code with the additional property that $\delta \geqslant 3$, so that one error will be corrected?

4   Professor McBrain has decided that each mathematics student at the University of Folornia will be allocated an 'identity number' in the form of a binary word.
   (i) If there are 53 students, find the least possible dimension of a *linear* code for this purpose. (It is assumed that some codewords will not be allocated.)
   (ii) If the code must allow for the detection and correction of one error, find the least possible length of the codewords.
   (iii) Write down a check matrix for a code which achieves the values found in (i) and (ii). Does the resulting code have the required error-correction property?

17.4   Correcting errors in linear codes

There is a very simple way of ensuring that a linear code defined by a check matrix will correct at least one error.

**Theorem
17.4**

If no column of H consists entirely of zeros, and no two columns are the same, then the code C defined by the check matrix H will correct one error.

Proof

By Theorem 17.1 we have to show that $\delta \geqslant 3$, and by Theorem 17.2 this is equivalent to $w_{\min} \geqslant 3$.

Suppose C contains a codeword **a** with $w(\mathbf{a}) = 1$. Then **a** has just one bit equal to 1: suppose it is the bit in position i. Since all the other bits are 0, $H\mathbf{a}'$ is equal to the ith column $\mathbf{h}^{(i)}$ of H, and the condition $H\mathbf{a}' = \mathbf{0}'$ means that $\mathbf{h}^{(i)}$ consists entirely of zeros, contrary to hypothesis. Hence C contains no words of weight 1.

Suppose C contains a codeword **b** with $w(\mathbf{b}) = 2$, so that **b** has 1's in bits i and j only. Then

$$H\mathbf{b}' = \mathbf{h}^{(i)} + \mathbf{h}^{(j)},$$

and so $H\mathbf{b}' = \mathbf{0}'$ implies that $\mathbf{h}^{(i)} = \mathbf{h}^{(j)}$, contrary to hypothesis. Hence C contains no words of weight 2 and $w_{min} \geqslant 3$, as required. □

Example (i) Let r be a given positive integer. What is the maximum possible number of columns in a binary matrix with r rows which satisfies the conditions of Theorem 17.4?

(ii) Write down such a matrix when $r = 3$, and find the corresponding code.

(iii) Find the parameters n, k, δ of the code for any given value of r.

Solution (i) Each column of the matrix has r entries, and each entry is 0 or 1. Hence the number of different columns is 2^r, and since one of them (the all-0 column) is not allowed, the maximum possible number is $2^r - 1$.

(ii) A natural way of writing down the matrix is to make the columns correspond to the binary representations of the integers $1, 2, \ldots, 7$ in order, thus

$$H = \begin{bmatrix} 0 & 0 & 0 & 1 & 1 & 1 & 1 \\ 0 & 1 & 1 & 0 & 0 & 1 & 1 \\ 1 & 0 & 1 & 0 & 1 & 0 & 1 \end{bmatrix}.$$

In this way we ensure that each of the 7 possible columns occurs just once. We remark that if the columns are rearranged so that columns 4, 2, 1 occur in the first three positions, then we have the *standard form* discussed in the previous section. However, it is quite possible to work with the matrix as given above, and we shall do this.

The equations arising from $H\mathbf{x}' = \mathbf{0}'$ are

$$x_4 = \quad x_5 + x_6 + x_7,$$
$$x_2 = x_3 \quad + x_6 + x_7,$$
$$x_1 = x_3 + x_5 \quad + x_7.$$

So if x_3, x_5, x_6, x_7 are given, then x_1, x_2, x_4 are determined. There are $2^4 = 16$ codewords, which may be obtained as in Table 17.4.1.

Table 17.4.1

x_3	x_5	x_6	x_7	x_1	x_2	x_4				x				$w(\mathbf{x})$
0	0	0	0	0	0	0	0	0	0	0	0	0	0	0
0	0	0	1	1	1	1	1	1	0	1	0	0	1	4
0	0	1	0	0	1	1	0	1	0	1	0	1	0	3
0	0	1	1	1	0	0	1	0	0	0	0	1	1	3
0	1	0	0	1	0	1	1	0	0	1	1	0	0	3
0	1	0	1	0	1	0	0	1	0	0	1	0	1	3
0	1	1	0	1	1	0	1	1	0	0	1	1	0	4
0	1	1	1	0	0	1	0	0	0	1	1	1	1	4
1	0	0	0	1	1	0	1	1	1	0	0	0	0	3
1	0	0	1	0	0	1	0	0	1	1	0	0	1	3
1	0	1	0	1	0	1	1	0	1	1	0	1	0	4
1	0	1	1	0	1	0	0	1	1	0	0	1	1	4
1	1	0	0	0	1	1	0	1	1	1	1	0	0	4
1	1	0	1	1	0	0	1	0	1	0	1	0	1	4
1	1	1	0	0	0	0	0	0	1	0	1	1	0	3
1	1	1	1	1	1	1	1	1	1	1	1	1	1	7

(iii) In part (i) we showed that $n = 2^r - 1$. For each codeword, the r bits $x_1, x_2, x_4, \ldots, x_{2^{r-1}}$ are determined by the remaining $2^r - 1 - r$ bits, as in part (ii). So the dimension k is $2^r - 1 - r$.

According to Theorem 17.4, $\delta \geq 3$. In fact, the word $11100 \cdots 0$ is a codeword for each $r \geq 2$ (why?) and it has weight 3, so $\delta = 3$. $\square$

The codes constructed in the *Example* are known as **Hamming codes** (after R. W. Hamming). Their parameters are

$$n = 2^r - 1, \qquad k = 2^r - 1 - r, \qquad \delta = 3,$$

and since $\delta = 3$ they will correct one error. In fact, the Hamming codes are the best possible codes with this property, in that they have the maximum possible number of codewords. In order to prove this, let C be any code of length n with $\delta = 3$. The set of words which can be obtained by making at most one error in a given codeword $\mathbf{c}$ is $S_1(\mathbf{c})$, in the notation of Section 17.2, and we have $|S_1(\mathbf{c})| = n + 1$. Since $\delta = 3$ the sets $S_1(\mathbf{c})$ do not overlap, and since the total number of words is 2^n we must have

$$|C| \times (n + 1) \leq 2^n.$$

For the Hamming codes $|C| = 2^k$, where $k = 2^r - 1 - r$, and $n + 1 = 2^r$, so we have equality and $|C|$ attains its maximum value.

The argument given in the previous paragraph is a special case of the one we used in the *Example* studied in Section 17.2. The general argument yields a similar bound for $|C|$ when the number of errors corrected takes any given value. Codes which attain this bound are said to be **perfect**; unfortunately, they are very rare.

Theorem 17.4 is a theoretical result. It tells us that if H satisfies some very simple conditions, the associated code C will correct one error by using the nearest-neighbour principle. In practice we should like to know how to correct the error without having to compare the received word with every codeword. Fortunately there is a simple way of doing this.

Suppose that the codeword **c** is sent, and an error is made in the ith bit. The received word is given by

$$\mathbf{z} = \mathbf{c} + \mathbf{e},$$

where **e** is the word with all bits equal to 0, except the ith bit, which is 1. Hence

$$H\mathbf{z}' = H(\mathbf{c} + \mathbf{e})' = H\mathbf{c}' + H\mathbf{e}'.$$

Now $H\mathbf{c}' = \mathbf{0}'$ since **c** is a codeword, and $H\mathbf{e}'$ is equal to $\mathbf{h}^{(i)}$, the ith column of H. To summarize, the following procedure will detect and correct single errors in the code C.

(i) Compute $H\mathbf{z}'$, where **z** is the received word.
(ii) If $H\mathbf{z}' = \mathbf{0}'$, **z** is a codeword.
(iii) If $H\mathbf{z}' \neq \mathbf{0}'$, locate the column $\mathbf{h}^{(i)}$ of H such that $H\mathbf{z}' = \mathbf{h}^{(i)}$, and change the ith bit of **z**.

A flowchart for the procedure is given in Fig. 17.3. Of course, the procedure will fail if more than one error has been made.

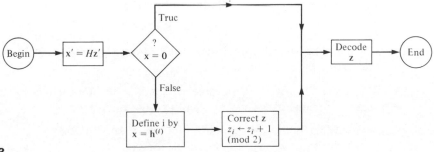

Fig. 17.3
How to correct a single error.

The Hamming codes provide a particularly neat application of the error-correcting procedure described above. When the check matrix for a Hamming code is written down in the natural way, the column $\mathbf{h}^{(i)}$ is just the binary representation of i. So if $\mathbf{z}$ is a received word containing one error, $H\mathbf{z}'$ is the binary representation of the integer i in which bit the error occurs. For instance, in the Hamming code of length 7 described in the *Example*, suppose the received word is $\mathbf{z} = 0\ 1\ 1\ 1\ 0\ 1\ 0$. We have

$$H\mathbf{z}' = \begin{bmatrix} 0 \\ 1 \\ 1 \end{bmatrix},$$

and since 0 1 1 is the binary representation of 3, we conclude that the third bit must be corrected. Thus the correct codeword is 0 1 0 1 0 1 0.

Exercises 17.4

1   Let C be the linear code defined by the check matrix

$$\begin{bmatrix} 1 & 1 & 0 & 1 & 0 & 1 \\ 1 & 1 & 0 & 0 & 1 & 0 \\ 1 & 0 & 1 & 1 & 0 & 0 \end{bmatrix}.$$

If the word 110110 is received, and only one error has been made, what is the intended codeword?

2   Write down a check matrix for the Hamming code of length 15. How many codewords are there? Assuming that the columns of your matrix have been written down in the natural order, as in part (ii) of the *Example*, which of the following are codewords?

011010110111000

100000000000011

110110110111111

Correct those words which are not codewords, assuming that only one error has been made.

3   Suppose we wish to be able to send 256 different messages by means of a linear code which will correct one error.

(i) What is the least possible length of such a code?
(ii) Write down a suitable check matrix.
(iii) Find a lower bound for the length of the code if two errors are to be corrected.

4 Show by arithmetical arguments that there cannot be a perfect code of length n which corrects e errors in the cases

(i) $n = 5, e = 1$; (ii) $n = 10, e = 2$.

17.5 Cyclic codes

In this section the bits of the word **a** of length n will be denoted by $a_0 a_1 a_2 \ldots a_{n-1}$. The reason for the rather strange choice of subscripts will appear very shortly.

A code C is said to be **cyclic** if it is a linear code and if

$$a_0 a_1 a_2 \ldots a_{n-1} \in C \quad \Rightarrow \quad a_{n-1} a_0 a_1 \ldots a_{n-2} \in C.$$

The word $\hat{\mathbf{a}} = a_{n-1} a_0 a_1 \ldots a_{n-2}$ is the first **cyclic shift** of **a**. If C is cyclic then the words obtained from **a** by any number of cyclic shifts, such as

$$a_i a_{i+1} \cdots a_{n-1} a_0 \cdots a_{i-1},$$

are also in C. For example, the code

$$C_2 = \{000, 110, 011, 101\}$$

discussed in Section 17.1 is a cyclic code.

Cyclic codes are useful for two reasons. On the practical side it is possible to implement them by simple devices known as shift-registers. This is an important point, but the details are not appropriate for a book like ours. On the theoretical side, which does concern us, cyclic codes can be constructed and investigated by means of the algebraic theory of rings and polynomials.

The key to the algebraic treatment of cyclic codes is the correspondence between

the word $\qquad$ $\mathbf{a} = a_0 a_1 \ldots a_{n-1}$ $\qquad$ in V^n
and
the polynomial $\quad a(x) = a_0 + a_1 + \cdots + a_{n-1} x^{n-1}$ in $\mathbb{Z}_2[x]$.

In this correspondence, the first cyclic shift $\hat{\mathbf{a}}$ of **a** is represented by the polynomial $\hat{a}(x)$, where

$$\hat{a}(x) = a_{n-1} + a_0 x + \cdots + a_{n-2} x^{n-1}$$
$$= x(a_0 + a_1 x + \cdots + a_{n-1} x^{n-1}) - a_{n-1}(x^n - 1)$$
$$= xa(x) - a_{n-1}(x^n - 1).$$

Since the coefficients belong to $\mathbb{Z}_2$ we could rewrite all the minus signs as plus signs, but we shall retain the minus signs for the sake of clarity. The result of the calculation can be expressed by saying that $\hat{a}(x)$ is equal to $xa(x)$ modulo x^n-1.

In Section 16.3 we explained that, when a polynomial $f(x)$ is given, addition and multiplication of polynomials modulo $f(x)$ can be formally regarded as addition and multiplication of equivalence classes of polynomials. The equivalence classes form a ring, and if $f(x)$ is irreducible, we even get a field. In this case the polynomial $f(x)$ is x^n-1, which is *not* irreducible except in the trivial case $n=1$. Indeed, the factorizations of $f(x)$ will play an important part in the following theory.

We shall denote by $V^n[x]$ the ring of polynomials modulo x^n-1, with coefficients in $\mathbb{Z}_2$. Working modulo x^n-1 is the same as replacing x^n by 1, x^{n+1} by x, x^{n+2} by x^2, and so on, whence it is clear that each class of polynomials modulo x^n-1 has a unique representative with degree less than n. We shall always use this representative to denote the equivalence class, since the correspondence between words in V^n and classes in $V^n[x]$ is manifestly obvious in this way. For example, in V^6,

$$110101 \quad \text{is represented by } 1+x+x^3+x^5$$
$$010110 \quad \text{is represented by } x+x^3+x^4,$$

where the polynomials are really classes in $V^6[x]$. In summary, we have a bijective correspondence between V^n and $V^n[x]$ such that, when $a(x)$ and $b(x)$ correspond to **a** and **b**, $a(x)+b(x)$ corresponds to $\mathbf{a}+\mathbf{b}$ and $xa(x)$ corresponds to the first cyclic shift $\hat{\mathbf{a}}$.

In Theorem 17.5 we shall prove that a cyclic code in V^n corresponds to a particular kind of subset of the ring $V^n[x]$. Subsets of this kind are studied by algebraists for other reasons as well, and they have a special name.

Definition Let R be a ring with commutative multiplication. A subset S of R is said to be an **ideal** if

(i) $a, b \in S \Rightarrow a+b \in S$
(ii) $r \in R$ and $a \in S \Rightarrow ra \in S.$

In other words, S is closed under addition, and under multiplication *by any element of R.*

Theorem 17.5 A code in V^n is cyclic if and only if it corresponds to an ideal in $V^n[x]$.

Proof Suppose C is a cyclic code, represented as a subset of $V^n[x]$. Since C is linear, if $a(x)$ and $b(x)$ are in C so is $a(x)+b(x)$, and the first condition for an ideal is satisfied. Since $xa(x)$ represents the first cyclic shift of $a(x)$, it follows that $xa(x)$ is in C whenever $a(x)$ is in C. By repeating the same argument $x^i a(x)$ is in C whenever $a(x)$ is in C, for any $i \geqslant 0$. Any polynomial $p(x)$ is the sum of a number of powers x^i and so, since C is linear, $p(x)a(x)$ is in C. Hence C is an ideal.

Conversely, if C is an ideal, condition (i) tells us that it represents a linear code. Condition (ii) tells us (in particular) that $xa(x)$ is in C whenever $a(x)$ is, so C is cyclic. □

It follows from the theorem that the construction of cyclic codes of length n is equivalent to the construction of ideals in $V^n[x]$. This is not just a piece of mathematical sophistry for its own sake, as might appear at first sight, because there is a very simple way of constructing ideals. Indeed, in the next section we shall show that the simple way is essentially the only way.

Let $f(x)$ be any polynomial in $\mathbb{Z}_2[x]$ with $\deg f(x) < n$, so that $f(x)$ is the natural representative of a class in $V^n[x]$. The set of all multiples of $f(x)$ in $V^n[x]$ is clearly an ideal, for if $a(x)$ and $b(x)$ are multiples of $f(x)$, so are $a(x)+b(x)$ and $p(x)a(x)$ for any $p(x)$. We denote this ideal by $\langle f(x) \rangle$ and refer to it as the ideal **generated** by $f(x)$.

For example, suppose

$$f(x) = 1 + x^2 \quad \text{in } V^3[x].$$

Multiplying $f(x)$ in turn by each element $p(x)$ of $V^3[x]$ (and remembering to reduce modulo $x^3 - 1$), we obtain Table 17.5.1.

Table 17.5.1

$p(x)$	$p(x)f(x) \bmod (x^3-1)$	Word		
0	0	0	0	0
1	$1+x^2$	1	0	1
x	$1+x$	1	1	0
$1+x$	$x+x^2$	0	1	1
x^2	$x+x^2$	0	1	1
$1+x^2$	$1+x$	1	1	0
$x+x^2$	$1+x^2$	1	0	1
$1+x+x^2$	0	0	0	0

So the ideal $\langle 1+x^2 \rangle$ has just four elements

$$0,\ 1+x,\ x+x^2,\ 1+x^2$$

and the corresponding code in V^3 is the code denoted by C_2 in Section 17.1:

$$C_2 = \{0\,0\,0,\ 1\,1\,0,\ 0\,1\,1,\ 1\,0\,1\}.$$

Exercises 17.5

1 Which of the following codes are cyclic?

     (i) $\{000, 100, 010\}$;   (ii) $\{000, 100, 010, 001\}$;

     (iii) $\{000, 111\}$;   (iv) $\{0000, 1010, 0101, 1111\}$.

2 Write down the codewords of the cyclic code corresponding to the ideal $\langle 1+x+x^2 \rangle$ in $V^3[x]$, and find a check matrix for this code.

3 Show that the ideal $\langle 1+x \rangle$ in $V^5[x]$ corresponds to the code in V^5 containing all the words of even weight. Does this result hold when 5 is replaced by an arbitrary integer $n \geqslant 2$?

17.6 Classification and properties of cyclic codes

The next theorem justifies the claim that every cyclic code can be constructed by the procedure outlined at the end of the previous section.

Theorem 17.6.1

Let C be a cyclic code (ideal) in $V^n[x]$. Then there is a polynomial $g(x)$ in C such that

$$C = \langle g(x) \rangle.$$

Proof

If C is the trivial code containing only the zero polynomial, then $C = \langle 0 \rangle$. If not, C contains a non-zero polynomial $g(x)$ of least degree. Suppose $f(x)$ is any element of C and note that by the division algorithm (Theorem 15.5) we have

$$f(x) = b(x)g(x) + r(x),$$

where either $\deg r(x) < \deg g(x)$ or $r(x) = 0$. (It will be recalled that the zero polynomial does not have a degree.) Because $f(x)$ and $g(x)$ are both in C and C is an ideal, it follows that $b(x)g(x)$ and

$$b(x)g(x) - f(x) = r(x)$$

are in C. This contradicts the definition of $g(x)$ as a polynomial of least degree in C, unless $r(x)=0$. Hence $f(x)=b(x)g(x)$, which means that $C=\langle g(x)\rangle$, as claimed. $\qquad\square$

The polynomial $g(x)$ obtained in the proof is uniquely determined by the property that it has the least degree in C. For if $g_1(x)$ and $g_2(x)$ both have this property, they have the same degree and their leading coefficients are both 1. (Since the coefficients are in $\mathbb{Z}_2$, 0 and 1 are the only possible coefficients.) Furthermore, since C is an ideal $g_1(x)-g_2(x)$ is also in C, and if it were not zero its degree would be strictly less than the degree of $g_1(x)$ and $g_2(x)$, contradicting our assumption. Hence we must have $g_1(x)=g_2(x)$.

In the *Example* discussed at the end of the previous section, we listed the ideal in $V^3[x]$ generated by $1+x^2$. Inspection of the list shows that in this case the unique non-zero polynomial of least degree is $1+x$, and Theorem 17.6.1 tells us that this polynomial is also a generator for the ideal. In general, a cyclic code C will have many generators, but only one of them will have the least degree in C. We shall refer to the unique polynomial with this property as the **canonical generator** of C.

Theorem 17.6.2   The canonical generator $g(x)$ of a cyclic code C in $V^n[x]$ is a divisor of x^n-1 in $\mathbb{Z}_2[x]$.

Proof   Using the division algorithm for $\mathbb{Z}_2[x]$ again, it follows that there are polynomials $h(x)$ and $s(x)$ such that

$$x^n-1=g(x)h(x)+s(x),$$

and either $\deg s(x)<\deg g(x)$ or $s(x)=0$. This equation implies that $s(x)=g(x)h(x)$ in the ring $V^n[x]$ of polynomials modulo x^n-1. Since C is the ideal in $V^n[x]$ generated by $g(x)$ it follows that $s(x)$ is in C. This contradicts the fact that $g(x)$ has the least degree in C unless $s(x)=0$. Hence

$$x^n-1=g(x)h(x)\quad\text{in }\mathbb{Z}_2[x],$$

which gives the required result. $\qquad\square$

Returning once again to our *Example* in $V^3[x]$, we verify that the canonical generator $1+x$ divides x^3-1, since

$$x^3-1=(1+x)(1+x+x^2)\quad\text{in }\mathbb{Z}_2[x].$$

Theorems 17.6.1 and 17.6.2 are not just pretty applications of the theory of rings and polynomials. We shall now show that, given the canonical generator $g(x)$ of C, the dimension of C and a check matrix for C can be easily determined.

Suppose $g(x)h(x) = x^n - 1$ in $\mathbb{Z}_2[x]$, where

$$g(x) = g_0 + g_1 x + \cdots + g_{n-k} x^{n-k}, \qquad h(x) = h_0 + h_1 x + \cdots + h_k x^k.$$

It is worth remarking that the 'extreme' coefficients g_0, h_0, g_{n-k}, h_k must all be 1, since the product of the polynomials is $x^n - 1$. The polynomial $g(x)$ corresponds to the word

$$\mathbf{g} = g_0 g_1 \ldots g_{n-k} 0\, 0 \ldots 0$$

in V^n, and the polynomials $x^i g(x)$ $(1 \leqslant i \leqslant k-1)$ correspond to cyclic shifts of $\mathbf{g}$; that is, the words

$$\mathbf{g}_{(i)} = 0\, 0 \cdots 0 g_0 g_1 \cdots g_{n-k} 0\, 0 \cdots 0,$$

where there are i zeros at the beginning and $k - 1 - i$ zeros at the end. We shall use the special notation $\tilde{\mathbf{h}}$ for the word whose first $k + 1$ bits are the coefficients of $h(x)$ in reverse order, followed by $n - k - 1$ zeros:

$$\tilde{\mathbf{h}} = h_k h_{k-1} \cdots h_0 0\, 0 \cdots 0.$$

The $(n-k) \times n$ matrix whose rows are $\tilde{\mathbf{h}}$ and the first $n - k - 1$ cyclic shifts of $\tilde{\mathbf{h}}$ will be denoted by H; thus

$$H = \begin{bmatrix} h_k & \cdot & \cdot & \cdot h_0 & & 0 & & 0 \\ 0 & h_k & \cdot & \cdot & \cdot & h_0 & & \\ & & \cdot & & & & \cdot & \\ 0 & & & \cdot & & & & \cdot \\ & & & & h_k & & & h_0 \end{bmatrix}$$

Theorem 17.6.3

The matrix H is a check matrix for the cyclic code $C = \langle g(x) \rangle$, and the dimension of C is k.

Proof

Let $c(x) = f(x)g(x)$ be any element of C, where

$$f(x) = f_0 + f_1 x + \cdots + f_{n-1} x^{n-1}.$$

Rearranging the product $f(x)g(x)$ we obtain

$$c(x) = f_0 g(x) + f_1 x g(x) + \cdots + f_{n-1} x^{n-1} g(x).$$

and so it follows that the word $\mathbf{c}$ corresponding to $c(x)$ is given by

$$\mathbf{c} = f_0 \mathbf{g} + f_1 \mathbf{g}_{(1)} + \cdots + f_{n-1} \mathbf{g}_{(n-1)}.$$

Hence, in order to show that H is a check matrix for C (that is, $H\mathbf{c}' = \mathbf{0}'$) it is sufficient to prove that $H\mathbf{g}'_{(i)} = \mathbf{0}'$ for $0 \leqslant i \leqslant n - 1$.

We are given that $g(x)h(x) = x^n - 1$. Equating coefficients of powers of x we obtain the equations

$$g_0h_1 + g_1h_0 = 0 \qquad \text{(coefficient of } x\text{)},$$
$$g_0h_2 + g_1h_1 + g_2h_0 = 0 \qquad \text{(coefficient of } x^2\text{)},$$
$$\cdots$$
$$g_{n-k-1}h_k + g_{n-k}h_{k-1} = 0 \qquad \text{(coefficient of } x^{n-1}\text{)}.$$

Also, since the coefficients of 1 and x^n are both 1, we have

$$g_0h_0 + g_{n-k}h_k = 0.$$

Recalling that $g_{n-k+1}, \ldots, g_{n-1}$ and $h_{k+1}, \ldots, h_{n-1}$ are all zero, these n equations can be written as

$$h_k g_{n-k+j} + h_{k-1}g_{n-k+j+1} + \cdots + h_0 g_{n+j} = 0$$

where $j = 0, 1, \ldots, n-1$, and the subscripts are taken modulo n. For suitable values of j the left-hand sides of these equations are precisely the expressions which occur in the evaluation of $H\mathbf{g}'_{(i)}$. Hence $H\mathbf{g}'_{(i)} = \mathbf{0}'$, as required.

Finally, suppose $\mathbf{y} = y_0 y_1 \cdots y_{n-1}$ is in C. The first equation arising from $H\mathbf{y}' = \mathbf{0}'$ is

$$y_k = h_k y_0 + h_{k-1}y_1 + \cdots + h_1 y_{k-1},$$

since $h_0 = 1$. So if the values of $y_0, y_1, \ldots, y_{k-1}$ are given, the value of y_k is determined. The next equation determines y_{k+1} in terms of $y_1, y_2, \ldots, y_k$, and so on. Since there are 2^k possible values for $y_0, y_1, \ldots, y_{k-1}$, we have $|C| = 2^k$, as claimed. $\square$

The foregoing theory indicates that in order to describe the cyclic codes of length n we must first find the factors of $x^n - 1$ in $\mathbb{Z}_2[x]$. For example, when $n = 7$, the factorization of $x^7 - 1$ into irreducible polynomials is

$$x^7 - 1 = (1+x)(1+x+x^3)(1+x^2+x^3).$$

(This factorization is related to the results obtained in Section 16.9: since $8 = 2^3$ the set of irreducible factors of $x^8 - x$ in $\mathbb{Z}_2[x]$ must consist of all the monic irreducible polynomials whose degree is a divisor of 3.) The equation shows that there are just eight divisors of $x^7 - 1$ in $\mathbb{Z}_2[x]$; they are the trivial divisors 1 and $x^7 - 1$, together with

$1 + x,$ $\qquad\qquad$ $1 + x + x^3,$ $\qquad\qquad$ $1 + x^2 + x^3,$
$(1+x)(1+x+x^3),$ $\quad$ $(1+x)(1+x^2+x^3),$ $\quad$ $(1+x+x^3)(1+x^2+x^3).$

Each of the divisors generates a cyclic code, and (by Theorems 17.6.1 and 17.6.2) these are the only cyclic codes of length 7. Clearly $\langle 1 \rangle$ is the code in which every word is a codeword, and $\langle x^7 - 1 \rangle = \langle 0 \rangle$ is the code in which the only codeword is $\mathbf{0}$. The other codes are more interesting. In particular, let C be the code with canonical generator $g(x) = 1 + x + x^3$, so that

$$h(x) = (1 + x)(1 + x^2 + x^3) = 1 + x + x^2 + x^4$$

in the notation introduced above. It follows from Theorem 17.6.3 that C has dimension 4, and a check matrix for C is

$$\begin{bmatrix} 1 & 0 & 1 & 1 & 1 & 0 & 0 \\ 0 & 1 & 0 & 1 & 1 & 1 & 0 \\ 0 & 0 & 1 & 0 & 1 & 1 & 1 \end{bmatrix}.$$

Inspection of this matrix reveals that its columns are the same as the columns of the check matrix for the Hamming code of length 7 (Section 17.4, *Example*), but in a different order. Hence the code C is essentially the same as the Hamming code.

The astute reader will have noticed that we have not said anything about the minimum distance δ of a cyclic code in general. For some classes of cyclic codes it is possible to obtain extremely useful results about δ; and these results, together with the theorems obtained in this section, mean that such codes are important both in theory and in practice. But these matters must be postponed to the course in coding theory which, it is hoped, the reader will be inspired to study after finishing this book.

Exercises 17.6

1 Write down the irreducible factors of $x^5 - 1$ in $\mathbb{Z}_2[x]$, and hence determine all cyclic codes of length 5. (There are four of them, all rather dull.)

2 Describe all the cyclic codes of length 7.

3 The factorization of $x^{15} - 1$ into irreducible polynomials in $\mathbb{Z}_2[x]$ is

$$x^{15} - 1 = (1 + x)(1 + x + x^2)(1 + x + x^4)(1 + x^3 + x^4)$$
$$\times (1 + x + x^2 + x^3 + x^4).$$

(i) Explain how this is related to the results obtained in Section 16.9.

(ii) What is the number of cyclic codes of length 15?

(iii) Find a canonical generator for a cyclic code equivalent to the Hamming code of length 15.

4   Use the results of Section 16.9 to find the number of irreducible factors of $x^{32} - x$ in $\mathbb{Z}_2[x]$, and hence determine the number of cyclic codes of length 31. Find a canonical generator for a cyclic code equivalent to the Hamming code of length 31.

7.7   Miscellaneous Exercises

1   Calculate the parameters (n, k, δ) for the linear code whose check matrix is

$$\begin{bmatrix} 0 & 1 & 1 & 0 & 0 \\ 1 & 0 & 0 & 1 & 0 \\ 1 & 1 & 0 & 0 & 1 \end{bmatrix}.$$

2   Decide whether or not the following words are codewords in the code defined in Ex. 1, and correct those which are not codewords, assuming that only one error has been made.

(i)   11111,.   (ii) 01101,   (iii) 01100.

How many words in all cannot be corrected by changing one bit?

3   Let C be a code with minimum weight $\delta = 3$. Show that there is a word x such that

$$\partial(x, c_1) = 1, \qquad \partial(x, c_2) = 2,$$

for some codewords c_1 and c_2 in C. What action should be taken if x is received?

4   Write down a check matrix for the linear code consisting of all the words of length 7 which have even weight.

5   Use the method of Section 17.2 to show that there is no code C in V^5 such that C has minimum distance $\delta = 3$ and $|C| = 6$.

6   By direct analysis of the possibilities show that the result given in the previous exercise remains true if the condition $|C| = 6$ if replaced by $|C| = 5$.

7   For which of the following values of the parameters (n, k, δ) is it possible that a linear code with those parameters exists?

(i) $(12, 7, 5)$;   (ii) $(11, 4, 5)$.

8   Let C be a linear code of length n and dimension k, and suppose i is any integer in the range $1 \le i \le n$. Show that

$$C_0 = \{c \in C \mid c_i = 0\}$$

is a linear code of length n and dimension $k - 1$.

9 Show by arithmetical means that a code C with length $n=6$ and minimum distance $\delta = 3$ must satisfy $|C| \leq 9$. Verify by direct means that there is no code with $|C| = 9$ and construct a code with $|C| = 8$.

10 Let C be a binary code of length n which will correct two errors and is perfect. Show that

(i) $n^2 + n + 2$ is a power of 2;
(ii) there is only one value of $n \leq 10$ which satisfies condition (i);
(iii) there is a code C with the required properties for the value of n obtained in (ii), but it is trivial.

Compute the next smallest value of n for which (i) is satisfied. (Unfortunately, it can be shown that there is no code in this case, or for any other value of n.)

11 Derive a condition analogous to (i) in the previous exercise for a perfect code which will correct three errors, and show that your condition is satisfied when $n = 7$ and $n = 23$. (In the next exercise we show that there is a non-trivial code when $n = 23$.)

12 Show that in $\mathbb{Z}_2[x]$ the polynomial $x^{23} - 1$ can be written as $(x-1)f(x)g(x)$, where $f(x)$ and $g(x)$ are polynomials of degree 11. Show that the cyclic code generated by $f(x)$ (or $g(x)$) is a perfect code which will correct three errors.

13 Let C_1 and C_2 be cyclic codes of length n, with generators $g_1(x)$ and $g_2(x)$ respectively. Show that

$$C_1 + C_2 = \{ \mathbf{x} \in V^n \mid \mathbf{x} = \mathbf{c}_1 + \mathbf{c}_2 \quad \text{for some} \quad \mathbf{c}_1 \in C_1, \mathbf{c}_2 \in C_2 \}$$

is a cyclic code with generator $\gcd(g_1(x), g_2(x))$.

14 Show that every Hamming code is a cyclic code.

15 Show that if a cyclic code contains a codeword whose weight is odd, then it contains the word $111 \cdots 1$.

16 Let $B_1, B_2, \ldots, B_7$ be the blocks of the Steiner triple system with seven varieties $1, 2, \ldots, 7$. Define a set C of seven words $\mathbf{w}^{(1)}, \mathbf{w}^{(2)}, \ldots, \mathbf{w}^{(7)}$ in V^7 by the rules

$$w_j^{(i)} = \begin{cases} 1 & \text{if } j \in B_i, \\ 0 & \text{otherwise.} \end{cases}$$

Show that C is a code with minimum distance 3.

17 Let C be the code constructed by the method given in the previous example, but using a projective plane over $\mathbb{F}_q$ in place of the Steiner triple system. What is the minimum distance of C?

18 Let S denote the set of codewords of weight 3 in a Hamming code of length $2^r - 1$. For each $\mathbf{w}$ in S define a subset B of $\{1, 2, \ldots, 2^r - 1\}$ by the rule that

$$i \in B \quad \Leftrightarrow \quad w_i = 1.$$

Show that the sets B are the blocks of a Steiner triple system.

Generating functions

18.1 Power series and their algebraic properties

In Chapter 15 we observed that a polynomial is really nothing more than its sequence of coefficients. The sequences which correspond to polynomials all have the characteristic property that they have only a finite number of terms; more precisely, they have only a finite number of non-zero terms. However, almost all the sequences which arise in practice do not share this property, and in order to study them by algebraic means we shall introduce a generalization of the polynomial concept.

The object which corresponds to an infinite sequence (a_n) in the same way as a polynomial corresponds to a finite sequence is known as a **power series** $A(x)$, written

$$A(x) = a_0 + a_1 x + a_2 x^2 + \cdots .$$

Just as with polynomials, we do not concern ourselves with the meaning of the symbol x and its powers; their function is simply to label the positions of the coefficients.

Power series can be added and multiplied in exactly the same way as polynomials. If we are given

$$A(x) = a_0 + a_1 x + a_2 x^2 + \cdots , \qquad B(x) = b_0 + b_1 x + b_2 x^2 + \cdots ,$$

then we have

$$A(x) + B(x) = (a_0 + b_0) + (a_1 + b_1)x + (a_2 + b_2)x^2 + \cdots ,$$
$$A(x)B(x) = a_0 b_0 + (a_0 b_1 + a_1 b_0)x + (a_0 b_2 + a_1 b_1 + a_2 b_0)x^2 + \cdots .$$

In other words, if $C(x) = A(x) + B(x)$ and $D(x) = A(x)B(x)$, then the coefficients c_i and d_i of $C(x)$ and $D(x)$ respectively are given by the equations

$$c_i = a_i + b_i, \qquad d_i = a_0 b_i + a_1 b_{i-1} + \cdots + a_i b_0 \qquad (i \geqslant 0).$$

With these definitions, the power series with coefficients in a commutative ring R are themselves the elements of another commutative ring, denoted by $R[\![x]\!]$. Checking the ring axioms is just as straightforward (and just as tedious) as for the ring of polynomials $R[x]$. We remark that $R[x]$ is a subset of $R[[x]]$, and it forms a ring with same algebraic operations; for that reason $R[x]$ is said to be a **subring** of $R[[x]]$.

From now on we shall frequently assume that the coefficients of our polynomials and power series belong to a field F. Both $F[x]$ and $F[[x]]$ are rings, rather than fields, but $F[[x]]$ is more like a field than $F[x]$, since it has many more invertible elements. It will be recalled (Ex. 15.5.4) that a polynomial is invertible in $F[x]$ only in trivial cases—when it is a non-zero constant. In $F[[x]]$ the situation is more promising.

Theorem 18.1

When F is a field, the power series
$$B(x) = b_0 + b_1 x + b_2 x^2 + \cdots$$
is invertible in $F[[x]]$ if and only if $b_0 \neq 0$.

Proof

If $B(x)$ is invertible there is a power series $U(x)$ in $F[[x]]$ such that $B(x)U(x) = 1$, that is
$$(b_0 + b_1 x + b_2 x^2 + \cdots)(u_0 + u_1 x + u_2 x^2 + \cdots) = 1.$$

Equating coefficients of the constant term we obtain $b_0 u_0 = 1$, so that $b_0 \neq 0$.

Conversely, suppose $b_0 \neq 0$. In order to determine $U(x)$ we must consider the equations obtained from the product formula:
$$b_0 u_0 = 1,$$
$$b_0 u_1 + b_1 u_0 = 0,$$
$$b_0 u_2 + b_1 u_1 + b_2 u_0 = 0,$$
$$\cdots \qquad .$$

Given that $b_0 \neq 0$, so that b_0^{-1} exists, we can determine the sequence (u_n) recursively, as follows:
$$u_0 = b_0^{-1},$$
$$u_1 = b_0^{-1}(-b_1 u_0),$$
$$u_2 = b_0^{-1}(-b_1 u_1 - b_2 u_0),$$
$$\cdots \qquad .$$

Hence $B(x)$ is invertible.   □

Naturally, we write $B(x)^{-1}$ for the inverse of $B(x)$: it is uniquely determined by the proof of the theorem.

The fact that so many power series have inverses in $F[[x]]$ is the basis for the methods to be discussed in the final part of this book. Almost always $B(x)$ itself will be a polynomial, but its inverse $B(x)^{-1}$ is necessarily a power series, rather than a polynomial.

Example   Show that the inverse of $1-x$ is the power series

$$1+x+x^2+\cdots.$$

Solution   Since the constant term of $1-x$ is 1, it has an inverse: suppose the inverse is $U(x)$. As in the proof of Theorem 18.1 we derive the following equations from $(1-x)(u_0+u_1x+u_2x^2+\cdots)=1$:

$$u_0=1,$$
$$u_1-u_0=0,$$
$$u_2-u_1=0,$$
$$\cdots$$

Hence $u_0=u_1=u_2=\cdots=1$, and we have

$$(1-x)^{-1}=1+x+x^2+\cdots.\qquad\square$$

In practice we usually resort to the high-school notation in which

$$B(x)^{-1}\quad\text{is written as}\quad\frac{1}{B(x)},$$

$$A(x)B(x)^{-1}\quad\text{is written as}\quad\frac{A(x)}{B(x)}.$$

But we should remember that $A(x)/B(x)$ simply means the product of the power series $A(x)$ and $B(x)^{-1}$. Most often we require the case when $A(x)$ and $B(x)$ themselves are polynomials, although of course $B(x)^{-1}$ is not. In that case we can compute the coefficients of $Q(x)=A(x)/B(x)$ by solving the equations arising from the consequent equation $B(x)Q(x)=A(x)$, that is

$$(b_0+b_1x+\cdots+b_mx^m)(q_0+q_1x+q_2x^2+\cdots)$$
$$=(a_0+a_1x+\cdots+a_nx^n).$$

Indeed, this method is precisely the same as the traditional method of dividing $B(x)$ into $A(x)$, provided we begin with the constant terms.

For example, in order to find the power series for

$$\frac{1+3x}{1-2x+x^2},$$

we proceed as follows.

$$
\begin{array}{r}
1+5x+\ 9x^2+13x^3+\cdots \\
1-2x+x^2\overline{\smash{\big)}\,1+3x} \\
\underline{1-2x+\ x^2} \\
5x-\ x^2 \\
\underline{5x-10x^2+\ 5x^3} \\
9x^2-\ 5x^3 \\
\underline{9x^2-18x^3+\ 9x^4} \\
13x^3-\ 9x^4 \\
\underline{13x^3-26x^4+13x^5} \\
17x^4-13x^5\ldots .
\end{array}
$$

In this example it is fairly easy to guess a formula for the coefficients, but that is not always so.

Exercises 18.1

1 Find the first four terms in the power series for

(i) $\dfrac{1+4x}{1+5x+x^2}$,   (ii) $\dfrac{2+6x+x^2}{3+x+5x^2+x^3}$.

(You may assume that the coefficients belong to the field $\mathbb{R}$ of real numbers.)

2 Show that the inverse of $1+x$ in $\mathbb{R}[\![x]\!]$ is

$$(1+x)^{-1}=1-x+x^2-x^3+\cdots,$$

where the coefficient of x^n is $(-1)^n$.

3 Using long division, show that the inverse of $1+x+x^2$ in $\mathbb{Z}_2[\![x]\!]$ is

$$(1+x+x^2)^{-1}=1+x+x^3+x^4+x^6+x^7+\cdots .$$

Guess the rule for the coefficients, and prove it.

4 What is the inverse of $1+x^2$ in $\mathbb{Z}_2[\![x]\!]$? What is its inverse in $\mathbb{Z}_3[\![x]\!]$?

18.2 Partial fractions

In this section we begin the task of finding an alternative way to compute the coefficients in the power series for $a(x)/b(x)$, when $a(x)$ and $b(x)$ are polynomials and $b_0 \neq 0$. Although the algorithmic method of long division is adequate for many purposes, it does not yield a formula for the coefficients, and there are circumstances in which such a formula is desirable.

The key to the problem is the decomposition of $a(x)/b(x)$ into *partial fractions*, a technique which may be familiar to some readers. Roughly speaking, if there is a factorization $b(x) = s(x)t(x)$, where $s(x)$ and $t(x)$ have no non-trivial common factor, then we can write

$$\frac{a(x)}{b(x)} = \frac{f(x)}{s(x)} + \frac{g(x)}{t(x)}$$

for some polynomials $f(x)$ and $g(x)$. For example, the polynomial $2 - 3x + x^2$ is equal to $(1-x)(2-x)$, and we have the equation

$$\frac{5-3x}{2-3x+x^2} = \frac{2}{1-x} + \frac{1}{2-x}.$$

It must be stressed that all the 'fractions' occurring in such equations are really power series, although in practice it is both convenient and harmless to retain the fractional notation. Note also that it is only necessary to discuss the case when the degree of $a(x)$ is strictly less than the degree of $b(x)$. This is because we have

$$a(x) = b(x)q(x) + r(x)$$

where either $r(x)$ is the zero polynomial or $\deg r(x) < \deg b(x)$, and consequently

$$\frac{a(x)}{b(x)} = q(x) + \frac{r(x)}{b(x)}.$$

Thus the problem is reduced to finding the partial fractions for $r(x)/b(x)$. For example,

$$\frac{7-2x-5x^2+2x^3}{2-3x+x^2} = \frac{(1+2x)(2-3x+x^2)+(5-3x)}{2-3x+x^2}$$

$$= (1+2x) + \frac{2}{1-x} + \frac{1}{2-x}.$$

We now establish the precise form of the result which yields partial fraction decompositions.

Theorem Let F be a field, and suppose $a(x)$ and $b(x)$ are polynomials in $F[x]$
18.2 such that

  (i) $\deg a(x) < \deg b(x)$,
  (ii) $b(x) = s(x)t(x)$ where $s(x)$ and $t(x)$ have no non-trivial common factor,
  (iii) $b_0 \neq 0$.

Then there are polynomials $f(x)$ and $g(x)$ such that

$$\deg f(x) < \deg s(x), \qquad \deg g(x) < \deg t(x),$$

and

$$\frac{a(x)}{b(x)} = \frac{f(x)}{s(x)} + \frac{g(x)}{t(x)},$$

where the equation holds in the ring $F[\![x]\!]$ of power series with coefficients in F.

Proof We remark first that $s(x)$ and $t(x)$ do have inverses in $F[\![x]\!]$, because the equation $s_0 t_0 = b_0$ and the condition $b_0 \neq 0$ imply that $s_0 \neq 0$ and $t_0 \neq 0$.

In the rest of the proof we shall, for convenience, use single letters to denote the polynomials. Since 1 is a gcd of t and s, it follows from Theorem 15.6 that there are polynomials λ and μ such that

$$1 = \lambda t + \mu s.$$

Multiplying by a and putting $\bar{f} = a\lambda$, $\bar{g} = a\mu$, we have

$$a = \bar{f}t + \bar{g}s.$$

In order to replace $\bar{f}$ and $\bar{g}$ by polynomials whose degrees satisfy the stated conditions, we proceed as follows. By the division algorithm, we have

$$\bar{f} = qs + f$$

where $\deg f < \deg s$ (or $f = 0$). Substituting for $\bar{f}$ in the equation $a = \bar{f}t + \bar{g}s$ we obtain

$$a = ft + gs,$$

where $g = \bar{g} + qt$. It remains to show that $\deg g < \deg t$.
We have

$$\deg a < \deg b, \qquad \deg ft < \deg st = \deg b.$$

Thus, since $gs = a - ft$,

$$\deg gs = \deg (a - ft) < \deg b = \deg st.$$

Hence $\deg g < \deg t$, as required.

Dividing the equation $a = ft + gs$ by $b = st$ we obtain the result. (Strictly speaking, we have to *multiply* by the *inverse* of b.) $\square$

It is easy to prove that the polynomials $f(x)$ and $g(x)$ which satisfy the conditions of Theorem 18.2 are unique.

Suppose that the factorization of $b(x)$ into irreducible polynomials is

$$b(x) = p_1(x)^{m_1} p_2(x)^{m_2} \cdots p_k(x)^{m_k}.$$

Repeated application of Theorem 18.2 tells us that if $\deg a(x) < \deg b(x)$ we can write

$$\frac{a(x)}{b(x)} = \frac{h_1(x)}{p_1(x)^{m_1}} + \frac{h_2(x)}{p_2(x)^{m_2}} + \cdots + \frac{h_k(x)}{p_k(x)^{m_k}},$$

where $\deg h_i(x) < \deg p_i(x)^{m_i}$ $(1 \le i \le k)$. In particular, when the irreducible factors are all linear, so that

$$b(x) = \beta(\alpha_1 - x)^{m_1}(\alpha_2 - x)^{m_2} \cdots (\alpha_k - x)^{m_k},$$

we have

$$\frac{a(x)}{b(x)} = \frac{h_1(x)}{(\alpha_1 - x)^{m_1}} + \frac{h_2(x)}{(\alpha_2 - x)^{m_2}} + \cdots + \frac{h_k(x)}{(\alpha_k - x)^{m_k}},$$

where $\deg h_i(x) < m_i$ $(1 \le i \le k)$.

Frequently, but not always, it is useful to carry the decomposition a stage further. Consider a typical fraction

$$h(x)/(\alpha - x)^m \quad \text{where} \quad \deg h(x) < m.$$

By successively applying the division algorithm (as in Ex. 18.2.4) we can determine coefficients $\gamma_1, \gamma_2, \ldots, \gamma_m$ such that

$$h(x) = \gamma_m + \gamma_{m-1}(\alpha - x) + \cdots + \gamma_1(\alpha - x)^{m-1},$$

whence

$$\frac{h(x)}{(\alpha - x)^m} = \frac{\gamma_1}{\alpha - x} + \frac{\gamma_2}{(\alpha - x)^2} + \cdots + \frac{\gamma_m}{(\alpha - x)^m}.$$

In this way we obtain an expression for $a(x)/b(x)$ in which the numerator of each partial fraction is a constant.

There are several ways of finding the constants in practice. One way is to 'multiply out' and equate coefficients in the resulting equation between two polynomials, as in the following *Example*.

Example   Find the partial fraction decomposition of

$$\frac{4+x-x^2}{3-5x+x^2+x^3}.$$

Solution   We have

$$3-5x+x^2+x^3 = (1-x)(3-2x-x^2)$$
$$= (1-x)(1-x)(3+x).$$

Thus the partial fraction decomposition takes the form

$$\frac{4+x-x^2}{3-5x+x^2+x^3} = \frac{A}{1-x} + \frac{B}{(1-x)^2} + \frac{C}{3+x}.$$

In order to determine A, B, C we multiply both sides by $(1-x)^2(3+x)$, obtaining

$$4+x-x^2 = A(1-x)(3+x) + B(3+x) + C(1-x)^2.$$

Equating coefficients of 1, x, x^2, we get

$$\begin{aligned} 4 &= & 3A + 3B + & C, \\ 1 &= & -2A + & B - 2C, \\ -1 &= & -A & + C, \end{aligned}$$

whence $A = \frac{1}{2}$, $B = 1$, $C = -\frac{1}{2}$.   □

In some circumstances it is easier to determine the coefficients by substituting values in the equations. For example, when we substitute $x = -3$ in the equation

$$4+x-x^2 = A(1-x)(3+x) + B(3+x) + C(1-x)^2$$

obtained above, we find that two of the terms on the right-hand side vanish and we get

$$4 + (-3) - (-3)^2 = C(1-(-3))^2.$$

This gives $C = -\frac{1}{2}$ as before. Similarly, substituting $x = 1$ yields $B = 1$. This method is the justification for the useful 'cover-up' rule, which can be formulated as follows.

Let $(x-\alpha)^r$ be the highest power of $x-\alpha$ which is a factor of

$b(x)$, so that we can write

$$\frac{a(x)}{b(x)}=\frac{a(x)}{(x-\alpha)^r c(x)}=\frac{K}{(x-\alpha)^r}+P(x),$$

where $P(x)$ is the sum of the remaining partial fractions. Then K is equal to the value obtained by 'covering up' the factor $(x-\alpha)^r$ in $a(x)/b(x)$ and putting $x=\alpha$ in what is left; that is, $K=a(\alpha)/c(\alpha)$. For example, the constant C in the partial fraction expansion of

$$\frac{4-x+x^2}{(3+x)(1-x)^2}$$

is obtained by covering up the factor $(3+x)$ and putting $x=-3$ in what is left:

$$C=\frac{4-(-3)+(-3)^2}{(1-(-3))^2}=-\tfrac{1}{2}.$$

The cover-up rule is very useful, but there are two points to remember. First, it only gives the numerator of the highest power of a linear factor, so the other coefficients must be obtained in some other way. Secondly, it should only be used when the coefficients are presumed to be real or complex numbers: this is because the technique of substituting values depends upon the assumption that a polynomial and the corresponding polynomial function are interchangeable, and this is false when the coefficients belong to a finite field (Section 15.8). Conventionally, we assume that we are working in the field $\mathbb{C}$ of complex numbers unless some other field is stated explicitly. One good reason for this convention is that any polynomial can be decomposed into linear factors in $\mathbb{C}[x]$, and so the complete decomposition of a 'quotient' into partial fractions with constant numerators can always be achieved.

Exercises 18.2

1 Use the cover-up rule to find the partial fraction decomposition of

(i) $\dfrac{3+4x}{(1-x)(2+x)}$; (ii) $\dfrac{2+x+x^2}{(1+x)(2+x)(3+x)}$.

2 Find the partial fraction decompositions of

(i) $\dfrac{1+3x}{1-3x^2+2x^3}$, (ii) $\dfrac{-5+3x}{6-11x+6x^2-x^3}$.

3   Find the partial fraction decomposition over $\mathbb{Z}_3$ of $1/(1+x^4)$. (Use the factorization obtained in Section 15.8.)

4   Let $h(x)$ be a polynomial with $\deg h(x) < m$. Show that if $q_i(x)$ and γ_i $(1 \le i \le m)$ are defined by

$$h(x) = (\alpha - x)q_1(x) + \gamma_m,$$
$$q_{i-1}(x) = (\alpha - x)q_i(x) + \gamma_{m-i+1} \quad (2 \le i \le m),$$

then

$$h(x) = \gamma_m + \gamma_{m-1}(\alpha - x) + \cdots + \gamma_1(\alpha - x)^{m-1}.$$

5   Find the partial fraction decomposition *over the complex field* $\mathbb{C}$ of $(1-x^4)^{-1}$.

18.3   The binomial theorem for negative exponents

The binomial theorem obtained in Section 4.3 can be formulated as a result about multiplication in the ring $\mathbb{Z}[x]$ of polynomials with integer coefficients. Specifically, it asserts that the coefficient of x^n in the polynomial $(1+x)^k$ is the binomial number $\binom{k}{n}$, that is,

$$(1+x)^k = \binom{k}{0} + \binom{k}{1}x + \binom{k}{2}x^2 + \cdots + \binom{k}{n}x^n + \cdots + \binom{k}{k}x^k.$$

In this chapter we have come to regard $(1+x)^{-1}$ as a power series with integer coefficients:

$$(1+x)^{-1} = 1 - x + x^2 - x^3 + \cdots,$$

and so in the ring of power series $\mathbb{Z}[\![x]\!]$ we can define $(1+x)^{-m}$ to be the product of m factors $(1+x)^{-1}$. Clearly this definition works for any integer $m \ge 1$, and naturally we should like to have a formula for the coefficient of x^n in the resulting power series. It turns out that the formula is remarkably simple.

Theorem 18.3   The coefficient of x^n in the power series $(1+x)^{-m}$ is

$$(-1)^n \binom{m+n-1}{n}.$$

Proof   In order to avoid the minus signs we shall consider $(1-x)^{-m}$ rather than $(1+x)^{-m}$. Since

$$(1-x)^{-1} = 1 + x + x^2 + \cdots,$$

it follows that $(1-x)^{-m}$ is the product of m factors, each of which is equal to the power series $1+x+x^2+\cdots$. We shall show that the coefficient of x^n in the product of these m factors is equal to the number of unordered selections of n of the m factors, with repetition allowed.

Suppose that each factor has a marker, initially positioned at the term 1, and that we make an unordered selection, with repetition, of size n from the m factors. Each time a particular factor is selected we move its marker to the next term, so that if that factor is selected i times in all the marker will finish up at x^i. Thus for each one of the $\binom{m+n-1}{n}$ possible selections we obtain a set of marked terms, one from each factor, with total exponent n. Each of these sets contributes 1 to the coefficient of x^n when the factors are multiplied, and so the coefficient of x^n is $\binom{m+n-1}{n}$. Replacing x by $-x$ we get the result as stated. □

For example, the coefficient of x^n in the power series $(1+x)^{-2}$ is

$$(-1)^n\binom{2+n-1}{n}=(-1)^n\binom{n+1}{n}=(-1)^n(n+1),$$

and so the power series is

$$(1+x)^{-2}=1-2x+3x^2-4x^3+\cdots.$$

It is possible to combine Theorem 18.3 and the binomial theorem for positive exponents obtained in Theorem 4.3 into a single formula. Given any integer α and any positive integer n define

$$\binom{\alpha}{n}=\frac{\alpha(\alpha-1)(\alpha-2)\cdots(\alpha-n+1)}{n!},$$

and let $\binom{\alpha}{0}=1$. This is legitimate extension of the meaning of binomial numbers, since when α is a positive integer k, we get the formula for $\binom{k}{n}$ obtained in Theorem 4.1.2. When α is a negative integer $-m$ we have

$$\binom{-m}{n}=\frac{(-m)(-m-1)(-m-2)\cdots(-m-n+1)}{n!}$$

$$=(-1)^n\frac{m(m+1)(m+2)\cdots(m+n-1)}{n!}$$

$$=(-1)^n\binom{m+n-1}{n}.$$

Thus, given the extended definition of $\binom{k}{n}$ for positive *and* negative integers k, we have shown that the coefficient of x^n in $(1+x)^k$ is $\binom{k}{n}$.

If we include the trivial case $(1+x)^0 = 1$, we have the following general form of the binomial theorem, which holds for all integers k:

$$(1+x)^k = \binom{k}{0} + \binom{k}{1}x + \binom{k}{2}x^2 + \cdots + \binom{k}{n}x^n + \cdots.$$

The general form is a power series, but when k is a positive integer we have

$$\binom{k}{n} = 0 \quad \text{whenever} \quad n > k,$$

so the right-hand side reduces to a polynomial.

There are several trivial extensions of the binomial theorem: formulae for $(a+x)^k$, $(a-x)^k$, and so on. In particular, we shall make considerable use of the formula for $(1-ax)^{-m}$, which is

$$(1-ax)^{-m} = 1 + max + \cdots + \binom{m+n-1}{n}a^n x^n + \cdots.$$

The binomial theorem has been presented here as a rule for finding the coefficients in the power series $(1+x)^k$. Many other power series can be transformed by algebraic manipulations into a form involving the basic power series, and so the binomial theorem can be used to find formulae for their coefficients. For example, in Section 18.1 we obtained the power series

$$\frac{1+3x}{1-2x+x^2} = 1 + 5x + 9x^2 + 13x^3 + \cdots.$$

Noting that the denominator is $(1-x)^2$ we can work out a formula for the general coefficient as follows.

$$\frac{1+3x}{1-2x+x^2} = (1+3x)(1-x)^{-2}$$

$$= (1+3x)(1+2x+\cdots+(n+1)x^n+\cdots).$$

The coefficient of x^n in the product is $(n+1)+3n$, which gives the answer $4n+1$ as one might guess from the first few values.

In later sections of this chapter we shall see how the technique of partial fractions enables us to reduce many power series to forms in which the binomial theorem can be applied.

Exercises 18.3

1 Write down, and simplify wherever possible, the coefficient of

(i) x^3 in $(1+2x)^7$,   (ii) x^n in $(1-x)^{-4}$,

(iii) x^{2r} in $(1-x)^{-r}$.

2   Write down the first four terms and the general term in the power series $(1-x)^{-3}$.

3   Let a_n be the coefficient of x^n in the power series $(1-x-x^2)^{-1}$. Show that

$$a_n = \binom{n}{0} + \binom{n-1}{1} + \binom{n-2}{2} + \cdots + \binom{n-r}{r},$$

where r is the greatest integer such that $0 \leqslant r \leqslant n-r$.

4   What is the coefficient of x^n in the power series

$$\frac{1+2x+2x^2}{1-3x+3x^2-x^3}?$$

5   If $a_0, a_1, \ldots, a_{6r}$ and $b_0, b_1, \ldots, b_{3r}$ are defined by

$$(1-x+x^2)^{3r} = a_0 + a_1 x + a_2 x^2 + \cdots + a_{6r} x^{6r},$$
$$(1+x)^{3r} = b_0 + b_1 x + b_2 x^2 + \cdots + b_{3r} x^{3r},$$

show that

$$\sum_{i=0}^{3r} a_i b_{3r-i} = \binom{3r}{r}.$$

18.4   Generating functions

Back in Chapter 12 we made the rather trivial observation that the solution of a combinatorial problem can often be expressed as a sequence (u_n). We now turn to methods based on the representation of (u_n) as a power series

$$U(x) = u_0 + u_1 x + u_2 x^2 + \cdots.$$

In this context $U(x)$ is traditionally referred to as the **generating function** for the sequence (u_n). (Strictly speaking it should be referred to as the *ordinary* generating function, since mathematicians often use other kinds of generating function. But in this book the ordinary generating function is the only kind needed, so we shall omit the word 'ordinary'.) Of course, we have gone to some lengths to emphasize that $U(x)$ is *not* a function; it is simply an alternative way of writing the sequence (u_n) so that certain algebraic manipulations can be carried out.

Perhaps the best example of a generating function arises from the binomial theorem. The formula

$$(1+x)^k = \binom{k}{0} + \binom{k}{1}x + \binom{k}{2}x^2 + \cdots + \binom{k}{n}x^n + \cdots,$$

can be regarded as saying that the generating function for the sequence defined by

$$u_n = \binom{k}{n},$$

for any given integer k, is

$$U(x) = (1+x)^k.$$

In the rest of this chapter we shall study the generating functions of sequences defined by recursion equations. The method we shall employ is in three stages:

(i) use the recursion for (u_n) to obtain an equation for $U(x)$;
(ii) solve the equation for $U(x)$;
(iii) use algebra (specifically, the partial fraction decomposition and the binomial theorem) to find a formula for the coefficients of $U(x)$.

The following *Example* gives a typical application of the method. The sequence considered was also discussed in Section 12.2.

Example   Joshua P. Stackenboom, the founder and benefactor of Folornia University has revealed how he became a millionaire. He began with nothing, and at the end of his first year of honest toil he had amassed net assets of one dollar. After a second year he had five dollars. Thereafter he made a rule that in the course of each year he would buy goods to the value of six times his net assets at the beginning of the previous year, and sell them for four times the value of his net assets at the beginning of the current year.

Find and solve a recursion for u_n, Mr Stackenboom's net assets at the end of the nth year. How many years did it take him to become a millionaire?

Solution   We are given that $u_0 = 0$, $u_1 = 1$, $u_2 = 5$. At the end of year $n+1$ Mr Stackenboom's net assets are equal to his net assets at the end of year n, less expenditure, plus income. That is,

$$u_{n+1} = u_n - 6u_{n-1} + 4u_n.$$

Rearranging and replacing n by $n+1$ we get the recursion

$$u_{n+2} - 5u_{n+1} + 6u_n = 0.$$

The argument given above shows that this holds for $n \geqslant 1$. By direct substitution it holds also when $n = 0$, and so we need assume only the initial conditions $u_0 = 0$, $u_1 = 1$. (There is a practical reason why we had to specify $u_2 = 5$ separately in the problem: can you spot it?)

Let $U(x)$ be the generating function for the sequence (u_n). Using the initial values of u_0 and u_1 and the equation for u_{n+2} $(n \geqslant 0)$ we have

$$U(x) = u_0 + u_1 x + u_2 x^2 + u_3 x^3 + \cdots$$
$$= 0 + x + (5u_1 - 6u_0)x^2 + (5u_2 - 6u_1)x^3 + \cdots$$
$$= x + 5(u_1 x^2 + u_2 x^3 + \cdots) - 6(u_0 x^2 + u_1 x^3 + \cdots).$$

The expression in the first bracket is $xU(x)$, since the missing term $u_0 x$ is 0, and in the second bracket we have $x^2 U(x)$. Hence

$$U(x) = x + 5xU(x) - 6x^2 U(x),$$

whence

$$U(x) = \frac{x}{1 - 5x + 6x^2}.$$

Since $1 - 5x + 6x^2 = (1 - 2x)(1 - 3x)$ we can decompose $U(x)$ into partial fractions as follows:

$$U(x) = \frac{-1}{1 - 2x} + \frac{1}{1 - 3x}.$$

Using the binomial expansions for $(1 - 2x)^{-1}$ and $(1 - 3x)^{-1}$ we obtain

$$U(x) - -(1 + 2x + (2x)^2 + \cdots) + (1 + 3x + (3x)^2 + \cdots),$$

so that

$$u_n = 3^n - 2^n.$$

A simple calculation shows that Mr Stackenboom was a millionaire at the end of thirteen years. $\square$

**Exercises
18.4**

1 Use the generating function method to find a formula for u_n when the sequence (u_n) is defined by

$$u_0 = 1, \qquad u_1 = 1, \qquad u_{n+2} - 4u_{n+1} + 4u_n = 0 \quad (n \geqslant 0).$$

2 Suppose $A(x)$ is the generating function for the sequence (a_n). What are the generating functions for the sequences (p_n), (q_n), and (r_n) defined as follows?

(i) $p_n = 5a_n$;　　(ii) $q_n = a_n + 5$;　　(iii) $r_n = a_{n+5}$.

3 Show that $x(1+x)/(1-x)^3$ is the generating function for the sequence whose nth term is n^2.

4   Let $A(x)$ be the generating function for the sequence (a_n) and define

$$s_n = a_0 + a_1 + \cdots + a_n \quad (n \geqslant 0).$$

Show that the generating function for (s_n) is

$$S(x) = \frac{A(x)}{1-x}.$$

Use this result in conjunction with Ex. 3 to find a formula for $\sum_{i=0}^{n} i^2$.

18.5   The homogeneous linear recursion

In Section 12.2 we found the explicit solution of the recursion

$$u_0 = c_0, \qquad u_1 = c_1, \qquad u_{n+2} + a_1 u_{n+1} + a_2 u_n = 0 \quad (n \geqslant 0).$$

This is the case $k = 2$ of the **homogeneous linear recursion** defined by the equations

[HLR] $\begin{cases} u_0 = c_0, \quad u_1 = c_1, \quad \ldots \quad, \quad u_{k-1} = c_{k-1}, \\ u_{n+k} + a_1 u_{n+k-1} + \cdots + a_k u_n = 0 \quad (n \geqslant 0). \end{cases}$

We shall now use the method of generating functions to derive the general solution to [HLR] for any value of k.

Theorem 18.5.1
The generating function for the sequence (u_n) defined by [HLR] is

$$U(x) = \frac{R(x)}{1 + a_1 x + \cdots + a_k x^k},$$

where $R(x)$ is a polynomial and $\deg R(x) < k$.

Proof
Consider the product

$(1 + a_1 x + \cdots + a_k x^k) U(x)$

$\qquad = (1 + a_1 x + \cdots + a_k x^k)(u_0 + u_1 x + \cdots + u_n x^n + \cdots).$

Using the multiplication rule, the coefficient of x^{n+k} is

$$u_{n+k} + a_1 u_{n+k-1} + \cdots + a_k u_n \quad (n \geqslant 0).$$

But since (u_n) satisfies [HLR], this is zero for all $n \geqslant 0$. The only coefficients which do not vanish are the coefficients of $1, x, \ldots, x^{k-1}$, and so the product is a polynomial $R(x)$ with $\deg R(x) < k$.

The non-vanishing coefficients of $R(x)$ may also be obtained by multiplying out the product given above:

$$R(x) = u_0 + (u_1 + a_1 u_0)x + \cdots + (u_{k-1} + a_1 u_{k-2} + \cdots + a_{k-1} u_0)x^{k-1}.$$

Since the values of $u_0, u_1, \ldots, u_{k-1}$ are specified explicitly by the equations [HLR] it follows that $R(x)$ may be determined by replacing u_i by c_i $(0 \leq i \leq k-1)$ in this formula. $\quad\square$

In Section 12.2 we showed that when $k = 2$ the form of the solution to a homogeneous linear recursion depends upon whether or not the roots α and β of the auxiliary equation

$$t^2 + a_1 t + a_2 = 0$$

are distinct. If $\alpha \neq \beta$ then the solution takes the form

$$u_n = A\alpha^n + B\beta^n,$$

whereas if $\alpha = \beta$ the solution takes the form

$$u_n = (Cn + D)\alpha^n.$$

In the general case we define the **auxiliary equation** for [HLR] to be

$$t^k + a_1 t^{k-1} + \cdots + a_k = 0,$$

and the form of the solution depends, as one would expect, on the behaviour of its roots.

We shall assume that the auxiliary equation has k roots, which will certainly be the case if we work in the field $\mathbb{C}$ of complex numbers. However, the k roots need not be distinct, and we shall suppose that the distinct values are $\alpha_1, \alpha_2, \ldots, \alpha_s$, occurring with multiplicities $m_1, m_2, \ldots, m_s$ respectively. In other words, the auxiliary equation can be rewritten as follows:

$$(t - \alpha_1)^{m_1}(t - \alpha_2)^{m_2} \cdots (t - \alpha_s)^{m_s} = 0,$$

where $m_1 + m_2 + \cdots + m_s = k$. Since the denominator of the formula for $U(x)$ given in Theorem 18.5.1 is obtained from the left-hand side of the auxiliary equation by dividing by t^k and replacing $1/t$ by x, it follows that $U(x)$ can be written as

$$U(x) = \frac{R(x)}{(1 - \alpha_1 x)^{m_1} \cdots (1 - \alpha_s x)^{m_s}}.$$

We can now apply the binomial theorem and obtain a general formula for u_n.

Theorem 18.5.2

Suppose (u_n) is defined by [HLR], and the auxiliary equation has roots $\alpha_1, \alpha_2, \ldots, \alpha_s$ with multiplicities $m_1, m_2, \ldots, m_s$. Then

$$u_n = P_1(n)\alpha_1^n + P_2(n)\alpha_2^n + \cdots + P_s(n)\alpha_s^n,$$

where, for $i = 1, 2, \ldots, s$, $P_i(n)$ is an expression of the form

$$A_0 + A_1 n + \cdots + A_{m_i-1} n^{m_i-1}.$$

In other words, $P_i(n)$ is a polynomial function of n with degree at most $m_i - 1$.

Proof

According to the theory of partial fractions developed in Section 18.2, $U(x)$ can be written as the sum of s expressions of the form

$$\frac{\gamma_1}{1-\alpha x} + \frac{\gamma_2}{(1-\alpha x)^2} + \cdots + \frac{\gamma_m}{(1-\alpha x)^m},$$

where, in each such expression, $\alpha = \alpha_i$ and $m = m_i$ for a specific value of i in the range $1 \leq i \leq s$. By the binomial theorem for negative integers, the coefficient of x^n in the corresponding power series is

$$\gamma_1 \binom{1+n-1}{n}\alpha^n + \gamma_2 \binom{2+n-1}{n}\alpha^n + \cdots + \gamma_m \binom{m+n-1}{n}\alpha^n.$$

Simplifying, and using the fact that $\binom{n+l}{n} = \binom{n+l}{l}$, we can write this in the form $P(n)\alpha^n$, where

$$P(n) = \gamma_1 \binom{n}{0} + \gamma_2 \binom{n+1}{1} + \cdots + \gamma_m \binom{n+m-1}{m-1}.$$

Now the formula

$$\binom{n+l}{l} = \frac{(n+l)(n+l-1)\cdots(n+1)}{l(l-1)\cdots 1}$$

shows that $\binom{n+l}{l}$ is a polynomial function of n with degree l. Hence $P(n)$ is a polynomial function of n with degree at most $m-1$, as required. $\qquad\square$

In practice, we use Theorems 18.5.1 and 18.5.2 as the basis for determining a formula for u_n. But we do not need to find the generating function or its partial fraction decomposition explicitly. We simply assume the final form of the result and obtain the coefficients of the polynomial $P_i(n)$ by substituting the initial values of $u_0, u_1, \ldots, u_{k-1}$.

Example

Find a formula for the nth term of the sequence defined by

$$u_0 = 0, \qquad u_1 = -9, \qquad u_2 = -1, \qquad u_3 = 21,$$

$$u_{n+4} - 5u_{n+3} + 6u_{n+2} + 4u_{n+1} - 8u_n = 0 \quad (n \geqslant 0).$$

Solution

The auxiliary equation is

$$t^4 - 5t^3 + 6t^2 + 4t - 8 = 0,$$

which is the same as

$$(t-2)^3(t+1) = 0.$$

Hence u_n is given by a formula

$$u_n = (An^2 + Bn + C)2^n + D(-1)^n,$$

where A, B, C, D are constants. Substituting the given values of u_0, u_1, u_2, u_3 we obtain the equations

$$\begin{aligned} C + D &= 0, \\ 2A + 2B + 2C - D &= -9, \\ 16A + 8B + 4C + D &= -1, \\ 72A + 24B + 8C - D &= 21. \end{aligned}$$

The standard methods of solving linear equations yield the solution $A = 1, B = -1, C = -3, D = 3$. Hence

$$u_n = (n^2 - n - 3)2^n + 3(-1)^n. \qquad \square$$

Exercises 18.5

1 Use the auxiliary equation method to find a formula for u_n when the sequence (u_n) is defined by:

(i) $u_0 = 1, \qquad u_1 = 3, \qquad u_{n+2} - 3u_{n+1} - 4u_n = 0 \quad (n \geqslant 0)$;

(ii) $u_0 = 2, \qquad u_1 = 0, \qquad u_2 = -2,$

$\qquad u_{n+3} - 6u_{n+2} + 11u_{n+1} - 6u_n = 0 \quad (n \geqslant 0)$;

(iii) $u_0 = 1, \qquad u_1 = 0, \qquad u_2 = 0,$

$\qquad u_{n+3} - 3u_{n+1} + 2u_n = 0 \quad (n \geqslant 0).$

2 Professor McBrain climbs stairs in an erratic fashion. Sometimes he takes two stairs in one stride, sometimes only one. Find a formula for b_n, the number of different ways in which he can climb n stairs.

3   Suppose the sequence (z_n) is defined by

$$z_0 = 1, \qquad z_{n+1} = \frac{z_n - a}{z_n - b} \quad (n \geqslant 0),$$

where a and b are real numbers and $b \neq 1$. Show that if the sequence (u_n) satisfies

$$u_{n+1}/u_n = z_n - b,$$

then

$$u_{n+2} + (b-1)u_{n+1} + (a-b)u_n = 0 \quad (n \geqslant 0).$$

Hence find a formula for z_n when $a = 0$ and $b = 2$.

4   Let (u_n), (v_n), (w_n) be the sequences defined by $u_0 = v_0 = w_0 = 1$ and

$$\begin{bmatrix} u_{n+1} \\ v_{n+1} \\ w_{n+1} \end{bmatrix} = \begin{bmatrix} 1 & 0 & 1 \\ 0 & -1 & 1 \\ 1 & -1 & 4 \end{bmatrix} \begin{bmatrix} u_n \\ v_n \\ w_n \end{bmatrix} \quad (n \geqslant 0).$$

Show that (u_n) satisfies a homogeneous linear recursion and hence find a formula for u_n.

18.6   Nonhomogeneous linear recursions

In some cases the method of generating functions can be used to solve the *nonhomogeneous* linear recursion

$$u_0 = c_0, \quad u_1 = c_1, \quad \ldots \quad , \quad u_{k-1} = c_{k-1},$$
$$u_{n+k} + a_1 u_{n+k-1} + \cdots + a_k u_n = f(n) \quad (n \geqslant 0).$$

The applicability of the method depends upon the particular form of $f(n)$. Roughly speaking, the technique is to work out

$$(1 + a_1 x + a_2 x^2 + \cdots + a_k x^k)u(x)$$

and hope that the terms involving $f(n)$ can be dealt with by some special device.

Example   Find an explicit formula for the terms of the sequence (u_n) defined by the recursion

$$u_0 = 0, \qquad u_1 = 1, \qquad u_{n+2} - u_{n+1} - 6u_n = n \quad (n \geqslant 0).$$

Solution   Using the technique proposed above, we calculate

$$(1-x-6x^2)(u_0+u_1x+u_2x^2+\cdots)$$

$$= u_0+(u_1-u_0)x+(u_2-u_1-6u_0)x^2+\cdots+(u_{n+2}-u_{n+1}-6u_n)x^{n+2}$$
$$+\cdots$$

$$= x+(x^3+2x^4+\cdots+nx^{n+2}+\cdots)$$

$$= x+x^3(1-x)^{-2}.$$

Since $1-x-6x^2=(1+2x)(1-3x)$, the generating function $U(x)$ satisfies

$$(1+2x)(1-3x)U(x)=\frac{x-2x^2+2x^3}{(1-x)^2},$$

and so

$$U(x)=\frac{x-2x^2+2x^3}{(1+2x)(1-3x)(1-x)^2}$$

$$=\frac{A}{1+2x}+\frac{B}{1-3x}+\frac{C}{1-x}+\frac{D}{(1-x)^2}.$$

for some constants A, B, C, D.

There are several ways of obtaining the constants: one way is to use the cover-up rule to find A, B, D by substituting $x=-\frac{1}{2}$, $x-\frac{1}{3}$, $x=1$ respectively. this gives

$$A=-\tfrac{2}{9}, \qquad B=\tfrac{1}{4}, \qquad D=-\tfrac{1}{6}.$$

Then C can be found by putting $x=0$ (equivalent to multiplying up and equating the constant terms), which gives

$$A+B+C+D=0,$$

and consequently $C=\tfrac{5}{36}$. Another way is to notice that the formula for $U(x)$ implies that

$$u_n = A(-2)^n + B(3^n)+C+D(n+1),$$

and use the values $u_0=0$, $u_1=1$, $u_2=1$, $u_3=8$ to determine A, B, C, D. The equations are:

$$0= \quad A+ \quad B+C+ \quad D,$$
$$1=-2A+ \ 3B+C+2D,$$
$$1= \ 4A+ \ 9B+C+3D,$$
$$8=-8A+27B+C+4D,$$

which yield the solution $A=-\tfrac{2}{9}$, $B=\tfrac{1}{4}$, $C=\tfrac{5}{36}$, $D=-\tfrac{1}{6}$. Rearranging, we obtain the formula

$$u_n = \tfrac{1}{36}[(-2)^{n+3} + 3^{n+2} - 6n - 1].$$ □

There are many other types of problem which can be dealt with by the generating function method. It is particularly enhanced when combined with analytic methods of dealing with power series, but that is beyond the scope of this book. Fortunately, some very important results can be obtained by the purely algebraic means already at our disposal, and several results of this kind will be discussed in the next two chapters.

Exercises 18.6

1 Show that the generating function for the sequence (u_n) defined by the recursion

$$u_0 = 1, \qquad u_{n+1} - 2u_n = 4^n \quad (n \geq 0)$$

is

$$U(x) = \frac{1 - 3x}{(1 - 2x)(1 - 4x)}.$$

Deduce that $u_n = 2^{2n-1} + 2^{n-1}$.

2 Let q_n be the number of words of length n in the alphabet $\{a, b, c, d\}$ which contain an odd number of b's. Show that

$$q_{n+1} = 4^n + 2q_n \quad (n \geq 1).$$

[Hint: partition the set of such words of length $n+1$ into those which begin with b and those which do not.] Hence find the generating function $Q(x)$ for (q_n), assuming $q_0 = 0$, and show that

$$q_n = \tfrac{1}{2}(4^n - 2^n).$$

3 Show that the generating function for the sequence defined by

$$u_0 = 1, \qquad u_{n+1} - 2u_n = n\alpha^n \quad (n \geq 0)$$

is

$$u(x) = \frac{1}{1 - 2x} + \frac{\alpha x^2}{(1 - 2x)(1 - \alpha x)^2},$$

provided $\alpha \neq 2$. Hence find a formula for u_n. What happens when $\alpha = 2$?

18.7 Miscellaneous Exercises

1   Which of the following are invertible in $\mathbb{R}[\![x]\!]$, and which of them are invertible in $\mathbb{Z}[\![x]\!]$?

$$\text{(i)}\ 1+x; \qquad \text{(ii)}\ x^2; \qquad \text{(iii)}\ 3+2x^3.$$

2   Write down the first four terms and the general term in the power series for $(1+x)^{-5}$.

3   Find the coefficient of x^n in the power series for

$$\frac{26-60x+25x^2}{(1-2x)(1-5x)^2}.$$

4   Find the first four terms and the coefficient of x^n in the power series for

$$\frac{1-x-x^2}{(1-2x)(1-x)^2}.$$

5   Find a formula for the coefficient of x^n in the power series for $(1+x)/(1+x+x^3)$ in $\mathbb{Z}_2[\![x]\!]$.

6   Find the partial fraction decomposition over $\mathbb{Z}_5$ of

$$\frac{4x+2}{x^3+2x^2+4x+3}.$$

7   Use the method of Ex. 18.4.4 to find formulae for $\sum i^3$ and $\sum i^4$, where the sums are taken over the range $1 \leqslant i \leqslant n$.

8   If $A(x)$ is the generating function for the sequence (a_n), define the *derivative* $A'(x)$ to be the generating function for the sequence (a_n') where $a_n' - (n+1)a_{n+1}$ $(n \geqslant 0)$. Show that

$$(AB)'(x) = A'(x)B(x) + A(x)B'(x).$$

9   By taking the derivative of the binomial expansion show that

$$\binom{n}{1} + 2\binom{n}{2} + 3\binom{n}{3} + \cdots + n\binom{n}{n} = 2^{n-1}n.$$

10   Use the formula $(1-x^2)^{-n} = (1-x)^{-n}(1+x)^{-n}$ to establish the identity

$$\sum_{i=0}^{r} (-1)^i \binom{n+r-i-1}{r-i}\binom{n+i-1}{i} = \begin{cases} 0 & \text{if } r \text{ is odd} \\ \binom{n+\frac{1}{2}r-1}{\frac{1}{2}r} & \text{if } r \text{ is even.} \end{cases}$$

11   Find an expression for $(1-x)^{-n}(1-x^k)^n$ as a polynomial of degree $n(k-1)$, and hence derive an identity

$$\binom{r}{n-1}\binom{n}{0} - \binom{r-k}{n-1}\binom{n}{1} + \binom{r-2k}{n-1}\binom{n}{2} - \cdots = 0,$$

valid when $r \geqslant nk$.

12   Write down the generating function for the sequence (u_n) whose terms represent the number of ways of distributing n different books to four people.

13   Let c_r be the number of ways in which the total r can be obtained when four dice are thrown. Show that the generating function for the sequence (c_r) is

$$C(x) = (x + x^2 + x^3 + x^4 + x^5 + x^6)^4.$$

14   Write down the generating function for the numbers b_r of integers n in the range $0 \leqslant n \leqslant 10^m - 1$ for which the sum of the digits in the base 10 representation is r.

15   Use the generating function method to solve the following recursion

$$u_0 = 2, \qquad u_1 = -6, \qquad u_{n+2} + 8u_{n+1} - 9u_n = 8(3^{n+1}) \quad (n \geqslant 0).$$

16   Find the general form of the solution to the recursion

$$y_{n+2} - 6y_{n+1} + 9y_n = 2^n + n \quad (n \geqslant 0).$$

17   When $k \geqslant 0$ let $f(n, k)$ be the number of k-subsets of $\{1, 2, \ldots, n\}$ which do not contain two successive integers. Show that

$$f(n, k) = f(n-2, k-1) + f(n-1, k).$$

Let $F_k(x)$ be the generating function of the numbers $f(n, k)$ for fixed k. Find a recursion for $F_k(x)$ and deduce that

$$f(n, k) = \binom{n-k+1}{k}.$$

18   Use the generating function method to solve the recursion

$$u_0 = 1, \qquad u_{n+1} = 3u_n + 2^{n-1} \quad (n \geqslant 0).$$

19   The *exponential* generating function for the sequence (u_n) is defined to be the power series

$$\tilde{u}(x) = u_0 + \frac{u_1}{1!} x + \frac{u_2}{2!} x^2 + \cdots + \frac{u_i}{i!} x^i + \cdots.$$

Use the recursion $d_n = nd_{n-1} + (-1)^n$ for the derangement numbers to show that the exponential generating function for (d_n) is $e^{-x}/(1-x)$. Hence derive the usual explicit formula for d_n.

20   Let $\tilde{Q}(x)$ be the exponential generating function for the numbers q_n, where q_n is the number of partitions of an n-set. Use the formula obtained in Ex. 5.7.10 to prove that

$$\tilde{Q}(x) = \exp(e^x - 1).$$

19
Partitions of a positive integer

19.1 Partitions and diagrams

Our study of the partitions of a positive integer began in Section 5.4, where we remarked that the problem of calculating the number $p(n)$ of partitions of n is not a simple one. Now that we have developed a range of techniques—in particular, the method of generating functions—we can attack the problem with renewed hope.

Let us begin by recalling the standard notation

$$[1^{\alpha_1}2^{\alpha_2}\ldots n^{\alpha_n}]$$

for a partition of n which has α_i parts of size i ($1 \le i \le n$). The partitions of 5, and their symbols in standard notation are:

5	$[5]$
$4+1$	$[14]$
$3+2$	$[23]$
$3+1+1$	$[1^23]$
$2+2+1$	$[12^2]$
$2+1+1+1$	$[1^32]$
$1+1+1+1+1$	$[1^5]$.

Hence the number of partitions of 5 is $p(5) = 7$. Some other values of $p(n)$ are

$$p(10) = 42$$
$$p(20) = 627$$
$$p(100) = 190\,569\,292$$
$$p(200) = 3\,972\,999\,029\,388.$$

Although there is no simple formula for $p(n)$, it is possible to obtain a very complicated exact formula by analytic techniques involving the generating function for the sequence $(p(n))$. But the generating function can also be used to set up an effective recursion for $p(n)$, and this is the approach which we shall adopt. The recursion will enable us to compute any given value, such as $p(200)$, by means of routine arithmetical operations.

We shall find it helpful to use a diagrammatic representation of partitions. In this representation the parts are arranged in order, with the largest part first, and each part is represented by a row of the appropriate number of marks. For example, the diagrams

$$
\begin{array}{ccccc}
\times & \times & \times & \times & \times \\
\times & \times & \times & & \\
\times & \times & & &
\end{array}
\qquad
\begin{array}{cccccc}
\times & \times & \times & \times & \times & \times \\
\times & \times & \times & \times & \times & \times \\
\times & \times & \times & & & \\
\times & & & & & \\
\times & & & & &
\end{array}
$$

represent the partitions

$$5+3+2 \qquad\qquad 6+6+3+1+1.$$

(Such a diagram is often called a *Ferrers diagram*, or a *Ferrers graph*, after its inventor N. M. Ferrers. His name is frequently misspelled.)

This simple idea is very useful for proving theorems about partitions. A good example is the following result.

Theorem 19.1 Let n and r be any positive integers. The number of partitions of n in which there are not more than r parts is equal to the number of partitions of $n+r$ in which there are exactly r parts.

Proof Suppose we are given a partition of $n+r$ with exactly r parts. Its diagram has exactly r marks in the first column, and if we delete this column we obtain the diagram of a partition of n into at most r parts. Conversely, if a partition of the second kind is given, we can add a new first column of r marks and obtain a partition of the first kind. In other words, there is bijection between the sets of partitions of the two kinds, and so they have the same cardinality. (The bijection is illustrated in Fig. 19.1.) ☐

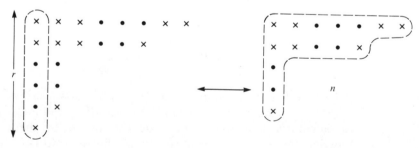

Fig. 19.1
Illustrating the proof of
Theorem 19.1.

If $\mathcal{P}$ is a property of partitions we shall use the notation $p(n \mid \mathcal{P})$ to denote the number of partitions of n which satisfy the property $\mathcal{P}$. With this notation, Theorem 19.1 can be expressed in the form

$$p(n \mid \text{number of parts} \leqslant r) = p(n+r \mid \text{number of parts} = r).$$

Exercises 19.1

1 Write down the diagrams representing the following partitions:

(i) $[1^2 3^3 5\ 7]$, (ii) $[2\ 4\ 6^2\ 7]$.

2 Find the values of $p(n)$, $1 \leqslant n \leqslant 7$, by listing all the possibilities.

3 Let

$$p_k(n) = p(n \mid \text{number of parts} = k).$$

Show that the following formula (obtained in Ex. 5.4.2) is a consequence of Theorem 19.1,

$$p_k(n) = p_k(n-k) + p_{k-1}(n-k) + \cdots + p_1(n-k),$$

and use it to calculate $p(8)$.

4 Use an argument based on diagrams to prove that the number of partitions of n which have at most m parts is equal to the number of partitions of $n + \frac{1}{2}m(m+1)$ in which there are m parts, all of them different. [Hint: add a 'triangle' with $\frac{1}{2}m(m+1)$ marks.]

19.2 Conjugate partitions

The diagrammatic representation of partitions suggests the following simple method of transforming one partition into another: switch the rows and columns of the diagram. For example, the partition $\lambda = [1^2 2 3^2 4]$ is transformed into the partition $\lambda' = [1\ 3\ 4\ 6]$, as the diagram indicates.

$$
\begin{array}{cccc}
\times & \times & \times & \times \\
\times & \times & \times & \\
\times & \times & \times & \\
\times & \times & & \\
\times & & & \\
\times & & & \\
\end{array}
\quad \rightarrow \quad
\begin{array}{cccccc}
\times & \times & \times & \times & \times & \times \\
\times & \times & \times & \times & & \\
\times & \times & \times & & & \\
\times & & & & & \\
\end{array}
$$

In general, partitions λ and λ' related in this way are said to be **conjugate**.

The transformation $\lambda \to \lambda'$ which takes a partition to its conjugate can be used to prove some useful results. A typical example is as follows. If the largest part of λ is m, then the first row of the diagram for λ and the first column of the diagram for λ' both contain m marks. Thus λ' is a partition with m parts. In other words, the transformation $\lambda \to \lambda'$ is a bijection from the set of partitions of n with largest part equal to m to the set of partitions of n with exactly m parts. So we have proved that

$$p(n \mid \text{largest part} = m) = p(n \mid \text{number of parts} = m).$$

A partition λ is said to be **self-conjugate** if $\lambda = \lambda'$.

Theorem 19.2 The number of self-conjugate partitions of n is equal to the number of partitions of n whose parts are distinct and odd.

Proof The first row and first column of the diagram for a self-conjugate partition of n contain the same number of marks, say k. Since they have one common mark, the total number of marks in the first row and column is the odd number $2k - 1$. In the same way, when the first row and column are removed, the remainder of the second row and column contains an odd number of marks, say $2l - 1$. Continuing, we obtain a partition of n in which each part is odd, and the parts are all distinct. The procedure may be visualized as a 'straightening' of the L-shaped sections of the given self-conjugate partition (Fig. 19.2)

Conversely, if we are given a partition of n with parts which are odd and distinct, then we can 'fold up' the diagram to get a self-conjugate partition of n. Hence the numbers of partitions of the two kinds are equal. $\square$

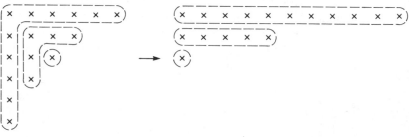

Fig. 19.2
Illustrating the proof of
Theorem 19.2.

Exercises 19.2

1 Write down the conjugates of the following partitions in standard notation:

$$\text{(i)} \quad [1^2356], \qquad \text{(ii)} \quad [2^23^358].$$

2 Which of the following partitions are self-conjugate?

$$\text{(i)} \quad [1^224], \qquad \text{(ii)} \quad [1^3456], \qquad \text{(iii)} \quad [2^235^2], \qquad \text{(iv)} \quad [1^42348].$$

3 Use the method described in Theorem 19.2 to list the self-conjugate partitions of 20.

4 Show that the number of self-conjugate partitions of n with largest part equal to k is the same as the number of self-conjugate partitions of $n - 2k + 1$ with largest part not exceeding $k - 1$. Hence find the number of self-conjugate partitions of 41 in which the largest part is 11.

19.3 Partitions and generating functions

In this section we shall obtain a formula for the generating function

$$P(x) = p(0) + p(1)x + p(2)x^2 + \cdots$$

where $p(n)$ is the number of partitions of n and, *by convention*, $p(0) = 1$.

The starting point is the formula

$$(1 - x^i)^{-1} = 1 + x^i + x^{2i} + x^{3i} + \cdots,$$

which should be regarded as an equation for the inverse of $1 - x^i$ in the ring of power series $\mathbb{C}[[x]]$. The formula can be obtained by direct calculation, or by replacing x by x^i ($i \geqslant 1$) in the standard formula for $(1 - x)^{-1}$. For our purposes, the key observation is that this power series is the generating function for the numbers f_n of partitions of a very simple kind:

$$f_n = p(n \mid \text{each part is equal to } i).$$

For clearly there are no such partitions unless n is a multiple of i, when there is just one partition $n = i + i + \cdots + i$. In other words

$$f_n = \begin{cases} 1 & \text{if } n = \alpha i \quad (\alpha = 0, 1, 2, \ldots), \\ 0 & \text{otherwise.} \end{cases}$$

Hence the formula tells us that the generating function for (f_n) is

$$F(x) = (1 - x^i)^{-1}.$$

Suppose we are given two different positive integers i and j, and let

$$h_n = p(n \mid \text{each part is equal to } i \text{ or } j).$$

What is the generating function $H(x)$ for the sequence (h_n)?

Let f_n and $F(x)$ be as above, and let g_n denote the number of partitions of n with each part equal to j. We have shown that $F(x) = (1 - x^i)^{-1}$, and by the same arguments the generating function $G(x)$ for (g_n) is equal to $(1 - x^j)^{-1}$. Now we remark that h_n is the number of ways of writing n as the sum of r and $n - r$, where r is partitioned into i's and $n - r$ is partitioned into j's. Thus h_n is a sum of terms $f_r g_{n-r}$, that is

$$h_n = f_0 g_n + f_1 g_{n-1} + \cdots + f_n g_0.$$

By the rule for multiplying power series,

$$H(x) = F(x)G(x) = (1 - x^i)^{-1}(1 - x^j)^{-1}.$$

For example, suppose $i = 2$ and $j = 3$. The product of the power series for $(1 - x^2)^{-1}$ and $(1 - x^3)^{-1}$ is

$$(1 + x^2 + x^4 + x^6 + x^8 + x^{10} + \cdots)(1 + x^3 + x^6 + x^9 + \cdots)$$
$$= 1 + x^2 + x^3 + x^4 + x^5 + 2x^6 + x^7 + x^8 + 2x^9 + 2x^{10} + \cdots.$$

The coefficient of x^{10} is 2, arising from the products $x^4 \times x^6$ amd $x^{10} \times 1$. These products correspond to the sums

$$10 = 4 + 6 = 2 + 2 + 3 + 3, \qquad 10 = 10 + 0 = 2 + 2 + 2 + 2 + 2,$$

which, in turn, correspond to the two ways of writing 10 as a partition with each part equal to 2 or 3.

It is now easy to see how to write down generating functions for the numbers of partitions in which each part is equal to one of a given set of numbers. For example, if

$$a_n = p(n \mid \text{each part is equal to } i \text{ or } j \text{ or } k)$$

then the corresponding generating function is

$$A(x) = (1 - x^i)^{-1}(1 - x^j)^{-1}(1 - x^k)^{-1}.$$

Example   In how many ways can a dollar (100 cents) be exchanged for quarters (25 cents), dimes (10 cents), and nickels (5 cents)?

Solution The question asks for the number of partitions of 100 into parts of size 5, 10, or 25. Thus the solution is the coefficient of x^{100} in

$$(1-x^5)^{-1}(1-x^{10})^{-1}(1-x^{25})^{-1}.$$

Clearly we can substitute $y = x^5$ and find the coefficient of y^{20} in $(1-y)^{-1}(1-y^2)^{-1}(1-y^5)^{-1}$. Also, we need never consider the terms beyond y^{20}. So first we calculate $(1-y^2)^{-1}(1-y^5)^{-1}$, that is

$$(1+y^2+y^4+y^6+y^8+y^{10}+y^{12}+y^{14}+y^{16}+y^{18}+y^{20}+\cdots)$$
$$\times(1+y^5+y^{10}+y^{15}+y^{20}+\cdots).$$

A good way of doing this is to 'detach' the coefficients, and lay out the working as in Table 19.3.1, where the first row contains the coefficients of the first factor and the other rows correspond to multiplication by the terms of the second factor.

Table 19.3.1

$1\ y\ y^2 y^3\ y^4$	$y^5\ \ldots$	$y^{10}\ \ldots$	$y^{15}\ \ldots$	y^{20}	
1 0 1 0 1	0 1 0 1 0	1 0 1 0 1	0 1 0 1 0	1	
	1 0 1 0 1	0 1 0 1 0	1 0 1 0 1	0	$(\times y^5)$
		1 0 1 0 1	0 1 0 1 0	1	$(\times y^{10})$
			1 0 1 0 1	0	$(\times y^{15})$
				1	$(\times y^{20})$
1 0 1 0 1	1 1 1 1 1	2 1 2 1 2	2 2 2 2 2	3	

Thus $(1-y^2)^{-1}(1-y^5)^{-1}$ is equal to

$$1+y^2+y^4+y^5+y^6+y^7+y^8+y^9+2y^{10}+y^{11}+2y^{12}+y^{13}+2y^{14}$$
$$+2y^{15}+2y^{16}+2y^{17}+2y^{18}+2y^{19}+3y^{20}+\cdots).$$

The term $(1-y)^{-1}$ is equal to $(1+y+y^2+\cdots+y^{20}+\cdots)$, so multiplying by it corresponds to adding all the coefficients obtained above. Hence the required number is 29. $\qquad\square$

Of course, the use of algebra in the *Example* is rather bogus, since we could have carried out the same calculations using the original, non-algebraic, terminology of the problem. However, the algebraic framework has the advantage of generality and the potential for further development. In particular, we can use it to obtain the generating function for the numbers $p(n)$ of unrestricted partitions.

**Theorem
19.3**

The generating function for the numbers $p(n)$ of partitions of n can be written as the infinite product

$$P(x) = \prod_{i=1}^{\infty} (1 - x^i)^{-1}.$$

Proof

Let m be a fixed positive integer and let $p^{(m)}(n)$ denote the number of partitions of n in which each part is one of the integers $1, 2, \ldots, m$. Then $p^{(m)}(n)$ is the number of ways of expressing n as a sum

$$n = s_1 + s_2 + \cdots + s_m,$$

where each s_i $(1 \leqslant i \leqslant m)$ is itself a sum of a number of i's. This is the same as the number of ways of choosing, for $i = 1, 2, \ldots, m$, a term x^{s_i} from each of the power series $(1 - x^i)^{-1}$ in such a way that the product of the terms is x^n. In other words, the generating function for the numbers $p^{(m)}(n)$ is

$$P^{(m)}(x) = (1 - x)^{-1}(1 - x^2)^{-1} \cdots (1 - x^m)^{-1}.$$

For any given value of n we have $p(n) = p^{(n)}(n)$, since a partition of n cannot contain any parts greater than n. This corresponds to the fact that the terms

$$(1 - x^i)^{-1} = 1 + x^i + x^{2i} + \cdots$$

with $i > n$ do not contribute to the coefficient of x^n in the infinite product $P(x)$. In other words, the coefficient of x^n in $P(x)$ is the same as the coefficient of x^n in $P^{(n)}(x)$, and this is just $p^{(n)}(n) = p(n)$. Thus $P(x)$ is the generating function for the sequence $(p(n))$. $\square$

As indicated in the last paragraph of the proof, the appearance of an infinite number of factors in the product $P(x)$ is not a real difficulty, because in order to calculate any given coefficient it is necessary to consider only a finite number of factors. We shall encounter some other infinite products of power series in the following sections, and they will all share this property. For that reason, we shall make no further reference to the supposed difficulties of dealing with expressions of this kind.

**Exercises
19.3**

1   Write down the generating functions for the sequences whose nth terms are:

(i) the number of partitions of n into parts equal to 3 or 5;

(ii) the number of partitions of n into parts equal to 2, 4, or 6.

2   By multiplying the relevant power series, determine the coefficient of x^9 in

$$\frac{1}{(1-x)(1-x^2)(1-x^3)}.$$

Interpret your result as the number of partitions of a certain kind, and check your answer by listing the partitions explicitly.

3   In how many ways can one pound (100 p) be exchanged for coins of the values 50p, 20p, 10p, and 5p?

4   In how many ways can a total weight of 15 kg be made up from weights of 1 kg, 2 kg, and 4 kg?

19.4   Generating functions for restricted partitions

Simple modifications of the arguments given in Section 19.3 enable us to write down generating functions for the numbers of partitions satisfying various restrictions. For instance, suppose we stipulate that each part must occur no more than k times. Then the contributions from the parts equal to i correspond to the terms of the polynomial

$$1+x^i+x^{2i}+\cdots+x^{ki},$$

rather than the power series $(1-x^i)^{-1}=1+x^i+x^{2i}+\cdots$, and so the generating function is

$$(1+x+\cdots+x^k)(1+x^2+\cdots+x^{2k})(1+x^3+\cdots+x^{3k})\cdots.$$

Since

$$1+x^i+x^{2i}+\cdots+x^{ki}=(1-x^{(k+1)i})/(1-x^i),$$

we can write this generating function in either of the forms

$$\prod_{i=1}^{\infty}(1+x^i+\cdots+x^{ki})=\prod_{i=1}^{\infty}\left(\frac{1-x^{(k+1)i}}{1-x^i}\right).$$

In particular when $k=1$, we have the generating function for the numbers of partitions with *distinct* parts. This is the first entry in Table 19.4.1, a short table of useful generating functions.

Table 19.4.1

u_n	$U(x)$
$p(n \mid$ each part is distinct)	$(1+x)(1+x^2)(1+x^3)\cdots$
$p(n \mid$ each part is odd)	$\dfrac{1}{(1-x)(1-x^3)(1-x^5)\cdots}$
$p(n \mid$ each part is even)	$\dfrac{1}{(1-x^2)(1-x^4)(1-x^6)\cdots}$
$p(n \mid$ each part is $\leqslant m$)	$\dfrac{1}{(1-x)(1-x^2)\cdots(1-x^m)}$

Elementary algebraic manipulations of such formulae can sometimes lead to unexpected results.

Example Use the method of generating functions to prove that

$$p(n \mid \text{each part is distinct}) = p(n \mid \text{each part is odd}).$$

Solution Let $D(x)$, $O(x)$ be the respective generating functions. We have

$$D(x) = (1+x)(1+x^2)(1+x^3)\cdots,$$
$$O(x) = (1-x)^{-1}(1-x^3)^{-1}(1-x^5)^{-1}\cdots.$$

In order to show that these two power series are the same, we use the identity

$$1+y = (1-y^2)/(1-y).$$

Thus

$$D(x) = \frac{(1-x^2)(1-x^4)(1-x^6)\cdots}{(1-x)(1-x^2)(1-x^3)\cdots}$$

$$= \frac{1}{(1-x)(1-x^3)(1-x^5)\cdots},$$

since each term of the form $1-x^{2i}$ can be cancelled. Hence $D(x) = O(x)$, and it follows that the two sequences of coefficients are the same. $\square$

Exercises 19.4

1 Write down formulae for the generating functions for the sequences whose nth terms are

(i) $p(n \mid$ each part occurs at most twice),

(ii) $p(n \mid$ each part is power of 2),

(iii) $p(n \mid$ the smallest part is 5).

2 Use the method of generating functions to find the number of partitions of 16 in which each part is an odd prime.

3 Write down the generating function for the sequence whose nth term is

$$p(n \mid \text{no even number occurs more than once as a part}).$$

Hence show that this number is equal to

$$p(n \mid \text{each part occurs at most 3 times}).$$

[Hint: $(1-y^4)=(1-y)(1+y+y^2+y^3)$.]

4 Show that

$$(1-x)(1+x)(1+x^2)(1+x^4) \cdots (1+x^{2^m}) = 1 - x^{2^{m+1}},$$

and derive the formula

$$(1-x)^{-1} = (1+x)(1+x^2)(1+x^4) \cdots ,$$

where the right-hand side is the product of the factors $1+x^{2^r}$ for all $r \geqslant 0$. Deduce that every positive integer has a unique partition whose parts are distinct powers of 2.

19.5 A mysterious identity

We have shown that the generating function for the numbers of partitions with distinct parts is

$$D(x) = (1+x)(1+x^2)(1+x^3) \cdots .$$

For example, the partition $7 = 1+2+4$ corresponds to the term $x \times x^2 \times x^4$ and contributes 1 to the coefficient of x^7. Now consider what happens when we change all the + signs to −, so that we get the power series

$$Q(x) = (1-x)(1-x^2)(1-x^3) \cdots$$
$$= 1 + q_1 x + q_2 x^2 + \cdots .$$

Here again each partition of n with distinct parts makes a contribution to the coefficient of x^n, but now the contribution is $(-1)^d$, where d is the number of parts. For example, the partition $7 = 1+2+4$ corresponds to the term

$$(-x) \times (-x^2) \times (-x^4) = (-1)^3 x^7$$

and contributes -1 to x^7. In general, each partition of n with an even number of distinct parts contributes 1 to q_n, and each partition with an odd number of distinct parts contributes -1. Thus

$$q_n = e_n - o_n,$$

where

$$e_n = p(n \mid \text{the parts are distinct and even}),$$
$$o_n = p(n \mid \text{the parts are distinct and odd}).$$

It is instructive to work out the first few terms of $Q(x)$ by multiplying the factors $1-x^i$ in turn, for $i = 1, 2, \ldots$. The reader who does this conscientiously (and correctly) up to $i = 26$ will find that

$$Q(x) = 1 - x - x^2 + x^5 + x^7 - x^{12} - x^{15} + x^{22} + x^{26} - \cdots.$$

There are several mysterious features. Many of the coefficients turn out to be 0, and the non-zero coefficients are all 1 or -1; furthermore, the signs appear to change regularly, two $+$ signs being followed by two $-$ signs, and so on. We shall use the combinatorial interpretation of q_n given above to explain this curious behaviour.

Theorem 19.5

If e_n and o_n are as above, then

$$e_n - o_n = \begin{cases} (-1)^m & \text{if } n = \tfrac{1}{2}m(3m \pm 1) \quad (m \geqslant 1), \\ 0 & \text{otherwise.} \end{cases}$$

Proof

We shall set up a transformation $\lambda \to \lambda^*$ which takes the partitions counted by e_n into those counted by o_n. For most values of n this transformation will be a bijection, so that $e_n = o_n$, but when $n = \tfrac{1}{2}m(3m - 1)$ or $n = \tfrac{1}{2}m(3m + 1)$ there will be one exceptional partition which causes the purported bijection to fail.

If λ is any partition with distinct parts, define $s(\lambda)$ to the smallest part of λ. Also, define $t(\lambda)$ to be the length of the sequence of parts which begins with the largest part and continues for as long as the parts decrease by 1 at each step. In the diagram for λ, $t(\lambda)$ is represented by the sequence of marks beginning at the top right-hand corner and continuing in a south-westerly direction. For example, when $\lambda = [13567]$ we have $s(\lambda) = 1$ and $t(\lambda) = 3$ (Figure 19.3).

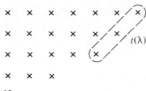

Fig. 19.3
When $\lambda = [13567]$, $s(\lambda) = 1$
and $t(\lambda) = 3$.

When λ is given, the definition of λ^* depends on the relative sizes of $s(\lambda)$ and $t(\lambda)$.

Case 1: $s(\lambda) \leqslant t(\lambda)$. In this case we obtain λ^* by removing the smallest part $s(\lambda)$ and adding 1 to each of the $s(\lambda)$ largest parts (Fig. 19.4).

Fig. 19.4
The transformation $\lambda \to \lambda^*$
when $s(\lambda) \leqslant t(\lambda)$.

Case 2: $s(\lambda) > t(\lambda)$. In this case we obtain λ^* by removing 1 from the $t(\lambda)$ largest parts and creating a new smallest part $t(\lambda)$, as illustrated in Fig. 19.5.

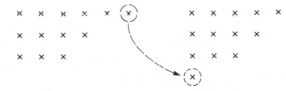

Fig. 19.5
The transformation $\lambda \to \lambda^*$
when $s(\lambda) > t(\lambda)$.

In both cases the transformation $\lambda \to \lambda^*$ changes the number of parts by 1, and so it takes a partition with an odd number of parts into one with an even number of parts, and vice versa. Thus, provided λ^* is always defined and is a partition with distinct parts, the transformation $\lambda \to \lambda^*$ is a bijection. There are just two situations where the definition breaks down.

In *Case 1* it may happen that the sections of the diagram representing $s(\lambda)$ and $t(\lambda)$ overlap (Fig. 19.6). In this case the rule yields a diagram for λ^* which is not a legal diagram for a partition. This happens precisely when $s(\lambda) = m$ and there are m parts: m, $m + 1, \ldots, 2m - 1$. Thus

$$n = m + (m + 1) + \cdots + (2m - 1) = \tfrac{1}{2}m(3m - 1).$$

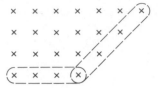

Fig. 19.6
The exceptional partition
in *Case 1*, when $s(\lambda) =$
$t(\lambda) = 4$.

In *Case 2* it may also happen that $s(\lambda)$ and $t(\lambda)$ overlap (Fig. 19.7). In this case, λ^* is not a partition with distinct parts, since the two smallest parts are equal.

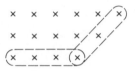

Fig. 19.7
The exceptional partition
in *Case 2*, when $s(\lambda) = 4$,
$t(\lambda) = 3$.

This happens precisely when $s(\lambda) = m + 1$ and there are m parts: $m + 1, m + 2, \ldots, 2m$. So we have

$$n = (m + 1) + (m + 2) + \cdots + 2m = \tfrac{1}{2}m(3m + 1).$$

In summary, we have a bijection $\lambda \to \lambda^*$ except when $n = \tfrac{1}{2}m(3m \pm 1)$. In the exceptional cases there is one spurious partition, which has m parts and contributes $(-1)^m$ to the difference $e_n - o_n$. Hence $e_n - o_n$ takes the values given in the statement of the theorem. $\qquad\square$

The theorem explains the mysterious behaviour of the coefficients in the power series expansion

$$(1 - x)(1 - x^2)(1 - x^3) \cdots = 1 - x - x^2 + x^5 + x^7 - x^{12} - x^{15} + \cdots.$$

We have shown that the coefficient of x^n is $e_n - o_n$, and that this is zero unless n is an integer of the form $\tfrac{1}{2}m(3m \pm 1)$. The first few integers of this form are

m	1	2	3	4	5	6
$\tfrac{1}{2}m(3m - 1)$	1	5	12	22	35	51
$\tfrac{1}{2}m(3m + 1)$	2	7	15	26	40	57 .

The result can also be expressed as an identity between an infinite product and an infinite series:

$$\prod_{i=1}^{\infty}(1 - x^i) = 1 + \sum_{m=1}^{\infty} (-1)^m (x^{\frac{1}{2}m(3m-1)} + x^{\frac{1}{2}m(3m+1)}).$$

In the next section we shall see that this is not just a pretty formula.

1 Show that the coefficient of x^n in the power series expansion of

$$(1+x)(1+x^2)(1+x^3)\cdots$$

is even unless n is a number of the form $\frac{1}{2}m(3m\pm1)$.

2 Let $q_{n,d}$ denote the number of partitions of n which have d parts, all distinct. Show that

$$q_{n,d} = q_{n-d,d} + q_{n-d,d-1} \quad (1\leqslant d\leqslant n).$$

3 Use the recursion equation of Ex. 2 to calculate $q_{n,d}$ for $1\leqslant n\leqslant 22$ and $1\leqslant d\leqslant6$, and verify that your answers are compatible with Ex. 1.

4 Show that the mysterious identity can also be written in the form

$$\prod_{i=1}^{\infty}(1-x^i) = \sum_{m=-\infty}^{\infty}(-1)^m x^{\frac{1}{2}m(3m-1)}.$$

19.6 The calculation of p(n)

It is possible to calculate $p(n)$ by simply listing all the partitions of n in some order (as in Ex. 19.7.13), but this is clearly a very inefficient procedure. We can do better by using a recursive approach, such as the one given in Ex. 19.1.3, but that particular method is also rather inefficient. A very good recursive procedure is based upon the mysterious identity established in Section 19.5, and this is the method which we shall study now.

The alert reader will have spotted that the generating function $P(x)$ for the numbers $p(n)$ is the inverse of the generating function $Q(x)$ for $q_n = e_n - o_n$; specifically,

$$P(x) = (1-x)^{-1}(1-x^2)^{-1}(1-x^3)^{-1}\cdots,$$
$$Q(x) = (1-x)(1-x^2)(1-x^3)\cdots.$$

It follows that

$$P(x)Q(x) = (1+p(1)x+p(2)x^2+\cdots+p(n)x^n+\cdots)$$
$$\times(1-x-x^2+x^5+x^7-x^{12}-x^{15}+\cdots)$$
$$= 1.$$

Thus the coefficient of x^n $(n \geqslant 1)$ in the product is zero. By the rule for multiplying power series, we obtain the equation

$$p(n) - p(n-1) - p(n-2) + p(n-5) + p(n-7) - p(n-12) - p(n-15)$$
$$+ \cdots = 0.$$

We can rewrite this in the form

$$p(n) = p(n-1) + p(n-2) - p(n-5) - p(n-7) + p(n-12) + p(n-15)$$
$$- \cdots,$$

where there are finitely many terms on the right-hand side, since only those terms $p(n-k)$ with $n-k \geqslant 0$ appear. For example,

$$p(13) = p(12) + p(11) - p(8) - p(6) + p(1).$$

These formulae provide an efficient recursive method for computing $p(n)$. The calculation may be set out as in Table 19.6.1, where we assume for convenience that the values $p(0) = 1$, $p(1) = 1$, $p(2) = 2$, $p(3) = 3$, $p(4) = 5$, $p(5) = 7$, $p(6) = 11$ are already known.

Table 19.6.1

n	7	8	9	10	11	12	13	14
$p(n-1)$	11	15	22	30	42	56	77	101
$p(n-2)$	7	11	15	22	30	42	56	77
$p(n-5)$	2	3	5	7	11	15	22	30
$p(n-7)$	1	1	2	3	5	7	11	15
$p(n-12)$	—	—	—	—	—	1	1	2
$p(n)$	15	22	30	42	56	77	101	135

Of course, extra rows of the table are required for the calculation of further values of $p(n)$.

So we have achieved our stated aim of finding an efficient method for the calculation of $p(n)$—just how efficient it is will be established in the following exercises. Remarkably, although the final form of the method is so simple, almost all the theory covered in this chapter is necessary to justify it.

Exercises 19.6

1 Extend the table as far as $p(20)$.

2 Find the least value of n for which $p(n) > 1000$.

3 Let $t(n)$ denote the number of terms on the right-hand side of the equation for $p(n)$. Show that

$$t(n) = \begin{cases} 2m-1 & \text{if } \tfrac{1}{2}m(3m-1) \le n < \tfrac{1}{2}m(3m+1), \\ 2m & \text{if } \tfrac{1}{2}m(3m+1) \le n < \tfrac{1}{2}(m+1)(3m+2). \end{cases}$$

Deduce that there is a constant K such that $t(n) \le K\sqrt{n}$ for all $n \ge 1$.

4 Show that the number of additions and subtractions required to compute $p(n)$ by the method of this section is $O(n^{3/2})$.

9.7 Miscellaneous Exercises

1 Calculate $p_5(9)$ and $p_5(10)$, where $p_k(n)$ is the number of partitions of n into k parts.

2 Use the Ferrers diagram to prove that the number of partitions of n in which each part is 1 or 2 is equal to the number of partitions of $n+3$ which have exactly two distinct parts.

3 Show that the number of partitions of n with at most two parts is $\lfloor \tfrac{1}{2}n \rfloor$.

4 Write down the generating functions for the numbers of:
(i) partitions of n in which each part is at most 5;
(ii) partitions of n in which only even parts can occur more than once;
(iii) partitions of n in which each part is a multiple of 3.

5 Given an alternative description of $p(n) - p(n-1)$ as the number of partitions of n which have a certain property.

6 Show that

$$p(n+2) - 2p(n+1) + p(n) \ge 0.$$

7 Show that the number of partitions of $2n$ into three parts, such that the sum of any two parts is greater than the third, is equal to the number of partitions of n with exactly three parts.

8 Show that the number of partitions of n into k parts satisfies

$$p_k(n) \ge \frac{1}{k!}\binom{n-1}{k-1}.$$

9 Verify that if λ is a partition with parts $\lambda_1, \lambda_2, \ldots, \lambda_r$ satisfying $\lambda_1 \ge \lambda_2 \ge \cdots \ge \lambda_r$, then the parts of the conjugate partition λ' are given by

$$\lambda'_i = \text{the number of parts of } \lambda \text{ not less than } i \; (1 \le i \le \lambda_1).$$

10 Suppose that λ and μ are partitions of n with parts $\lambda_1 \ge \lambda_2 \ge \cdots \ge \lambda_r$ and $\mu_1 \ge \mu_2 \ge \cdots \ge \mu_s$ respectively. Show that λ and μ are conjugate if and

only if

$$\lambda_1 = s, \qquad \mu_1 = r, \qquad \lambda_i + \mu_j \neq i + j - 1 \quad (1 \leq i \leq r, \ 1 \leq j \leq s).$$

11 Show that the number of partitions of $n - m$ into exactly $k - 1$ parts, none of the parts being greater than m, is equal to the number of partitions of $n - k$ into $m - 1$ parts, none of the parts being greater than k.

12 Show that the number of partitions of n in which no part occurs more than $2^{k+1} - 1$ times is equal to the number of partitions of n in which no multiple of 2^k can occur more than once as a part.

13 The following rule is the basis for a method of listing all partitions of n in **lexicographic order**. The first partition is $[n]$. Suppose the current partition λ has parts $\lambda_1 \geq \lambda_2 \geq \cdots \geq \lambda_r$. Then the next partition is found as follows:

(i) if $\lambda_r \neq 1$, then the parts of the next partition are $\lambda_1, \lambda_2, \ldots, \lambda_{r-1}, \lambda_r - 1, 1$;

(ii) if $\lambda_r = \lambda_{r-1} = \cdots = \lambda_{r-s+1} = 1$ but $\lambda_{r-s} = x \neq 1$, then the parts of the next partition are obtained by replacing $\lambda_{r-s}, \lambda_{r-s+1}, \ldots, \lambda_r$ by $x - 1$, $x - 1, x - 1, \ldots, x - 1, y$, where $1 \leq y \leq x - 1$ and the number of parts $x - 1$ is chosen so that the result is a partition of n.

Use this rule to list the partitions of 8.

14 Write a program based on the algorithm outlined in the previous exercise.

15 Write a program to calculate $p(n)$ for $n \leq 100$ by the method of Section 19.6.

16 In a Ferrers diagram the largest possible square of marks in the top left-hand corner is called a **Durfee square**. Show that for each self-conjugate partition of n there is an associated integer k and a partition of n into one part of size k^2 (the Durfee square) and some parts of even size not greater than $2k$. Deduce that

$$(1 + x)(1 + x^3)(1 + x^5) \cdots = 1 + \sum_{k=1}^{\infty} \frac{x^{k^2}}{(1 - x^2)(1 - x^4) \cdots (1 - x^{2k})}.$$

17 Use the Durfee square construction to show that the partition generating function $P(x)$ satisfies the identity

$$P(x) = 1 + \sum_{k=1}^{\infty} \frac{x^{k^2}}{(1 - x)^2 (1 - x^2)^2 \cdots (1 - x^k)^2}.$$

18 Show that

$$(1 + x^2)(1 + x^4)(1 + x^6) \cdots = 1 + \sum_{k=1}^{\infty} \frac{x^{k(k+1)}}{(1 - x^2)(1 - x^4) \cdots (1 - x^{2k})}.$$

20
Symmetry and counting

20.1 The cycle index of a group of permutations

In the final chapter of this book we return to the study of counting problems in which a group of permutations is involved. We shall augment the basic techniques developed in Chapter 14 by some formal algebra based on the theory of generating functions, and thereby obtain an elegant piece of mathematics with important practical applications.

As an example of the kind of problem which we have in mind, let us consider the number of ways of assigning one of the colours black or white to each corner of a square. Since there are two colours and four corners there are basically $2^4 = 16$ possibilities. But when we take account of the symmetry of the square we see that some of the possibilities are essentially the same. For example, the first colouring depicted in Fig. 20.1 is the same as the second one after rotation through $180°$.

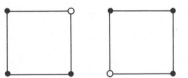

Fig. 20.1
Two indistinguishable colourings.

In this light, we regard two colourings as being indistinguishable if one is transformed into the other by a symmetry of the square. It is easy to find the distinguishable colourings (in this example) by trial and error: there are just six of them, as shown in Fig. 20.2.

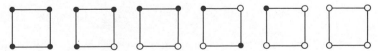

Fig. 20.2
The six distinguishable
colourings.

In the study of general problems of this kind, the most important tool is a compact notation which records information about the cycle structures of permutations in a group. Let G be a group of permutations of a set X, where frequently we take X to be the set $\{1, 2, \ldots, n\}$. Each element g in G can be written in cycle notation with α_i cycles of length i $(1 \leqslant i \leqslant n)$, and we recall that the *type* of g is the corresponding partition

$$[1^{\alpha_1}2^{\alpha_2}\ldots n^{\alpha_n}]$$

of n. Of course, we have $\alpha_1 + 2\alpha_2 + \cdots + n\alpha_n = n$. We shall associate with g an expression

$$\zeta_g(x_1, x_2, \ldots, x_n) = x_1^{\alpha_1}x_2^{\alpha_2}\cdots x_n^{\alpha_n},$$

where the x_i $(1 \leqslant i \leqslant n)$ are, for the moment, simply formal symbols like the x in a polynomial. For example, if G is the group of symmetries of a square, regarded as permutations of the corners 1, 2, 3, 4, then the expressions ζ_g are given in Table 20.1.1. Note that although the 1-cycles are conventionally omitted in the notation for g it is important to include them in ζ_g.

Table 20.1.1

g	α_1	α_2	α_3	α_4	ζ_g
id	4	—	—	—	x_1^4
(1234)	—	—	—	1	x_4
(13)(24)	—	2	—	—	x_2^2
(1432)	—	—	—	1	x_4
(12)(34)	—	2	—	—	x_2^2
(14)(23)	—	2	—	—	x_2^2
(13)	2	1	—	—	$x_1^2x_2$
(24)	2	1	—	—	$x_1^2x_2$

The formal sum of the ζ_g, taken over all g in G, is a 'polynomial' in $x_1, x_2, \ldots, x_n$. Dividing by $|G|$ we obtain the **cycle index** of the group of permutations:

$$\zeta_G(x_1, x_2, \ldots, x_n) = \frac{1}{|G|}\sum_{g \in G} \zeta_g(x_1, x_2, \ldots, x_n).$$

For example, the cycle index of the group of the square, as considered above, is

$$\tfrac{1}{8}(x_1^4 + 2x_1^2x_2 + 3x_2^2 + 2x_4);$$

where we have collected the terms corresponding to permutations of the same type. In general, we can write

$$\zeta_G(x_1, x_2, \ldots, x_n) = \frac{1}{|G|} \sum c(\alpha_1, \alpha_2, \ldots, \alpha_n) x_1^{\alpha_1} x_2^{\alpha_2} \cdots x_n^{\alpha_n},$$

where $c(\alpha_1, \alpha_2, \ldots, \alpha_n)$ is the number of permutations in G which have type $[1^{\alpha_1} 2^{\alpha_2} \ldots n^{\alpha_n}]$, and the sum is taken over all types (that is, over all partitions of n). Thus we see that the cycle index is like the index of a good book: it is a list which tells us something about the permutations contained in G.

Exercises 20.1

1 Use the method of trial and error to find all the distinguishable colourings of the corners of an equilateral triangle, assuming that three colours are available. (Regard the triangle as a flat piece of card, and assume that both sides are coloured.)

2 A uniform rod one metre in length is to be decorated by dividing its surface into five 20 cm bands and colouring each band red or blue. In how many essentially different ways can this be done?

3 Write down the cycle index of

(i) the group of symmetries of an equilateral triangle regarded as permutations of the corners,
(ii) the alternating group A_4,
(iii) the symmetric group S_5.

4 The graphs illustrated in Fig. 20.3 are to be regarded as plane frameworks lying in three-dimensional space. In each case find the cycle index of the group of symmetries, regarded as a group of permutations of the vertices.

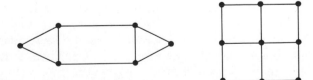

Fig. 20.3
Find the cycle indexes.

20.2 Cyclic and dihedral symmetry

In order to deploy our mathematical theories we shall need a small collection of useful cycle indexes, and so in this section and the next one we shall work out some important examples.

The most frequently occurring kind of symmetry is that associated with a circular object, such as a coloured disc or a necklace. When faced with a problem involving an object of this kind, we must settle one crucial question at the outset: do the conditions of the problem allow us to overturn (or take a mirror-image of) the object concerned? For example, if we have a disc divided into equal sectors on one side only, then it is clear that overturning is not allowed. In Fig. 20.4 there are two black-and-white colourings of one side of a disc which are not equivalent under these conditions. On the other hand,

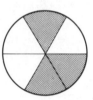

Fig. 20.4
Two inequivalent colourings of a disc.

if we have a circular string of coloured beads, then we normally assume that overturning is allowed, and consequently we should regard the two necklaces shown in Fig. 20.5 as the same.

Fig. 20.5
Two equivalent necklaces.

We begin by considering the case when overturning is not allowed, in which case we are dealing with a *cyclic* group of symmetries. Suppose we label the corners of a regular n-gon in clockwise order with the symbols $1, 2, \ldots, n$. Then a rotation through $2\pi/n$ about the centre of the n-gon corresponds to the permutation

$$\pi = (123 \ldots n)$$

of the corners. The group of rotations of the n-gon is the cyclic group of order n generated by π, that is

$$C_n = \{\mathrm{id}, \pi, \pi^2, \ldots, \pi^{n-1}\}.$$

For example, when $n = 6$ the permutations and the associated expressions ζ_{π^i} are as follows:

id	(six	1-cycles)	x_1^6
(123456)	(one	6-cycle)	x_6
(135)(246)	(two	3-cycles)	x_3^2
(14)(25)(36)	(three	2-cycles)	x_2^3
(153)(264)	(two	3-cycles)	x_3^2
(165432)	(one	6-cycle)	x_6.

In this case, the cycle index is $\frac{1}{6}(x_1^6 + x_2^3 + 2x_3^2 + 2x_6)$. The general case follows the same pattern, and the details are given in the next theorem.

Theorem 20.2.1

Let C_n be the cyclic group of permutations generated by $\pi = (123 \ldots n)$. Then for each divisor d of n there are $\phi(d)$ permutations in C_n which have n/d cycles of length d, and hence the cycle index of C_n is

$$\xi_{C_n}(x_1, x_2, \ldots, x_n) = \frac{1}{n} \sum_{d|n} \phi(d) x_d^{n/d}.$$

Proof

According to Theorem 13.9 a cyclic group of order n contains exactly $\phi(d)$ elements of order d, for each divisor d of n. In this case, the $\phi(d)$ permutations are those of the form $\pi^{kn/d}$, where $1 \le k \le d$ and $\gcd(k, d) = 1$. It remains only to show that these permutations have n/d cycles of length d.

Let m be the length of a shortest cycle of the permutation π^i $(1 \le i \le n-1)$, and suppose x is in a cycle of length m. Then

$$\pi^{im}(x) = (\pi^i)^m(x) = x.$$

For any y in $\{1, 2, \ldots, n\}$ both x and y are in the single cycle of π, so $y = \pi^r(x)$ for some r. Now

$$(\pi^i)^m(y) = \pi^{im}\pi^r(x) = \pi^r\pi^{im}(x) = \pi^r(x) = y,$$

so that y is in a cycle of π^i whose length divides m. But m is the minimum length of a cycle, so this cycle has length m. Thus all cycles of π^i have the same length m. If the order of π^i is d we must have $d = m$, and so there are n/d cycles of length d as claimed. $\square$

It is worth remarking that if we substitute $x_i = 1$ $(1 \le i \le n)$ in this result we regain a version of the classic formula first obtained in Section 3.3:

$$\frac{1}{n} \sum_{d|n} \phi(d) = 1.$$

We now turn to the case in which mirror-image symmetry is allowed. Suppose first that n is even and $n \geqslant 4$, and let $n' = \frac{1}{2}n$, so that the corners of our n-gon are labelled $1, 2, \ldots, n', n'+1, \ldots, n$. Overturning the n-gon about the perpendicular bisector of the side $1n$ is the same as taking its 'reflection' in that axis (Fig. 20.6), and

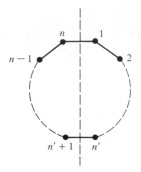

Fig. 20.6
Symmetry of an even polygon.

the corresponding permutation is

$$\sigma = (1\ n)(2\ n-1) \ldots (n'\ n'+1).$$

Taking $\pi = (1\ 2\ 3 \ldots n)$ as usual, we find that

$$\sigma\pi = (1\ n-1)(2\ n-2) \ldots (n'-1\ n'+1)(n')(n),$$

which represents a reflection in the axis nn'.

Clearly, there are $n' = \frac{1}{2}n$ reflections in the perpendicular bisectors of the sides; these correspond to the permutations

$$\sigma, \sigma\pi^2, \sigma\pi^4, \ldots, \sigma\pi^{n-2}.$$

Also, there are $n' = \frac{1}{2}n$ reflections in the axes joining opposite corners, and these correspond to the permutations

$$\sigma\pi, \sigma\pi^3, \sigma\pi^5, \ldots, \sigma\pi^{n-1}.$$

Thus we have a group of $2n$ permutations, the n rotations π^i and the n reflections $\sigma\pi^i$ $(0 \leqslant i \leqslant n-1)$. It is called the **dihedral group** of order $2n$, and we shall denote it by D_{2n}. (*Caution*: some authors write D_n instead of D_{2n}.)

When n is odd, say $n = 2n' + 1$, there are again n reflections, but now they are all of the same type, since each one is a reflection in an axis joining a corner to the mid-point of the opposite side. For example, choosing the corner n' as in Fig. 20.7, we get

$$\sigma = (1\,n)(2\,n-1)\ldots(n'-1\ \ n'+1)(n').$$

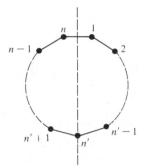

Fig. 20.7
Symmetry of an odd
polygon.

Here again we have a dihedral group of order $2n$ consisting of the n rotations π^i and the n reflections $\sigma\pi^i$ ($0 \leqslant i \leqslant n-1$).

**Theorem
20.2.2**

The cycle index of D_{2n} is

$$\tfrac{1}{2}\zeta_{C_n}(x_1, x_2, \ldots, x_n) + \begin{cases} \tfrac{1}{4}(x_2^{n/2} + x_1^2 x_2^{n/2-1}) & \text{if } n \text{ is even,} \\ \tfrac{1}{2}x_1 x_2^{(n-1)/2} & \text{if } n \text{ is odd.} \end{cases}$$

Proof

In the even case, D_{2n} contains the n elements of C_n, together with $\tfrac{1}{2}n$ permutations (like σ) of type $[2^{n/2}]$ and $\tfrac{1}{2}n$ permutations (like $\sigma\pi$) of type $[1^2 2^{n/2-1}]$. Hence $\zeta_{D_{2n}}(x_1, \ldots, x_n)$ is equal to

$$\frac{1}{2n}\left(\sum_{d\,|\,n} \phi(d) x_d^{n/d} + \frac{n}{2} x_2^{n/2} + \frac{n}{2} x_1^2 x_2^{n/2-1}\right)$$

which reduces to the form given. In the odd case we have the n permutations of C_n together with n permutations of type $[1\,2^{n/2-1}]$, and the result follows as before.   □

**Exercises
20.2**

1   Write out in full the cycle indexes for C_{12}, D_{12}, and D_{14}.

2   Write down the cycle index for D_{2p} when p is an odd prime, and prove that

$$\zeta_{D_{2p}}(r, r, \ldots, r)$$

is an integer whenever r is a positive integer.

3   Find all the subgroups of D_{12} and sketch the lattice of subgroups.

4   Each of the groups C_{12}, D_{12}, A_4 has order 12. By finding the number of elements of each order in each group show that no two of them are isomorphic.

5   One of the groups $C_4 \times C_3$, $C_6 \times C_2$, $D_6 \times D_2$ is not isomorphic to any of the groups studied in Ex. 4. Which?

20.3   Symmetry in three dimensions

Most people become familiar with the symmetry properties of an ordinary cube (Fig. 20.8) in early infancy, but the mathematical description of these properties is often omitted from their formal education. In this section we shall attempt to remedy this deficiency.

We begin with an important general point: in our discussion of the symmetry of three-dimensional objects we shall consider only *rotational* symmetries. The reader may well ask why reflectional symmetries are excluded. After all, when we discuss necklace problems we must take account of the reflections if we are to set up a correct model of the situation. This is because we can achieve a

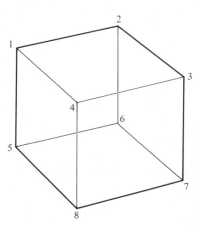

Fig. 20.8
A cube.

reflection of a two-dimensional necklace by using a rotation (over-turning) in three dimensions. On the other hand, there is no way in which we can achieve a reflection of a solid three-dimensional object by means of an actual movement of the object. For example, in Fig. 20.8 the reflection of the cube in the horizontal plane through its centre corresponds to the permutation (15)(26)(37)(48) of its corners, but there is no physical motion of the solid cube which realizes this permutation.

With this in mind, we shall tabulate (see Table 20.3.1) the rotational symmetries of the cube; according to Ex. 14.3.2 there are 24 of them. We shall classify the rotations by specifying the axis and the order of each class, and in each case we shall give a typical permutation of the corners (labelled as in Fig. 20.8). Note that a rotation of order m is the same as a rotation through the angle $2\pi/m$. Hence the cycle index of the group, regarded as a group of permutations of the corners is

$$\tfrac{1}{24}(x_1^8 + 8x_1^2x_3^2 + 9x_2^4 + 6x_4^2).$$

Table 20.3.1

Axis	Order	Permutation	Number
Line joining mid-points of opposite faces	4	(1234)(5678)	6
The same	2	(13)(24)(57)(68)	3
Line joining mid-points of opposite sides	2	(15)(28)(37)(46)	6
Line joining opposite corners	3	(245)(386)	8
–	1	id	1

In practice, we often encounter problems about the cube which involve the faces, rather than the corners. (Children's bricks have coloured faces, not coloured corners!) One way of dealing with such problems is by working out the permutations of the faces which correspond to the various symmetry transformations: for example, the first symmetry in the table above induces a permutation (A) (B) $(CDEF)$ of the six faces, where A and B are the top and bottom faces in Fig. 20.8, C is the front face, and so on. Alternatively, we can think of the problem in a different light, by using the construction illustrated in Fig. 20.9.

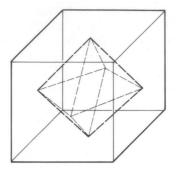

Fig. 20.9
A cube and the dual
octahedron.

We put a new corner in the centre of each face of the cube, and join two new corners by an edge whenever they correspond to adjacent faces of the cube. In this way we obtain a regular octahedron inscribed in the cube; it is known as the *dual* of the cube. Each symmetry transformation of the cube is a symmetry transformation of the octahedron, and vice versa. Since the faces of the cube correspond to the corners of the octahedron, any problem about the former can be translated into a problem about the latter. In particular, the cycle index of the symmetry group acting on the faces of the cube is the same as the cycle index of the symmetry group acting on the corners of the octahedron. This cycle index is given in the table at the end of the section.

Thus far we have discussed only one group of rotational symmetries in three dimensions. Clearly, the cyclic and dihedral groups C_n and D_{2n} can also occur in this context, for example, as groups of rotations of an n-sided prism (Fig. 20.10). One might expect that there would be many other kinds of rotational symmetry groups, but in fact it is not so. Apart from the cyclic, dihedral, and cubic-octahedral types, there are only two other rotational symmetry groups in three dimensions. This important result can be deduced from the theory developed in Chapter 14 and the proof is indicated in Ex. 14.7.19–21. However, for our purposes it is sufficient to point out that the two remaining groups are also associated with regular polyhedra.

Fig. 20.10
An n-sided prism.

  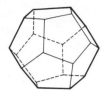

(a) Tetrahedron (b) Icosahedron (c) Dodecahedron

Fig. 20.11
Regular polyhedra.

First, there is the group of rotations of the regular tetrahedron (Fig. 20.11a), which we discussed in Section 14.3. As a group of permutations of the corners it contains the identity, three permutations of type $[2^2]$, and eight permutations of type $[13]$. As a group of permutations of the faces it has the same form, corresponding to the geometrical fact that the dual of the tetrahedron is another tetrahedron.

Secondly, we have the group of rotations of the regular icosahedron (Fig. 20.11b) and its dual, the regular dodecahedron (Fig. 20.11c). This fascinating group has many remarkable properties, but we shall be content to record the relevant cycle indexes. Table 20.3.2 gives the cycle index of each rotational symmetry group G which acts as a group of permutations of the set x of corners of a regular polyhedron in three-dimensional space.

Table 20.3.2

Polyhedron	$|X|$	$|G|$	Cycle index
Tetrahedron	4	12	$\frac{1}{12}(x_1^4 + 8x_1x_3 + 3x_2^2)$
Octahedron	6	24	$\frac{1}{24}(x_1^6 + 6x_1^2x_4 + 3x_1^2x_2^2 + 6x_2^3 + 8x_3^2)$
Cube	8	24	$\frac{1}{24}(x_1^8 + 8x_1^2x_3^2 + 9x_2^4 + 6x_4^2)$
Icosahedron	12	60	$\frac{1}{60}(x_1^{12} + 24x_1^2x_5^2 + 15x_2^6 + 20x_3^4)$
Dodecahedron	20	60	$\frac{1}{60}(x_1^{20} + 20x_1^2x_3^6 + 15x_2^{10} + 24x_5^4)$

Exercises 20.3

1   In Ex. 14.3.1 we showed that the rotational symmetry group of the regular tetrahedron is isomorphic to the alternating group A_4, that is, the group of even permutations of the corners. Show that every odd permutation of the corners corresponds either to a reflection of the tetrahedron, or to the composite of a reflection and a rotation.

2   Write down the cycle index of the rotational symmetry group of the tetrahedron, regarded as a group of permutations of the edges.

3   The types of permutations which belong to the rotational symmetry group of a dodecahedron are specified in the cycle index given above. Label the corners of a dodecahedron (using Fig. 20.11c) and write down one permutation of each type.

4   Write down a permutation which corresponds to a reflectional symmetry of the dodecahedron.

20.4   The number of inequivalent colourings

In this section we return to the general problem of finding the number of distinguishable colourings when a group of permutations is involved. A simple example is the problem of black-and-white colourings of the corners of a square, discussed in Section 20.1: in that case there are 16 colourings in all, but only six essentially different ones when the symmetry is taken into account (Fig. 20.2).

In general, suppose we have a group G of permutations of an n-set X, and each element of X can be assigned one of r colours. If we denote the set of colours by K, then a *colouring* is simply a function ω from X to K. There are r^n colourings in all, and we shall denote the set of them by Ω.

Now each permutation g in G induces a permutation $\hat{g}$ of Ω in the following way. Given a colouring ω, we define $\hat{g}(\omega)$ to be the colouring in which the colour assigned to x is the colour ω assigns to $g(x)$; that is,

$$(\hat{g}(\omega))(x) = \omega(g(x)).$$

The definition is illustrated in Fig. 20.12, where g is the clockwise rotation through 90°, and ω is the colouring on the right-hand side.

The function taking g to $\hat{g}$ is a representation of G as a group $\hat{G}$

Fig. 20.12
Illustrating the definition
of $\hat{g}(w)$.

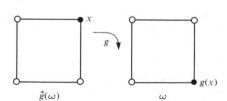

of permutations of Ω. Two colourings are indistinguishable if one of them can be transformed into the other by some permutation $\hat{g}$: that is, if they belong to the same orbit of $\hat{G}$ on Ω. So the number of distinguishable (or as we shall say, *inequivalent*) colourings is just the number of orbits in the action of $\hat{G}$ on Ω.

It will be recalled that Theorem 14.4 gives a general formula for the number of orbits. But before we apply it we shall check that the representation $g \to \hat{g}$ is faithful, so that the image group $\hat{G}$ is isomorphic to G. Suppose then that $\hat{g}_1 = \hat{g}_2$, so that

$$(\hat{g}_1(\omega))(x) = (\hat{g}_2(\omega))(x)$$

and consequently

$$\omega(g_1(x)) = \omega(g_2(x)) \quad (\omega \in \Omega, \ x \in X).$$

Since the equation is true for all ω, it is true in particular for the colouring which assigns a specified colour to $g_1(x)$ and another colour to every other member of X. In this case the equation implies that $g_1(x) = g_2(x)$, and because the same argument works for each x in X, we conclude that $g_1 = g_2$. Hence the representation $g \to \hat{g}$ is faithful, and we are ready to apply Theorem 14.4. The elegance of the result justifies our preparations.

Theorem 20.4

If G is a group of permutations of X, and $\zeta_G(x_1, \ldots, x_n)$ is its cycle index, then the number of inequivalent colourings of X with r colours available is

$$\zeta_G(r, r, \ldots, r).$$

Proof

We have shown that the representation $g \to \hat{g}$ is a bijection, so that $|G| = |\hat{G}|$. Furthermore, the formula giving the number of orbits of $\hat{G}$ on Ω is

$$\frac{1}{|\hat{G}|} \sum_{\hat{g} \in \hat{G}} |F(\hat{g})| = \frac{1}{|G|} \sum_{g \in G} |F(\hat{g})|,$$

where $F(\hat{g})$ is the set of colourings fixed by $\hat{g}$.

Suppose ω is a colouring fixed by $\hat{g}$, so that $\hat{g}(\omega) = \omega$, and let $(xyz \ldots)$ be any cycle of g. We have

$$\omega(y) = \omega(g(x)) = (\hat{g}(\omega))(x) = \omega(x),$$

and so ω assigns the same colour to x and y. By repeating the same argument, we can show that y and z have the same colour, and so on. Thus ω is constant on each cycle of g. If g has k cycles in all, then the number of colourings which have this property is just r^k, since one element of each cycle can be coloured arbitrarily, and the colours of the rest are then determined. So if g has α_i cycles of length i $(1 \leq i \leq n)$, we have $\alpha_1 + \cdots + \alpha_n = k$, and

$$|F(\hat{g})| = r^k = r^{\alpha_1 + \cdots + \alpha_n} = \zeta_g(r, r, \ldots, r).$$

Substituting in the formula for the number of orbits given above, we obtain the result $\zeta_G(r, r, \ldots, r)$ as claimed. $\square$

It follows from this theorem that the problem of finding the number of inequivalent colourings when r colours are available can be reduced to the problem of calculating the cycle index. When the cycle index is known, we have only to replace each of $x_1, \ldots, x_n$ by r in order to get the result. For example, the number of inequivalent colourings of the corners of a square is

$$\tfrac{1}{8}(r^4 + 2r^3 + 3r^2 + 2r),$$

obtained by putting $x_1 = x_2 = x_3 = x_4 = r$ in the cycle index computed in Section 20.1. When $r = 2$ we obtain

$$\tfrac{1}{8}(16 + 16 + 24 + 4) = 6,$$

in agreement with Fig. 20.2. Similar formula for other regular polygons and polyhedra can be derived from the cycle indexes computed in Sections 20.2 and 20.3.

For the sake of variety, we give an example involving the symmetry of an irregular polygon—a rectangle.

Example The competitors in the University of Folornia Classic Marathon Race (founded last year) are distinguished not, as is usual, by the wearing of numbers, but by rectangular badges made from four pieces of coloured material, as in Fig. 20.13. Unfortunately, many of the runners are eminent academics, likely to pin on their badges upside-down, or back-to-front, or (in the case of the Medical Faculty) both. If 160 runners intend to compete this year, what is

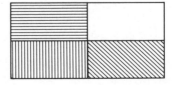

Fig. 20.13
The latest technological breakthrough from Folornia.

the smallest number of coloured materials required to make the badges?

Solution   The segments of the badge can be identified with the corners of the rectangle, which we shall label 1, 2, 3, 4 in clockwise order with 1 at the top left. The relevant group consists of the permutations

id	(correct position)
(13)(24)	(upside-down)
(12)(34)	(back-to-front)
(14)(23)	(Medical position).

Hence the cycle index for the group of the rectangle is

$$\tfrac{1}{4}(x_1^4 + 3x_2^2),$$

and the number of badges with r colours available is $\tfrac{1}{4}(r^4 + 3r^2)$. The smallest integer r for which

$$\tfrac{1}{4}(r^4 + 3r^2) \geqslant 160$$

is 5, so this is the number of colours required.   □

Exercises 20.4

1   Use the table of cycle indexes given in Section 20.3 to write down formulae for the number of ways of colouring with r colours

(i) the corners of a tetrahedron;
(ii) the faces of a tetrahedron;
(iii) the faces of a cube.

2   Find the cycle index for the group of symmetries of a uniform straight rod divided along its length into m equal sections. (You will have to treat the cases m odd and m even separately.) Hence write down the number of ways of colouring the sections when r colours are available.

3   In how many inequivalent ways can the faces of a regular dodecahedron be coloured red, white, or blue?

4   A circular disc is divided into eight equal sectors on one side, and each sector is painted red, green, or yellow. How many different discs can be constructed in this way?

20.5   Sets of colourings and their generating functions

Theorem 20.4 is only a special case of an even more elegant and comprehensive result. This result will enable us to calculate not just the total number of inequivalent colourings, but also the number of them in which the individual colours are used a specified number of times. For example, the six inequivalent black-and-white colourings of the corners of a square can be listed (see Table 20.5.1) according to the numbers of black and white corners.

Table 20.5.1

Black	White	Number of colourings
4	0	1
3	1	1
2	2	2
1	3	1
0	4	1

We can represent these numbers in an obvious way as coefficients of the *generating function*

$$U(b, w) = b^4 + b^3 w + 2b^2 w^2 + bw^3 + w^4,$$

and our aim is to show how $U(b, w)$ can be derived from the appropriate cycle index.

Suppose that the set X is to be coloured, and that the set of available colours is $K = \{a, b, \ldots, h\}$. Associated with each colouring $\omega: X \to K$ is a formal expression, the **indicator** of ω, defined by

$$\text{ind} (\omega) = a^{n_a} b^{n_b} \cdots h^{n_h},$$

where $n_a, n_b, \ldots, n_h$ are the numbers of members of X which receive colours $a, b, \ldots, h$ respectively. Of course, $n_a + n_b + \cdots + n_h = n$, where $n = |X|$.

Given any subset A of the set Ω of all colourings, we define the generating function U_A to be the formal sum

$$U_A(a, b, \ldots, h) = \sum_{\omega \in A} \text{ind} (\omega).$$

Clearly, when the terms of U_A are collected in the usual way, the coefficient of the term $a^s b^t \ldots$ is just the number of colourings in A in which colour a is used s times, colour b is used t times, and so on.

We shall require an explicit formula for one particular generating function of this kind. Suppose we have a partition

$$X = X_1 \cup X_2 \cup \cdots \cup X_k,$$

in which $|X_i| = m_i$ $(1 \le i \le k)$, and $m_1 + m_2 + \cdots + m_k = n$.

Theorem 20.5

Let X be partitioned as above, and let B denote the set of colourings $\omega: X \to \{a, b, \ldots, h\}$ which assign the same colour to each member of X_i $(1 \le i \le k)$. Then

$$U_B(a, b, \ldots, h) = (a^{m_1} + b^{m_1} + \cdots + h^{m_1})(a^{m_2} + b^{m_2} + \cdots + h^{m_2}) \times \cdots$$
$$\cdots \times (a^{m_k} + b^{m_k} + \cdots + h^{m_k}).$$

Proof

We construct a bijective correspondence between the colourings in B and the terms obtained by multiplying out the right-hand side. Given a colouring ω, suppose that ω assigns colour c_1 to the whole of X_1, c_2 to the whole of X_2, and so on; then we choose the terms

$$c_1^{m_1} \quad \text{from the first bracket,}$$
$$c_2^{m_2} \quad \text{from the second bracket,}$$
$$\cdots$$
$$c_k^{m_k} \quad \text{from the last bracket.}$$

Conversely, any such choice of terms determines a unique colouring satisfying the conditions for membership of B.

Now the product of the terms chosen to correspond to ω is just the indicator ind (ω), and so when we multiply out the right-hand side we obtain

$$\sum_{\omega \in B} \text{ind}(\omega).$$

But this is the definition of U_B, so we have the result. $\square$

When we put $a = b = \cdots = h = 1$ in the formula for $U_B(a, b, \ldots, h)$ each factor on the right-hand side reduces to r, the number of colours. Hence the total number of colourings in B is r^k, in agreement with the fact that we must assign one of r colours to each one of the k sets $X_1, \ldots, X_k$.

The primary reason for proving Theorem 20.5 is that it forms an important step in the proof of the main theorem, to be given in the next section. The result is, however, sometimes useful in other ways.

Example

Five families arrange a party for their children. Each child will get a prize, either a balloon or a candy-bar, and children from the same family must get the same prize. Two of the families have four children and the others have two each. Write down the generating function for the numbers of ways of distributing the prizes, and find the number of ways in which six balloons and eight candy-bars can be used.

Solution The set X is the set of children, partitioned into two subsets of size 4 and three of size 2. The 'colours' are b (for balloon) and c (for candy-bar). According to Theorem 20.5 the generating function is

$$(b^4 + c^4)^2(b^2 + c^2)^3 = (b^8 + 2b^4c^4 + c^8)(b^6 + 3b^4c^2 + 3b^2c^4 + c^6).$$

We could multiply out and obtain all the coefficients, but only the coefficient of b^6c^8 is asked for, and by inspection this is $(2 \times 3) + (1 \times 1) = 7$. □

Exercise 20.5

1   Write down the generating function for the set of colourings which assign the same colour to each member of a part of X, when $|X| = 10$ and

(i) there are three colours available, and X is partitioned into two equal subsets;

(ii) there are two colours available, and X is partitioned into subsets of size 4, 2, 2, 2.

2   In the *Example*, suppose there were six families, three of them with four children, one with three, and two with two. In how many ways can 11 balloons and 8 candy-bars be distributed?

20.6 Pólya's theorem

The culmination of our efforts is now at hand, in the form of a theorem discovered by George Pólya in 1935. It is an elegant combination of the theory of groups of permutations and the powerful method of generating functions.

Suppose G is group of permutations of a set X, and let $\hat{G}$ be the induced group of permutations of the set Ω of colourings of X, defined by the rule $(\hat{g}(\omega))(x) = \omega(g(x))$. We require the generating function $U_D(a, b, \ldots, h)$, where D is a set of colourings containing one representative of each orbit of $\hat{G}$ on Ω. The coefficient of $a^s b^t \cdots$ in U_D will be the number of distinguishable colourings in which colour a is used s times, colour b is used t times, and so on.

Pólya's theorem asserts that U_D is obtained from the cycle index $\zeta_G(x_1, \ldots, x_n)$ by substituting

$$a^i + b^i + \cdots + h^i$$

for x_i $(1 \le i \le n)$. Before embarking on the proof, let us see how this works in the simple case of the black-and-white colourings of the corners of a square. The cycle index is

$$\tfrac{1}{8}(x_1^4 + 2x_1^2 + 3x_2^2 + 2x_4),$$

and we have to substitute

$$x_1 = b + w, \qquad x_2 = b^2 + w^2, \qquad x_3 = b^3 + w^3, \qquad x_4 = b^4 + w^4.$$

We get

$$U_D(b, w) = \tfrac{1}{8}[(b + w)^4 + 2(b + w)^2(b^2 + w^2) + 3(b^2 + w^2)^2 + 2(b^4 + w^4)]$$
$$= b^4 + b^3 w + 2b^2 w^2 + bw^3 + w^4,$$

which agrees with Table 20.5.1. With this small comfort, we shall attack the theorem.

Theorem 20.6

Let $\zeta_G(x_1, x_2, \ldots, x_n)$ be the cycle index for a group G of permutations of X. The generating function $U_D(a, b, \ldots, h)$ for the numbers of inequivalent colourings of X, when the colours available are $a, b, \ldots, h$, is given by

$$U_D(a, b, \ldots, h) = \zeta_G(\sigma_1, \sigma_2, \ldots, \sigma_n),$$

where

$$\sigma_i = a^i + b^i + \cdots + h^i \quad (1 \le i \le n).$$

Proof

We shall begin by finding an alternative formula for

$$U_D(a, b, \ldots, h) = \sum_{\omega \in D} \text{ind}(\omega),$$

where D is a set of colourings containing one representative of each orbit of $\hat{G}$ on Ω. We do this by invoking the 'weighted' form of Theorem 14.4, as given in Ex. 14.4.5. Applying this result to the action of $\hat{G}$ on Ω, we obtain

$$\sum_{\omega \in D} \text{ind}(\omega) = \frac{1}{|\hat{G}|} \sum_{\hat{g} \in \hat{G}} \left[\sum_{\omega \in F(\hat{g})} \text{ind}(\omega) \right].$$

Now the sum in the bracket is just $U_{F(\hat{g})}$, by definition. Furthermore, a colouring ω is in $F(\hat{g})$ if and only if it is constant on each cycle of g, as we showed in the proof of Theorem 20.4. Hence the explicit form of $U_{F(\hat{g})}$ is given by Theorem 20.5:

$$U_{F(\hat{g})}(a, b, \ldots, h) = (a^{m_1} + \cdots + h^{m_1}) \times \cdots \times (a^{m_k} + \cdots + h^{m_k})$$

$$= \sigma_{m_1} \cdots \sigma_{m_k},$$

where $m_1, m_2, \ldots, m_k$ are the lengths of the cycles of g. In other words, if g has α_i cycles of length i $(1 \le i \le n)$ then

$$U_{F(\hat{g})}(a, b, \ldots, h) = \sigma_1^{\alpha_1} \sigma_2^{\alpha_2} \cdots \sigma_n^{\alpha_n}$$

$$= \zeta_g(\sigma_1, \sigma_2, \ldots, \sigma_n).$$

Since the representation $g \to \hat{g}$ is a bijection, we have $|G| = |\hat{G}|$, and substituting for $U_{F(\hat{g})}$ above we get

$$U_D(a, b, \ldots, h) = \zeta_G(\sigma_1, \sigma_2, \ldots, \sigma_n),$$

as required. $\qquad\square$

Pólya's theorem provides a mechanical way of computing numbers of inequivalent colourings of various types. In general, the major task is to calculate the cycle index for the relevant group, and it is for this reason that we have armed ourselves with a small list of useful cycle indexes. The secondary task is to expand the expression obtained by substituting for x_i in the cycle index, and hence find the required coefficients. This can be messy, but often it is not necessary to work out all the details.

Example (Section 14.4 revisited.) How many necklaces can be made from a string of 3 black beads and 13 white beads?

Solution The group is the dihedral group D_{32} acting on the corners of a 16-gon, and according to Theorem 20.2.2 the appropriate cycle index is

$$\tfrac{1}{32}(x_1^{16} + 9x_2^8 + 2x_4^4 + 4x_8^2 + 8x_{16} + 8x_1^2 x_2^7).$$

In order to find the required number we have to substitute $x_i = b^i + w^i$ $(1 \le i \le 16)$, and calculate the coefficient of $b^3 w^{13}$. By inspection we notice that odd powers of b and w can arise only from the terms x_1^{16} and $8x_1^2 x_2^7$, so we require the coefficient of $b^3 w^{13}$ in

$$\tfrac{1}{32}[(b+w)^{16} + \cdots + 8(b+w)^2(b^2+w^2)^7]$$

$$= \tfrac{1}{32}[(b^{16} + \cdots + \tbinom{16}{13}b^3 w^{13} + \cdots + w^{16}) + \cdots$$

$$\cdots + 8(b^2 + 2bw + w^2) \times (b^{14} + \cdots + 7b^2 w^{12} + w^{14})].$$

Hence the answer is

$$\frac{1}{32}\left(\frac{16\times15\times14}{3\times2\times1}+8\times2\times7\right)=21. \qquad \square$$

In the *Example* the terms which contribute to the answer are precisely those which contribute to the 'fixed-point sum' when we use the more basic method of Section 14.4. Of course, that is to be expected, since the new method is based on the old one.

Exercises 20.6

1 In how many ways can the faces of a cube be coloured so that there are two red faces, one white face, and three blue faces?

2 Work out explicitly the generating function $U_D(b, g)$ for the number of ways of colouring the faces of a regular octahedron when the colours available are blue and green.

3 (i) How may coloured discs can be constructed by dividing one side of the disc into five equal sectors and colouring two sectors red, two sectors white, and one sector blue?

(ii) How many necklaces can be constructed using two red beads, two white beads, and one blue bead?

4 When only two colours are involved it is often convenient to use the symbols 1 and x instead of b and w (or whatever names the actual colours may have). In this way we obtain a generating function of the form

$$f_0+f_1x+f_2x^2+\cdots+f_nx^n,$$

where f_i is the number of configurations with i black objects and $n-i$ white ones. Work out this generating function explicitly for the necklace problem with 16 black or white beads, as discussed in the *Example*.

5 Chemists consider the kinds of molecule which can be formed from one C (carbon) atom linked to four radicals, each of which may be $HOCH_2$ (hydroxymethyl), C_2H_5 (ethyl), Cl (chlorine), or H (hydrogen). There are good reasons for using a picture of this situation in which the C atom is located at the centre of a regular tetrahedron, and the radicals occupy the corners.

(i) Show that there are 36 possible molecules.
(ii) Show that there are 15 molecules which contain on H radical.

(iii) Compute the generating function

$$H(x) = h_0 + h_1 x + h_2 x^2 + h_3 x^3 + h_4 x^4,$$

where h_i $(0 \leqslant i \leqslant 4)$ is the number of molecules which contain exactly i H radicals.

20.7 Miscellaneous Exercises

1   Show that the cycle index of the rotational symmetry group of the cube, regarded as a group of permutations of the edges, is

$$\tfrac{1}{24}(x_1^{12} + 3x_2^6 + 6x_4^3 + 6x_1^2 x_5^2 + 8x_3^4).$$

2   Let Δ be an abstract group containing elements π and σ such that

(i) π has order n,   (ii) σ has order 2,   (iii) $\sigma\pi = \pi^{-1}\sigma$.

Suppose also that any element of Δ can be expressed as a product of terms π, π^{-1}, and σ. (Note that $\sigma = \sigma^{-1}$.) Prove that every element of Δ can be expressed uniquely in the form $\sigma^e \pi^i$ ($e = 0$ or 1, $0 \leqslant i \leqslant n-1$), and deduce that $\Delta \approx D_{2n}$.

3   Show that, for any positive integer n, $D_{2n} \times C_2 \approx D_{4n}$.

4   Describe the centre $Z(D_{2n})$, distinguishing between the cases when n is odd and n is even.

5   Calculate the number of ways of colouring a disc divided on one side into p sectors, when three colours are available for each sector and p is an odd prime.

6   Calculate the number of necklaces which can be constructed with p beads when three colours are available and p is an odd prime.

7   Suppose six distinguishable dice are thrown. Use Theorem 20.5 to construct a generating function $U^*(a_1, a_2, \ldots, a_6)$ for the numbers of ways of obtaining a given result when the first four dice show the same number and the last two dice show the same number. By substituting x^i for a_i $(1 \leqslant i \leqslant 6)$ find the number of ways in which a total of 18 can be obtained in this manner.

8   Calculate the number of different ways of colouring the faces of a dodecahedron with three colours, subject to the condition that each colour is used at least once.

9   Suppose that three red balls, two blue balls, and one yellow ball are placed at the corners of an octahedron. In how many ways can this be done?

10   Show that when $n_1 + n_2$ is an odd prime the number of necklaces which

can be made with n_1 black beads and n_2 white beads is

$$\frac{1}{2(n_1+n_2)}\binom{n_1+n_2}{n_1}+\frac{1}{2}\binom{\frac{1}{2}(n_1+n_2-1)}{\lfloor\frac{1}{2}n_1\rfloor}.$$

11 An unsharpened pencil is to be painted with seven bands of equal size, the colours available being red, green, and yellow. In how many ways can this be done if there are

(i) two red, two green, and three yellow bands;
(ii) four red, one green, and two yellow bands?

12 Calculate the number of ways of colouring the faces of a regular dodecahedron in such a way that there are six red faces, four yellow faces, and two blue faces.

13 In Ex. 14.7.21 we obtained three 'special' finite groups of rotations, having orders 12, 24, and 60. These are the rotational symmetry groups of the regular tetrahedron, cube, and regular dodecahedron. Describe the rotations geometrically in each case, specifying a typical axis for a rotation of each order.

14 Compute the centre of each of the three groups discussed in the preceding exercise.

15 Show that the group of the regular dodecahedron is isomorphic to the alternating group A_5.

Answers to selected exercises

1.1   3   Take $c = a - b = a + (-b)$.

1.2   4   (i) Yes, -4.
             (ii) No.
             (iii) Yes, 0.

       6

1	2	3	4	25	$s = 10,\ t = 5.$
6	7	8	9	10	
11	12	13	14	15	
16	17	18	19	20	
21	22	23	24	5	

1.3   2 $t_1 = 2,\ t_n = 2t_{n-1}$ $(n \geq 2)$.
        4   (i) $u_n = 3n - 2$.
             (ii) $u_n = (n!)^2$.

1.4   2 $S_n = (\tfrac{1}{2}n(n+1))^2$.
        5   (i) $n_0 = -2$.
             (ii) $n_0 = 6$.

1.5   2 $11\,111\,000\,001$, $30\,420$, 1545.
        3   (i) 221.
             (ii) 1468.

1.6   2   If $a = pc$, $b = qc$, then $xa + yb = (xp + yq)c$.

1.7   1   gcd $(721, 448) = 7$.
        5   gcd $(966, 686) = 14$, a solution is $x = -110$, $y = 155$.

1.8   2 $201 = 3 \times 67$, $1001 = 7 \times 11 \times 13$, $201\,000 = 2^3 \times 3 \times 5^3 \times 67$.
        9   11.

2.1   2   Take $g(i) = i$ for $1 \leq i \leq 5$.
        3   (i) Yes, if U includes dead citizens.
             (ii) No, x may have no daughters or several daughters.
             (iii) No, women do not have wives.

2.2 1   (i) Injection.
         (ii) Bijection.
         (iii) Injection.
         (iv) Not an injection or a surjection.
    2   Note $u(n) = u(n-7)$ $n \geqslant 8$.

2.3 1   (i) $n = 5$, $f(i) = 2i$.
         (ii) $n = 6$, $f(i) = 2 - 5i$.
         (iii) $n = 8$, $f(i) = (i+2)^2 + 1$.

2.4 1   3, 12.
    5   The set $\{n+1, n+2, \ldots, 2n\}$ will do.

2.5 1   (i) A suitable injection is the inclusion $i : \mathbb{N} \to \mathbb{Z}$.
         (ii) $i(n) = -n$.
         (iii) $i(n) = n + 10^6 - 1$.
    3   Consider $N = 6p_1 p_2 \cdots p_k - 1$ where $p_1, p_2, \ldots, p_k$ are assumed to be all the primes of the form $6m + 5$.

3.1 1   49.
    2   The multiples of 2 and multiples of 3 are not disjoint.

3.2 1   20.
    3   No, since 3 is not a divisor of 40.
    5   26^4, 25^4.

3.3 1   18, 8, 12
    4   $\phi(a)\phi(b) = \phi(ab)$ if $\gcd(a, b) = 1$.

3.4 1 4^n.
    3   7.

3.5 1   14 529 715 200.
    2   5040.

3.6 1   (137)(2548)(6)(9).
    3   9.
    4   $\alpha_1 = (12)(34)$, $\alpha_2 = (13)(24)$, $\alpha_3 = (14)(23)$.

4.1 3   The complement of an r-subset is an $(n-r)$-subset.
    5   1820, 6188.
    6   The r zeros can be in any of the n places.

4.2 2 $\dbinom{n+5}{5}$.
    3   $\frac{1}{2}(n+1)(n+2)$.

4.3 2   (i) 462.
         (ii) 45.

(iii) 10.

(iv) 34 560.

4 Put $x = 1$ and $x = -1$.

4.4 1 8, 6.

2 582.

3 9.

4.5 1

n	95	96	97	98	99	100
$\phi(n)$	72	32	96	42	60	40
$\mu(n)$	1	0	−1	0	0	0

4 $d \mid x$ and $d \mid n$ if and only if $d \mid n - x$ and $d \mid n$.

4.6 1 (i) $4v = 7k$.

(ii) Four at least.

2 (i) 123, 456.

(ii) Impossible.

(iii) 123, 234, 345, 456, 567, 167, 127.

4 $v' = v$, $k' = v - k$, $r' = vr/k - r$.

4.7 1 4, 12, 30, 66, 132.

2 (i) Impossible.

(ii) Possible.

5.1 1 1 127 966 1701 1050 266 28 1.

5.2 2 $\{1, 11, 6\}, \{2, 7\}, \{5\}, \{9\}$.

3 (i) 24.

(iii) 6.

4 $n!$, $(n - 1)!$.

6 Given a, there may not be a b such that $a \sim b$.

5.3 1 34 650.

2 126 000, 756.

5 (i) 2520.

(ii) 2520.

5.4 1 There are 15 partitions.

2 Subtract 1 from each part.

5.5 2 A suitable permutation is $\sigma = (1)(268\,974)(35)$.

3 $\pi\tau = \tau^{-1}(\tau\pi)\tau$.

5 265.

6 15.

5.6 1 $\text{sgn } \alpha = 1$, $\text{sgn } \beta = -1$, $\text{sgn } \gamma = 1$.

4   (i) Possible.

    (ii) Impossible.

6.1 2   (i) False.

    (ii) False.

    (iii) Possibly true, but false in fact.

3 6, 5.

6.2 3 $x = 2$, $y = 1$. No solution in $\mathbb{Z}_5$.

4 3, 5.

6.3 4 6, 13, 7, 8.

5 4.

6.4 1   (i) Yes.

    (ii) No.

    (iii) Yes.

2 $v = 4n + 3$, $k = 2n + 1$, $r_2 = n$.

6.5 1

AS	KH	QC	JD
KC	AD	JS	QH
QD	JC	AH	KS
JH	QS	KD	AC

3 Because 4 is not prime.

7.1 1 $c_1 = t(x_0 y)$, $p_0 = u(x_0 y)$; $c_{i+1} = t(x_i y + c_i)$, $p_i = u(x_i y + c_i)$ $(1 \leqslant i \leqslant n - 1)$; $p_n = c_n$.

7.2 1 Program calculates $\gcd(2406, 654) = 6$.

2 $b = \min\{x_1, x_2, \ldots, x_n\}$.

7.4 3 $z + xy$.

4 $z = nm$.

7.5 1   (i) n.

    (ii) $n - 1$.

    (iii) $x_1 > x_2 > \cdots > x_n$.

3   (iii) $(a_k, b_k) = (f_{k+1}, f_k)$, where f_k is defined by $f_1 = 1$, $f_2 = 2$, $f_r = f_{r-1} + f_{r-2}$ $(r \leqslant 2)$.

7.6 1   (i) n^3.

    (ii) n^2.

(iii) $(3/2)^n$.

(iv) n^n.

4 Because $\log_a n = \log_b n/\log_b a = C \log_b n$.

7.7 1 54, 9.

3 $ax + bz = m_1 + m_4 - m_5 + m_7;$ $ay + bt = m_3 + m_5;$ $cx + dz = m_2 + m_4;$
$cy + dt = m_1 - m_2 + m_3 + m_6.$

7.8 1 3 4 5 1 2 7 8 6 9

3 4 1 2 5 7 6 8 9

3 1 2 4 5 6 7 8 9

1 2 3 4 5 6 7 8 9

Remaining passes effect no changes.

8.1 1 No.

3 $\frac{1}{2}n(n-1)$ edges; $n \geq 5$ impossible.

8.2 1 Second graph has no 3-cycles.

8.3 1 (i) No.

(ii) Yes.

(iii) No.

(iv) No.

2 $n - 1 - d_1, n - 1 - d_2, \ldots, n - 1 - d_n$.

3 There are only two possibilities.

5 There cannot be vertices of valency 0 and $n - 1$ in the same graph.

8.4 1 3.

4 There is an Eulerian path but no Hamiltonian cycle.

5 No.

8.5 4 Calculate componentwise.

8.6 1 (i) n.

(ii) 2.

(iii) 3.

2 3, 4, 4.

9.1 2 3.

4 $\lceil \log_2 20 \rceil = 5$.

5 12, 13.

7 2.

9.2 1 5.

2 (i) 6.

(ii) 6.

(iii) 5.

3 (i) 16, 55, 33, 63, 81, 76.

    (ii) 12, 21, 17, 28, 32, 51, 19, 84, 38, 49, 77, 73, 56.

9.3 3 The MST is unique.

      4 Four MST's, weight 20.

9.4 1 G is connected.

      2 3.

9.5 2 Not connected.

      3 $n, 1; n, \lfloor n/2 \rfloor$.

      5 The number of vertices at level i is $\leqslant k(k-1)^{i-1}$.

9.6 1 $v\ a\ d\ e\ b\ g\ h\ w$.

      2 A D C F.

10.1 2 (i) s.

        (ii) r.

        (iii) rs.

        (iv) Every x is related to every y.

      3 $\frac{1}{4}rs(r-1)(s-1)$.

10.2 1 (i) 3.

        (ii) 5.

        (iii) 3.

      3 Give $x_i y_j$ colour $i-j$ (mod n).

10.3 2 $Q = E$.

10.4 1 $\Gamma\{x_2, x_3, x_4\} = \{y_2, y_4\}$.

      2 (i) $x_2 y_5 x_5 y_1$ will do.

        (ii) $M' = \{x_2 y_5, x_3 y_2, x_4 y_4, x_5 y_1\}$.

        (iv) Yes.

10.5 2 All 3-subsets except $\{x_1, x_2, x_3\}$.

10.6 1 Let a, l, b, e, t, s represent the sets in the given order.

      3 There are five sets containing only the four elements a, e, m, r.

11.1 1 $c, b, a, d, e, f; d, e, f, a, d$.

      2 1, 2, 4, 5, 6, 7, 8, 3, 9.

11.2 1 Critical path s, p, q, z, t.

      2 Critical path s, a, e, f, t.

11.3 1 The cut $\{s, b\}, \{a, c, d, t\}$ has capacity 10.

      2 Flow: 4, 5, 1, 3, 4, 2, 5, 2.

        Cut: $\{s, a, b, d\}, \{c, t\}$.

11.4 1 (i) 11.
   (ii) s, c, b, d, t; 12.
   (iii) $\{s, b, c, e\}, \{a, d, t\}$.
   (iv) $f^{\star}$ is max flow.
   2 Max flow $= 38$.

11.5 1 Max flow $= 55$.
   2 Max flow $= 39$.

12.1 1 In the recursive method, number of multiplications is $O(n)$.
   3 Same number of multiplications, less store needed.

12.2 1 (i) $(4^n + 4(-1)^n)/5$.
   (ii) $3n - 2$.
   4 (i) $F_n = q_{n-1}$.

12.3 3 $a_n = 2n - \sqrt{n}$.

12.4 1 9; there are four optimal policies.
   2 21; allocate $2, 2, 1$ or $0, 2, 3$.

12.5 2 $x_1 = 1, x_2 = 0, x_3 = 5$; profit 184.
   $x_1 = 4, x_2 = 0, x_3 = 0$; profit 166.

12.6 1 s, b, f, i, m, t or s, d, h, k, o, t.
   2 $x_1 = 2, x_2 = 1, x_3 = 1$; optimum 34.

13.1 1 $+$ has all four properties, $-$ has all except Associativity, and $\times$ has all except Inverse.

13.2 3 (i) All $n \geqslant 2$.
   (ii) No such n.
   (iii) n prime.

13.3 2 (i) $xy = 1 \Rightarrow x = y^{-1}$.
   3 Use part (ii) of previous question.
   5 $(ab)c = c$ but $a(bc) = b$.

13.4 1 $10, 4; 6, 6$.
   3 Orders of $A, B = 7, 6$.
   4 The matrix $\begin{bmatrix} -1 & b \\ 0 & 1 \end{bmatrix}$ has order 2 for any b.

13.5 2 (iii) Either every element has order 1 or 2, or there is an element x for which $x^2 \neq 1$.

13.6 1 $U \approx \langle 3 \rangle = \{3, 3^2, 3^3, 3^4, 3^5, 3^6 = 1\}$.
   2 The function taking $\begin{bmatrix} 1 & m \\ 0 & 1 \end{bmatrix}$ to m is an isomorphism from M to $\mathbb{Z}$.

13.7   1 K_1 yes, K_2 no, K_3 no.

          2    $H = \{i, x\}$, $K = \{i, y\}$ will do.

          4    $Z(G)$ is the intersection of all $C(g)$, $g \in G$.

13.8   2   Cosets: $\{i, x\}$, $\{r, z\}$, $\{s, y\}$.

          3   Order 1: {identity};
              Order 2: {identity and a reflection};
              Order 5: {identity and 4 rotations};
              Order 10: the whole group.

13.9   1 $\langle z^7 \rangle = C_{24}$, $\langle z^8 \rangle = C_3$, $\langle z^9 \rangle = C_8$.

          2    $\phi(60) = 16$.

14.1   1   (i) No.
              (ii) Yes.
              (iii) Yes.
              (iv) No.

          2   (i) 4.
              (ii) 2.
              (iii) 15.

          3   1, 2, 3, 4, 5, 6, 7, 8, 10, 12, 15.

          5   Group has order 12.

14.2   1   id, (15)(24).

          2    $\{1, 6, 7\}$, $\{2, 4, 5\}$, $\{3\}$.

          3    $\{1, 3\}$, $\{2\}$, $\{4\}$, $\{5, 7\}$, $\{6, 8\}$.

14.3   3 $\{a, f\}$, $\{b, c, d, e\}$; $|G_a| = 4$, $|G_b| = 2$, $|G| = 8$.

14.4   2   21.

          4   42.

14.5   2   (i) Yes.
              (ii) Yes.
              (iii) No.
              (iv) No.

14.6   1   24.

          3   (13)(24); g_3.

          5    $\sigma = (78)$.

15.1   1   (iv) m^4.
              (vi) There are six of them.

15.2   1 $\{1, 3, 7, 9\}$, cyclic, generated by 3; $\{1, 2, \ldots, 10\}$, cyclic, generated by 2; $\{1, 5, 7, 11\}$, not cyclic, $C_2 \times C_2$.

          3    $U(\Gamma) = \{1, i, -1, -i\}$, cyclic, generated by i.

15.3   2   5 is a generator.

15.4   1   (i) x^3+4x^2+x+3, $x^5+2x^4+x^3+3x+2$.
  (ii) $x^4+x^3+x^2+4x+3$, $x^7+x^4+x^3+x^2+3x+2$.

  4   True when m is a prime.

15.5   1 $q=x$, $r=x+1$.
  2   $q=x^3+4x^2+x+2$, $r=2x+1$.

15.6   1 $x+1$; $\lambda(x)=1$, $\mu(x)=2x+2$.
  2   $x+4$.
  3   (i) x^2+1.
    (ii) $x+1$.
    (iii) x^3+1.

15.7   6 $\mathbb{Z}_{15}$ is not a field, factorization in $\mathbb{Z}_{15}[x]$ is not unique.

15.8   1   (i) $(x+2)(x+3)$.
    (ii) $(x+8)(x+9)(x+10)$.
    (iii) $(x^2+2)(x^2+3x+3)$.
  2   x^2+1, x^2+x+2, x^2+2x+2.

16.1   1   If $c=2x+1$, generators are c, $c^3=x+1$, $c^5=x+2$, $c^7=2x+2$.
  2   x, 2, $2x$, 1.

16.2   1   (i) $0=w$, $1=y$.
    (iii) 2.

16.3   2   Take $0\to w$, $1\to y$, $x\to z$, $x+1\to t$.
  3   3, 7, 11, 19, 23; groups are cyclic.
  4   The number of monic irreducible polynomials is >0.

16.4   1   (i) 3.
    (ii) 2.
    (iii) 5.
  3   (i) Reducible.
    (ii) Irreducible, primitive.
    (iii) Reducible.
  4   $\phi(31)=30$; that is, all except 0 and 1.
  5   4, 5, 6, 7.

16.5   1

66	23	50	41
03	79	88	97
10	87	99	78
31	98	77	89

16.6   2   (i) $(0, 2)$, $(1, 4)$, $(2, 4)$.
          (ii) $x = 1$, $x + 3y = 1$, $x + y = 1$.
          (iii) $(1, 0)$.
       3   (i) 3.
          (ii) 2.
          (iii) 4.
          (iv) 2.

16.7   1   13 weeks, scheme is a projective plane over $\mathbb{F}_3$.
       2   Complement of each quadrangle is a line.

16.8   1 $\{1, 4, 5, 6, 7, 9, 11, 16, 17\}$.

16.9   1 $(p^4 - p^2)/4$, $(p^5 - p)/5$, $(p^6 - p^3 - p^2 + p)/6$.
       2   $x^4 + x^3 + 1$, $x^4 + x + 1$, $x^4 + x^3 + x^2 + x + 1$.
       3   x, $x + 1$, $x^2 + x + 1$, and the polynomials obtained in question 2.
       5   (iii) 2, 4, 8, 64.

17.1   1   (i) $\delta = 2$, detects 1, corrects 0.
          (ii) $\delta = 2$, detects 1, corrects 0.
          (iii) $\delta = 3$, detects 2, corrects 1.
       2   (ii) and (iii) can be extended.

17.2   1 $k = 1$, $\delta = n$.
       3   2; $\{00\,000\,000, 11\,111\,000, 00\,011\,111, 11\,100\,111\}$.

17.3   1   8 codewords.
       2   $n = 7$, $k = 3$, $\delta = 3$.
       4   (i) 6.
          (ii) 10.

17.4   1   100 110.
       2   2048; only the second one is a codeword.
       3   (i) 12.
          (iii) 15.

17.5   1   (i) No.
          (ii) No (not linear).
          (iii) Yes.
          (iv) Yes.
       2   000, 111; $H = \begin{bmatrix} 1 & 1 & 0 \\ 0 & 1 & 1 \end{bmatrix}$ will do.

17.6   1 $(x + 1)(x^4 + x^3 + x^2 + x + 1)$.
       3   (ii) 32.
          (iii) $x^4 + x + 1$.

18.1 1 (i) $1-x+4x^2-19x^3$.
  (ii) $\frac{2}{3}+\frac{16}{9}x-\frac{37}{27}x^2-\frac{221}{81}x^3$.
  4 $1+x^2+x^4+x^6+\cdots;\ 1+2x^2+x^4+2x^6+\cdots$.

18.2 1 (i) $\frac{7}{3}/(1-x)+(-\frac{5}{3})/(2+x)$.
  (ii) $1/(1+x)+(-4)/(2+x)+4/(3+x)$.
  3 $(2+2x)/(1+2x+2x^2)+(2+x)/(1+x+2x^2)$.
  5 $\frac{1}{4}\{(1+x)^{-1}+(1-x)^{-1}+i(i-x)^{-1}+i(i+x)^{-1}\}$.

18.3 1 (i) 280.
  (ii) $(n+1)(n+2)(n+3)/6$.
  (iii) $\binom{3r-1}{r-1}$.
  4 $\frac{1}{2}(5n^2+3n+2)$.

18.4 1 $(1-\frac{1}{2}n)2^n$.
  2 (i) $5A(x)$.
  (ii) $A(x)+5(1-x)^{-1}$.
  (iii) $x^{-5}(A(x)-a_0-a_1x-a_2x^2-a_3x^3-a_4x^4)$.

18.5 1 (i) $(4^{n+1}+(-1)^n)/5$.
  (ii) $5-2^{n+2}+3^n$
  (iii) $(8-6n+(-2)^n)/9$.
  2 $b_n=F_n$ (as in Ex. 12.2.2).
  3 When $a=0$ and $b=2$, $z_n=(-1)^n3(2^{n+1}+(-1)^n)^{-1}$.

18.6 2 $Q(x)=x/(1-2x)(1-4x)$.

19.1 2 1, 2, 3, 5, 7, 11, 15.

19.2 1 (i) $[12^23^25]$.
  (ii) $[1^32^257^2]$.
  2 (i) No.
  (ii) No.
  (iii) Yes.
  (iv) Yes.
  3 There are seven of them.
  4 7.

19.3 1 (i) $(1-x^3)^{-1}(1-x^5)^{-1}$.
  (ii) $(1-x^2)^{-1}(1-x^4)^{-1}(1-x^6)^{-1}$.
  3 49.
  4 25.

19.4 1 (i) $(1+x+x^2)(1+x^2+x^4)(1+x^3+x^6)\cdots$.
  (ii) $(1-x)^{-1}(1-x^2)^{-1}(1-x^4)^{-1}\cdots$.
  (iii) $x^5(1-x^5)^{-1}(1-x^6)^{-1}(1-x^7)^{-1}\cdots$.

19.5   2   Subtract 1 from each part.

19.6   1 $p(15) = 176$, $p(16) = 231$, $p(17) = 297$, $p(18) = 385$, $p(19) = 490$, $p(20) = 627$.

      2   22.

20.1   1   10 distinguishable colourings.

      2   20.

      3   (i) $(x_1^3 + 3x_1x_2 + 2x_3)/6$.

          (ii) $(x_1^4 + 8x_1x_3 + 3x_2^2)/12$.

          (iii) $(x_5^1 + 10x_1^3x_2 + 20x_1^2x_3 + 15x_1x_2^2 + 30x_1x_4 + 20x_2x_3 + 24x_5)/120$.

      4   $(x^6 + x_1^2x_2^2 + 2x_2^3)/4$; $(x_1^9 + 2x_1x_4^2 + x_1x_2^4 + 4x_1^3x_2^3)/8$.

20.2   2 $(x_1^p + (p-1)x_p + px_1x_2^{p-1})/2p$; use Fermat's theorem.

      5   $C_6 \times C_2$.

20.3   2 $(x_1^6 + 8x_3^2 + 3x_1^2x_2^2)/12$.

20.4   1   (i) $(r^4 + 11r^2)/12$.

          (ii) $(r^4 + 11r^2)/12$.

          (iii) $(r^6 + 3r^4 + 12r^3 + 8r^2)/24$.

      3   9099.

      4   834.

20.5   1   (i) $(a^5 + b^5 + c^5)^2$.

          (ii) $(a^4 + b^4)(a^2 + b^2)^3$.

      2   6.

20.6   1   3.

      3   (i) 6.

          (ii) 8.

Index